Civil Engineering Hydraulics and Engineering Hydrology

Civil Engineering Hydraulics and Engineering Hydrology

Editor:

Dr. Tanjina Nur, Ph.D.

ArclerPress
www.arclerpress.com

Civil Engineering Hydraulics and Engineering Hydrology

Editor: Dr. Tanjina Nur, Ph.D.

708 3rd Avenue, 6th Floor
New York
NY 10017
United States of America
www.arclerpress.com

ISBN: 978-1-68094-381-8

Library of Congress Control Number: 2016951069

About the Editor

Dr. Tanjina Nur, Ph.D.

Tanjina finished her PhD in Civil and Environmental Engineering in 2014 from University of Technology Sydney (UTS). Now she is working as Post-Doctoral Researcher in the Centre for Technology in Water and Wastewater (CTWW) and published about eight International journal papers with 80 citations. Her research interest is wastewater treatment technology using adsorption process.

Contents

List of Contributors

Arindam Singha
Mechanical Engineering Department, Texas A&M University at Qatar, Doha 77844, Qatar

Ramaswami Balachandar
Civil Engineering Department, University of Windsor, Windsor, ON, Canada N9B 3P4

Thorsten Stoesser
School of Engineering, Cardiff University, Cardiff CF24 3AA, UK

Richard McSherry
School of Engineering, Cardiff University, Cardiff CF24 3AA, UK

Bruno Fraga
School of Engineering, Cardiff University, Cardiff CF24 3AA, UK

Laurent Schindfessel
Hydraulics Laboratory, Department of Civil Engineering, Ghent University, St-Pietersnieuwstraat 41 B5, Ghent B-9000, Belgium

Stéphan Creëlle
Hydraulics Laboratory, Department of Civil Engineering, Ghent University, St-Pietersnieuwstraat 41 B5, Ghent B-9000, Belgium

Tom De Mulder
Hydraulics Laboratory, Department of Civil Engineering, Ghent University, St-Pietersnieuwstraat 41 B5, Ghent B-9000, Belgium

Andy Ward
Department of Food, Agricultural and Biological Engineering, The Ohio State University, Columbus, OH 43210, USA

Molly Moran
Department of Civil, Environmental and Geodetic Engineering, The Ohio State University, Columbus, OH 43210, USA

María L. Merino-Solís
Environmental Technology, Research Center and Assistance in Technology and Design of Jalisco, N.P.O., Normalistas 800, Guadalajara 44270, Jalisco, Mexico

Edgardo Villegas
Environmental Technology, Research Center and Assistance in Technology and Design of Jalisco, N.P.O., Normalistas 800, Guadalajara 44270, Jalisco, Mexico

José de Anda
Environmental Technology, Research Center and Assistance in Technology and Design of Jalisco, N.P.O., Normalistas 800, Guadalajara 44270, Jalisco, Mexico

Alberto López-López
Environmental Technology, Research Center and Assistance in Technology and Design of Jalisco, N.P.O., Normalistas 800, Guadalajara 44270, Jalisco, Mexico

Do Guen Yoo
Research Center for Disaster Prevention Science and Technology, Korea University, Seoul 136-713, Korea

Doosun Kang
Department of Civil Engineering, Kyung Hee University, 1732 Deogyeong-daero, Giheung-gu, Yongin-si 446-701, Gyeonggi-do, Korea

Hwandon Jun
Department of Civil Engineering, Seoul National University of Science and Technology, 172, Gongreung 2-dong, Nowon-Gu, Seoul 139-743, Korea

Joong Hoon Kim
School of Civil, Environmental and Architectural Engineering, Korea University, Anam-ro 145, Seongbuk-gu, Seoul 136-713, Korea

Seung Oh Lee
Department of Urban & Civil Engineering, Hongik University, 72-1 Sangsu-dong Mapo-gu, Seoul 121-791, Korea

Sooyoung Kim
Department of Urban & Civil Engineering, Hongik University, 72-1 Sangsu-dong Mapo-gu, Seoul 121-791, Korea

Moonil Kim
Department of Civil and Environmental Engineering, Hanyang University, Ansan 426-791, Korea

Kyoung Jae Lim
Department of Regional Infrastructural Engineering, Kangwon National University, Chuncheon 200-701, Korea

Younghun Jung
Department of Regional Infrastructural Engineering, Kangwon National University, Chuncheon 200-701, Korea

José A. Amador
Department of Natural Resources Science, Coastal Institute-Kingston, University of Rhode Island, Kingston, RI 02881, USA

David A. Potts
Geomatrix, LLC, 114 Mill Rock Road East, Old Saybrook, CT 06475, USA

George W. Loomis
New England Onsite Wastewater Training Program, Coastal Institute-Kingston, University of Rhode Island, Kingston, RI 02881, USA

David V. Kalen
New England Onsite Wastewater Training Program, Coastal Institute-Kingston, University of Rhode Island, Kingston, RI 02881, USA

Erika L. Patenaude
Department of Natural Resources Science, Coastal Institute-Kingston, University of Rhode Island, Kingston, RI 02881, USA

Josef H. Görres
Department of Natural Resources Science, Coastal Institute-Kingston, University of Rhode Island, Kingston, RI 02881, USA

Mark E. Grismer
Hydrologic Sciences and Biological & Agricultural Engineering, UC Davis, Davis, CA 95616, USA

Hua Zhong
Nanjing Hydraulic Research Institute, Nanjing 210029, China
State Key Laboratory of Hydrology–Water Resources and Hydraulic Engineering, Nanjing 210098, China

Pieter van Gelder
Safety and Security Science Group, Faculty of Technology, Policy and Management, Delft University of Technology, Delft 2628 BX, The Netherlands

Wen Wang
State Key Laboratory of Hydrology–Water Resources and Hydraulic Engineering, Nanjing 210098, China
College of Hydrology and Water Resources, Hohai University, Nanjing 210098, China

Gaoxu Wang
Nanjing Hydraulic Research Institute, Nanjing 210029, China

State Key Laboratory of Hydrology–Water Resources and Hydraulic Engineering, Nanjing 210098, China

Yongzhi Liu
Nanjing Hydraulic Research Institute, Nanjing 210029, China
State Key Laboratory of Hydrology–Water Resources and Hydraulic Engineering, Nanjing 210098, China

Shuai Niu
Nanjing Hydraulic Research Institute, Nanjing 210029, China
State Key Laboratory of Hydrology–Water Resources and Hydraulic Engineering, Nanjing 210098, China

Terry Lucke
Stormwater Research Group, School of Engineering, Faculty of Science and Engineering, University of the Sunshine Coast, Warana, QLD 4558, Australia

Mohamed Ansaf Kachchu Mohamed
Stormwater Research Group, School of Engineering, Faculty of Science and Engineering, University of the Sunshine Coast, Warana, QLD 4558, Australia

Neil Tindale
Stormwater Research Group, School of Engineering, Faculty of Science and Engineering, University of the Sunshine Coast, Warana, QLD 4558, Australia

Younghun Jung
Department of Civil Engineering, Inha University, 100 Inha-ro, Nam-gu, Incheon 402-751, Korea

Venkatesh Merwade
School of Civil Engineering, Purdue University, 550 Stadium Mall Drive, West Lafayette, IN 47907, USA

Kyudong Yeo
Department of Civil Engineering, Inha University, 100 Inha-ro, Nam-gu, Incheon 402-751, Korea

Yongchul Shin
Department of Biological & Agricultural Engineering, Texas A&M University, College Station, TX 77843, USA

Seung Oh Lee
School of Urban and Civil Engineering, Hongik University, Seoul 121-791, Korea

Thelma Dede Baddoo
State Key Laboratory of Hydrology-Water Resources and Hydraulic Engineering, Hohai University, Nanjing 210098, China

Yiqing Guan
State Key Laboratory of Hydrology-Water Resources and Hydraulic Engineering, Hohai University, Nanjing 210098, China
College of Hydrology and Water Resources, Hohai University, Nanjing 210098, China

Danrong Zhang
State Key Laboratory of Hydrology-Water Resources and Hydraulic Engineering, Hohai University, Nanjing 210098, China
College of Hydrology and Water Resources, Hohai University, Nanjing 210098, China

Samuel A. Andam-Akorful
School of Earth Sciences and Engineering, Hohai University, Nanjing 210098, China
Department of Geomatic Engineering, Kwame Nkrumah University of Science and Technology, Private Mail Box, Kumasi, Ghana

Jeng-Wen Lin
Department of Civil Engineering, Feng Chia University, Taichung 40724, Taiwan

Chih-Wei Huang
Ph.D. Program in Civil and Hydraulic Engineering & Research Center for Science and Technology across the Taiwan Strait, Feng Chia University, Taichung 40724, Taiwan

Yin-Sung Hsu
Department of Water Resources Engineering and Conservation, Feng Chia University, Taichung 40724, Taiwan

Adama Toure
Graduate Research Program on Climate Change and Water Resources, West African Science Service Center on Climate Change and Adapted Land Use (WASCAL), Department of Applied Hydrology, University of Abomey-Calavi, Cotonou, Abomey-Calavi BP 2008, Benin

Bernd Diekkrüger
Department of Geography, University of Bonn, Bonn 53115, Germany

Adama Mariko
Department of Geology, National Engineer School, Bamako BP 242, Mali

Bruno Mazzorana
Department of Hydraulic Engineering—Abteilung Wasser Schutzbauten, Autonomous Province of Bolzano, Italy, Cesare Battisti Street 23, Bolzano 39100, Italy
Faculty of Science and Technology, Free University of Bolzano, Universitätsplatz 5, Bolzano 39100, Italy

Laura Levaggi
Faculty of Science and Technology, Free University of Bolzano, Universitätsplatz 5, Bolzano 39100, Italy

Omar Formaggioni
Department of Hydraulic Engineering—Abteilung Wasser Schutzbauten, Autonomous Province of Bolzano, Italy, Cesare Battisti Street 23, Bolzano 39100, Italy

Claudio Volcan
Department of Hydraulic Engineering—Abteilung Wasser Schutzbauten, Autonomous Province of Bolzano, Italy, Cesare Battisti Street 23, Bolzano 39100, Italy

Marta Kiraga
Water Engineering Division, Department of Hydraulic Engineering, Warsaw University of Life Sciences—SGGW, 159 Nowoursynowska Street, Warsaw 02-787, Poland

Zbigniew Popek
Water Engineering Division, Department of Hydraulic Engineering, Warsaw University of Life Sciences—SGGW, 159 Nowoursynowska Street, Warsaw 02-787, Poland

Lihwa Lin
Coastal and Hydraulics Laboratory, U.S. Army Engineer Research and Development Center, 3909 Halls Ferry Road, Vicksburg, MS 39180, USA

Zeki Demirbilek
Coastal and Hydraulics Laboratory, U.S. Army Engineer Research and Development Center, 3909 Halls Ferry Road, Vicksburg, MS 39180, USA

Donald Ward
Coastal and Hydraulics Laboratory, U.S. Army Engineer Research and Development Center, 3909 Halls Ferry Road, Vicksburg, MS 39180, USA

David King
Coastal and Hydraulics Laboratory, U.S. Army Engineer Research and Development Center, 3909 Halls Ferry Road, Vicksburg, MS 39180, USA

Preface

The objective of this book is to give knowledge of open channel hydraulics and hydrology, leading to understanding of the scientific foundations and basic principles of these fields, and the ability to apply hydraulic and hydrological methods to engineering applications in an integrated way. The first part of the book describes an assessment of the literature and recent developments on the shallow wake in an open channel flow and secondary currents and turbulence over a non-uniformly roughened open-channel bed. Then we present flow patterns in an open channel confluence with increasingly dominant tributary inflow and a novel approach for estimating the recurrence intervals of channel-forming discharges. After that, we try to find the effect of the hydraulic retention time on the performance of an ecological wastewater treatment system with a constructed wetland and rehabilitation priority determination of water pipes based on hydraulic importance.

In the middle part of the book, we focus on the effect of hydraulic characteristics on algal bloom in an artificial seawater canal of South Korea and improvement of hydraulic and water quality renovation functions by intermittent aeration of soil treatment areas in onsite wastewater treatment systems. Determination of watershed infiltration and erosion parameters from field rainfall simulation analyses and the Influence of statistical uncertainty in the hydraulic boundary conditions on the probabilistically computed high water level frequency curve are also discussed. Then we present pollutant removal and hydraulic reduction performance of field grassed swales during runoff simulation experiments and an approach using a 1D hydraulic model, landsat imaging and generalized likelihood uncertainty estimation for an approximation of flood discharge.

At the end, we try to describe rainfall variability in the watershed and its relationship with ENSO and micro radar surface velocimetry for hydrologic signal processing using a filtering approach. Also impact of climate change on groundwater resources in the Southern Mali and physical vulnerability assessment based on fluid and classical mechanics to support cost-benefit analysis of flood risk mitigation strategies are also discussed. Then modified lane's relation in local bed scouring studies in the laboratory channel and wave and hydrodynamic modelling for engineering design of Jetties at Tangier Island is presented briefly.

Editor
Dr. Tanjina Nur, Ph.D.

INTRODUCTION

Civil engineering is one of the oldest engineering disciplines, since civil engineers of one form or another have been around ever since humans started building major public works such as roads, bridges, tunnels, and large public buildings. It is also an incredibly broad discipline, spanning treatment of environmental issues, transportation, power generation, and major structures. To become a civil engineer, a person must typically study engineering at a university and then participate in field work for practical training. Many nations also require that students pass a competency exam to ensure that they will be able to design and build safe, stable structures.

There are many branches of civil engineering, and a wide range of specialties. Some engineers focus on conception and initial design of a project, analyzing the site, the needs, and the resources to come up with a workable project plan. Others specialize in contracting, physically building the structure, managing the site crew, and handling materials and supply. In other cases, civil engineers focus on maintenance of the project after it is completed, to make sure that it is safe and useful.

Most people pick a focus while they are receiving an education. Engineers who focus on transportation, for example, might choose to specialize in building bridges, tunnels, and roads. Others might lean towards power generation facilities, water treatment, waste management, construction of light railways and subways, or other disciplines. In all cases, extensive training is undertaken so that the prospective civil engineer understands his or her chosen field in depth. Behind every major public works is a team of civil engineers.

One of the primary concerns of civil engineering is public safety and health. A value is also placed on building structures that are functional, efficient, and also aesthetically pleasing. Structural soundness, conformity with local codes, and functionality are all issues which are faced in the discipline. Some civil engineers work directly for the public in the form of government agencies, while others find employment with public firms.

Education does not end with a degree and a course of fieldwork with trained and experienced civil engineers. Continuing education is also an important part of this discipline. As advances are made in the field, engineers are expected to keep pace with them, especially when the advances improve safety for workers and the public. There are many trade journals and annual conferences in the field to keep engineers updated.

Hydraulic engineering

Hydraulic engineering as a sub-discipline of civil engineering is concerned with the flow and conveyance of fluids, principally water and sewage. One feature of these systems is the extensive use of gravity as the motive force to cause the movement of the fluids. This area of civil engineering is intimately related to the design of bridges, dams, channels, canals, and levees, and to both sanitary and environmental engineering. Hydraulic engineering is the application of fluid mechanics principles to problems dealing with the collection, storage, control, transport, regulation, measurement, and use of water. Before beginning a hydraulic engineering project, one must figure out how much water is involved. The hydraulic engineer is concerned with the transport of sediment by the river, the interaction of the water with its alluvial boundary, and the occurrence of scour and deposition. "The hydraulic engineer actually develops conceptual designs for the various features which interact with water such as spillways and outlet works for dams, culverts for highways, canals and related structures for irrigation projects, and cooling-water facilities for thermal power plants."

What is Hydrology?

Hydrology is the scientific study of the movement, distribution, and quality of water on Earth and other planets, including the hydrologic cycle, water resources and environmental watershed sustainability. A practitioner of hydrology is a hydrologist, working within the fields of earth or environmental science, physical geography, geology or civil and environmental engineering.

Hydrology subdivides into surface water hydrology, groundwater hydrology (hydrogeology), and marine hydrology. Domains of hydrology include hydrometeorology, surface hydrology, hydrogeology, drainage-basin management and water quality, where water plays the central role.

Oceanography and meteorology are not included because water is only one of many important aspects within those fields.

Infiltration (hydrology)

Infiltration is the process by which water enters the soil. Some of the water is absorbed, and the rest percolates down to the water table. The infiltration capacity, the maximum rate at which the soil can absorb water, depends on several factors. The layer that is already saturated provides a resistance that is proportional to its thickness, while that plus the depth of water above the soil provides the driving force (hydraulic head). Dry soil can allow rapid infiltration by capillary action; this force diminishes as the soil becomes wet. Compaction reduces the porosity and the pore sizes. Surface cover increases capacity by retarding runoff, reducing compaction and other processes. Higher temperatures reduce viscosity, increasing infiltration.

What Does a Hydraulic Engineer Do?

A hydraulic engineer specializes in the effects of water in one or many disciplines. The nature of water in regard to flow, runoff, flooding, and how it is affected by weather can be used from an engineering perspective to help in the design of civil engineering projects. Hydraulic engineers help design dams and bridges, analyze the path of waste water in the ground, and study the effects of the ocean on the shoreline and offshore structures like oil platforms. They can also assist with human development or disaster planning in flood zones. Often assisted by powerful computer programs, hydraulic engineer duties sometimes involve working with other engineers and scientists as well as public officials.

The physical nature of the environment needs to be generally understood by a hydraulic engineer. Water flow and the forces that go along with it are often studied so that structures can be designed accordingly. Supports for bridges, and the capacity for a dam to support the weight of water in a reservoir, usually require the ingenuity of an engineer to be designed correctly. In agriculture, irrigation systems driven by gravity, pumps, or valves are typically designed according to engineering principles involving water.

Hydraulic engineer jobs can also be at industrial facilities. Industrial machines such as turbines, pipes, and pumps often operate on the flow of water. The design of car engines typically requires some analysis of the flow of gasoline. Engineers sometimes analyze the runoff of sewage in urban areas and the operation of hydraulic systems in water treatment plants. Knowledge of how to manage water resources is usually required to operate such systems, so a hydraulic engineer may work as part of the plant staff.

As part of the general workday, the hydraulic engineer job description often includes the need for computer skills. Software programs that specialize in data analysis and hydrologic modeling are used in many cases. Business meetings are sometimes a part of the routine, both with teams and external customers. Engineers specializing in hydraulics may need experience in aerodynamics if they are working for an aerospace company. They usually need to know how the hydraulic power components and actuators interact with other systems.

For many jobs, hydraulic engineer requirements include a degree in civil or agricultural engineering. Coursework in concepts related to a specialized field, such as oceanography, geophysics, or meteorology, may also be required. Skills in using computer aided design software or geographical information system programs are sometimes necessary as well.

Hydraulic engineer jobs can also be at industrial facilities. Industrial machines such as turbines, pipes, and pumps often operate on the flow of water. The design of car engines typically requires some analysis of the flow of gasoline. Engineers sometimes analyze the runoff of sewage in urban areas and the operation of hydraulic systems in water treatment plants. Knowledge of how to manage water resources is usually required to operate such systems, so a hydraulic engineer may work as part of the plant staff.

As part of the general workday, the hydraulic engineer job description often includes the need for computer skills. Software programs that specialize in data analysis and hydrologic modeling are used in many cases. Business meetings are sometimes a part of the routine, both with teams and external customers. Engineers specializing in hydraulics may need experience in aerodynamics if they are working for an aerospace company. They usually need to know how the hydraulic power components and actuators interact with other systems.

For many jobs, hydraulic engineer requirements include a degree in civil or agricultural engineering. Coursework in concepts related to a specialized field, such as oceanography, geophysics, or meteorology, may also be required. Skills in using computer aided design software or geographical information system programs are sometimes necessary as well.

Hydraulic Intensifier

A hydraulic intensifier is a device that is attached to a normal hydraulic machine to increase the amount of pressure the machine can produce. This is done by using two pistons, which usually are the same size but can be different sizes, to increase the pressure on the internal fluid. One problem with using a hydraulic intensifier is that it can only increase the pressure for one stroke, so this device cannot be used in machines that require more than one stroke. This can help reduce the hydraulic cylinder's size, so this device is most often used in machines that do not have room for large cylinders. This intensifier normally is not used for very small jobs, because there rarely is a need to increase the pressure on small tasks.

Hydraulic intensifier devices are made with two pistons that push into the hydraulic fluid to increase the overall pressure. The two pistons normally are the same size, and they increase the force pushing on the fluid. There also are versions that have differently sized pistons, but this exerts less force, so these are used for intensifiers that do not need the highest force. While a hydraulic intensifier can greatly increase the pressure of hydraulic machines — usually from 25 percent to 50 percent — there is one problem with using an intensifier. This device can only be made for single-stroke applications, meaning the pistons can only push down on the hydraulic fluid once with any effect. The pistons can go up and down, but they cannot build up pressure beyond the first stroke. The result is that hydraulic systems that need more than one stroke cannot use this device.

Aside from increasing the pressure of hydraulic systems, a hydraulic intensifier also can decrease the size of the cylinder. This is useful for machines that cannot use a very large cylinder without detriment, such as a jack. At the same time, large intensifiers can be used to increase the force of large cylinders.

Small hydraulic systems generally do not have hydraulic intensifier devices, because small jobs rarely need the extra force. For example, a car hood that uses hydraulics to keep the hood up would not use an intensifier, because the increased force would either do no good or actually be detrimental by damaging the hood. An intensifier normally is only used for medium or large hydraulic systems that are used for lifting or crushing heavy items.

Hydraulic and Water Resources Engineering

Water resources engineering is the quantitative study of the hydrologic cycle -- the distribution and circulation of water linking the earth's atmosphere, land and oceans. Surface runoff is measured as the difference between precipitation and abstractions, such as infiltration (which replenishes groundwater flow), surface storage and evaporation. Applications include the management of the urban water supply, the design of urban storm-sewer systems, and flood forecasting.

Hydraulic engineering consists of the application of fluid mechanics to water flowing in an isolated environment (pipe, pump) or in an open channel (river, lake, ocean). Civil engineers are primarily concerned with open channel flow, which is governed by the interdependent interaction between the water and the channel.

Applications include the design of hydraulic structures, such as sewage conduits, dams and breakwaters, the management of waterways, such as erosion protection and flood protection, and environmental management, such as prediction of the mixing and transport of pollutants in surface water. Hydroelectric-power development, water supply, irrigation and navigation are some familiar applications of water resources engineering involving the utilization of water for beneficial purposes. More recently, concern for preserving our natural environment and meeting the needs of developing countries has increased the importance of water resources engineering.

Civil engineers play a vital role in the optimal planning, design and operation of water resource systems. Job opportunities in hydrology and water resources are quite varied.

Hydrology Engineering

Hydrology engineering is the design and construction of equipment and structures to monitor, control and store water. The term "hydrology" refers to the relationship of water to the environment. A hydrology engineer uses engineering principles to design dams for water containment, pumping stations for distribution and levees and bulkheads for flood control.

Early applications of hydrology engineering can be traced back thousands of years. Dams were built on the Nile River to control flooding. The Greeks and Romans built aqueducts and canals to manage water supplies to their cities. Earthworks were built by early civilizations to control river flooding. Water control gates and valves were used to flood or irrigate fields when river levels were adequate, and pumps were designed to move water from rivers or lakes to fields that needed irrigation.

The industrial revolution brought a more structured approach to hydrology engineering. Water was used to power grain mills and simple mechanical devices. Dams and piping were needed to supply growing cities with fresh water, and electricity was generated from these same storage systems. Sanitary systems were needed to manage a growing waste problem. Growing cities expanded into areas that were undeveloped because of flooding or high water tables, resulting in the need for flood and drainage control.

The roles of the hydrologist and hydrology engineer diverged to a certain extent during the 20th century. Hydrologists typically are more focused on water resource management, flood planning and the impacts of weather on the balance of water in a given region. Hydrology engineers use the research and estimates of hydrologists to design and build the systems that provide water, remove waste, control flooding and provide storage for current and projected water needs.

A growing interest in hydrology has resulted in some universities recognizing hydrology and hydrology engineering as degree programs. Many hydrologists and engineers who have graduated since the 1930s might have studied civil engineering and pursued their interest in hydrology after graduation. Hydrology engineering also is critical for proper design of agricultural irrigation and flood control and has resulted in university studies connecting agricultural engineering and hydrology.

Hydrology engineering affects many other areas around the world. Flood control and river flow management affects bridge design. Harbor and port designs constantly change to accommodate new cargo and passenger ships. Changing global climate in the 21st century and beyond might require massive flood control redesign if proposed global warming estimates become a reality and sea levels rise as predicted. As food production becomes critical in many parts of the world, hydrology engineers and water resources engineers might have a growing responsibility to provide adequate water supplies.

What does a Hydrologist do?

A hydrologist studies the physical properties of the earth's water systems by performing extensive field and laboratory research. He or she may study the role of water in an ecosystem, measure the amount of rainfall in a certain area, or test water samples to determine the presence of pollutants. Most hydrologists work for universities, private research institutions, government agencies, environmental protection organizations, and engineering consultation firms.

Research hydrologists frequently dedicate months or even years to specific projects which require tracking changes in water systems over time. A scientist might determine the amount of ground or surface water in a given area, the rate of precipitation, and the ways in which water is used by people, plants, and animals. He or she might also gather data about the negative or positive effects of human activity on a marine ecosystem.

A hydrologist might take samples directly from lakes, streams, or other water sources to analyze water content and check for pollutants. He or she might also retrieve nearby soil and plant samples to determine the effects of water pollution on living organisms. Hydrologists often conduct several laboratory trials to verify their results and write detailed scientific papers based on their findings.

Outside of research institutions, scientists may work for engineering firms, federal and state governments, and nonprofit environmental groups. Those who are employed by engineering and consulting firms ensure that buildings and industries do as little harm as possible to nearby water sources. They may be responsible for designing efficient, clean water systems for use within a company, or devising waste management programs that do not have negative impacts on the environment. Government and nonprofit hydrologists are often involved in conservation and awareness efforts, teaching people about the dangers of polluting water systems and wasting valuable water resources.

To become a hydrologist, a person must typically have at least a master's degree in hydrology, environmental science, or a related scientific discipline. Many government employees, university professors, and individuals who conduct independent research hold doctoral degrees. Additional licensing or certification is not typically required, though a hydrologist may choose to take a certifying exam administered by a nationally recognized organization, such as the American Institute of Hydrology in the United States.

Most new scientists begin their careers as field and laboratory assistants, learning about practical research techniques firsthand from experienced hydrologists. Assistants may be responsible for setting up experiments, entering data, and interpreting results. After a certain period of time working as an assistant, a hydrologist who proves his or her competency can begin designing new projects and conducting independent research.

Hydrologic Cycle

The Earth as we know it is made up of primarily water, considered the most precious of our natural resources. Snow storms, rain showers and thunderstorms play just a small part of what is known as the hydrologic cycle, or water cycle. The hydrologic cycle is made up of water in all three of its forms: solid, liquid and gas. The hydrosphere is the area which contains all the water in the atmosphere. Every cycle occurs in five different stages.

The hydrologic cycle starts in the evaporation stage. All water on the ground — including lakes, rivers and oceans — is at the mercy of the sun. Sunlight aids in the process to raise the water temperature and eventually change it from a liquid to a gas. As a gas, the water vapor rises into the atmosphere and starts to condense. During the condensation stage, the water vapor condenses on particles in the atmosphere and forms clouds. A smaller version of this process is dew on the grass in the morning.

From condensation to precipitation, the water can come in a variety of forms. Even though this is known as the precipitation stage, hail is one of the forms which can fall. The other likely scenarios are snowfall and rain showers. Once clouds form, upper atmosphere winds spreads the clouds across the entire globe until the clouds cannot hold all the moisture. Weather systems over lakes in the winter can actually stall out and lead to a longer more intense system as it sits over a larger body of water which speeds up the hydrologic cycle.

The infiltration and runoff stages are the final two parts of the hydrologic cycle. Both stages occur simultaneously. When the snow and rain hit the ground, it begins to seep into the ground during the infiltration process. During a flood, the ground has already absorbed the amount of water it can hold and begins to accumulate on the ground. The runoff stage goes from the ground, either above or below ground and flows into rivers and streams. The streams and rivers flow back into lakes and oceans which lead back to the start of the process, evaporation.

CHAPTER 1

RECENT DEVELOPMENTS ON THE SHALLOW WAKE IN AN OPEN CHANNEL FLOW

Arindam Singha and Ramaswami Balachandar

INTRODUCTION

Numerous flows occurring in nature can be regarded as shallow [1]. In these flows, the length scale in the horizontal direction is much greater than that in the vertical direction [2]. Examples of shallow flow can be found in rivers, headlands, estuaries, stratified lakes, and coastal seas [1]. Heat dissipated from the earth can bring forth density variation in the vertical direction which can induce shallow flow over mountains and hills [3]. Knowledge of the characteristics of shallow flow and its transport capacity is crucial in modelling and predicting the flow, besides assisting in analyzing the dispersion of heat, pollutants, and biological species. Furthermore, it would also aid in weather modelling as the waterbeds and marshlands play a crucial role in controlling the local weather condition [2].

From the perspective of fluid dynamics, shallow flows display important features due to the presence of the bounding surfaces in the form of the bed and the free surface. The bed imparts vertical shear; while the free surface acts as a stress-free weak boundary. A typical shallow wake can be generated by introducing a disturbance in the form of a bluff body in an otherwise plane flow. The resulting separation of flow from the sides of the body though similar to a deep wake would be modified due to the effect of the imposed boundary conditions. In the subsequent sections, the effects of all these are introduced step by step. Initially, some of the key features of the shallow open channel flow are discussed. This is very crucial since the open channel flow is the background flow for the shallow wake. Then the cause and effect of various factors are discussed parametrically.

Citation: Arindam Singha and Ramaswami Balachandar, "Recent Developments on the Shallow Wake in an Open Channel Flow," Advances in Civil Engineering, vol. 2010, Article ID 969407, 12 pages, 2010. doi:10.1155/2010/969407.

SHALLOW OPEN CHANNEL FLOW

Open channel flows can be classified as shallow when the vertical length scale of the flow (usually the depth, H) is significantly smaller than the horizontal/transverse length scales [2]. This definition is based on the kinematic condition, but in turn leads to certain dynamical aspects of the flow. Jirka and Uijttewaal [2] had listed a number of dynamic events that can be observed because of the aforementioned constraint. The main complexity of the shallow channel flow arises from the fact that disturbances from the bed may reach the free surface, and vice versa, as a result of vertical confinement. For a conventional deep flow, this interaction is either absent or minimal. Separate quantification (mainly in deep channel flow) of events occurring at bed or at free surface is fairly mature, but the nature of the interaction between these distinct events is not clear enough. The key point of understanding shallow open channel flow may lie in analyzing the nature of this interaction in greater detail.

The near-bed region of a typical channel flow is mainly governed by the wall turbulence produced by the effect of bed shear. This region is highly chaotic, three-dimensional, and a number of dynamical events are found to take place at this region. Numerous studies have been devoted towards exploration of the effect of boundary layer in an effort to understand the associated dynamics of the near-bed mechanisms (e.g., [4–7]). Mainly, the bed region is occupied by negative streamwise fluctuations with fluid particles being lifted away from the bed by positive bed-normal fluctuations (ejection) and positive streamwise fluctuations moving towards the bed (sweep). Above the near-bed region, turbulent bulges of a length scale of the order of two-to-three wall units can be observed [8]. In addition, structures inclined about 45° to the bed, resembling a horseshoe, appear in the bed, named as hairpin vortices by Theodorsen [9]. Another picture of the near-bed region was put forward by Kline and Robinson [10], which postulates three separate layers of flow close to the bed: low speed streamwise streaks submerged in the viscous sublayer, various flow interaction events corresponding to ejection and sweeps, and above all, three-dimensional turbulent bulges and hairpin vortices. Recent PIV measurements attempt to resolve this discrepancy and found evidence of fluid structures using analysis based on swirling strength, conditional averaging, and two point auto-correlation functions [11, 12]. Lin et al. [13] used proper orthogonal decomposition technique to prove the role of the large-scale structures in the flow features and turbulent structures in the channel flow. Most recently, Roussinova et al. [14] used the same technique on PIV data to expose signature of horseshoe vortex, sweep, and ejections for a typical shallow channel flow case.

The effect of free surface on shallow channel flow was investigated by a number of researchers [15–17], and the general consensus is that the free surface plays an important role in redistributing the turbulent kinetic energy of the flow. Close to the surface, the surface-normal velocity has to diminish to zero, except for the case of waves or flow with inclined surface. The free surface, being stress-free (assuming zero wind-induced stress on the top) boundary, plays a major role in this redistribution process. But a wide range of opinion exists about where the redistribution takes place. Hunt and Graham [18] indicated that the redistribution takes place in two separate regions directly beneath the free surface: the outer source region and the inner viscous region. According to them, the normal to surface component of turbulence decreases and the surface-parallel component increases inside the outer layer. The thickness of the outer layer is of the order of the integral length scale. The inner layer lies in between the outer layer and the free surface and the vertical velocity rapidly decays to zero to the free surface within this layer. Walker et al. [19] estimated the thickness of these two layers as approximately one and one-tenth of the turbulent integral length scale. Contrary to the opinion of Hunt and Graham [18], they indicated that the energy redistribution takes place only in the large-scale structures of the flow, but the small-scales remain unaffected. However, experimental evidence put forward by Balachandar and Patel [20] shows that for low turbulent intensity, the whole flow field can be affected by the boundary condition imposed by the presence of the free surface. They concluded a change in the Cole's Π parameter as well as rate of production of turbulent kinetic energy due to the presence of free surface disturbance.

Like the near-bed region, it is equally important to identify and characterize the fluid structures associated with the free-surface dynamics, and a number of studies are directed towards this objective. At this point, it must be noted that investigation of near-surface structures can be very difficult experimentally (appearance of surface waves, bubbles) as well as computationally (lack of proper model). Generally speaking, close to the free surface mainly two types of structures can be observed: surface-normal vortex tubes with predominantly vertical velocity component and surface-parallel vortex tubes buried inside the surface layer. Physically, it appears relevant to assume a direct linkage between the surface normal vorticity, and the surface undulations via the "pressure" exerted by the vorticity patch. Dommermuth [21] measured the free surface deformation, and vorticity magnitude simultaneously to conclude the existence of a rather weak correlation between them, and it was supported experimentally by Zhang et al. [22]. But Weigard and Gharib [23] used simultaneous shadowgraph and PIV results to show that the location of the peaks of vertical vorticity matches well within range with the location of free surface deformation. Along the same line, efforts were made to computationally examine the free surface flow structures, and the first direct simulation study was reported by Lam and Banerjee [24]. Worth mentioning are the studies of Komori et al. [25] and Handler et al. [26], which used a full free surface model unlike the case of Lam and Banerjee [24] who used a rigid-lid assumption. The most important and comprehensive study was carried out by Pan and Banerjee [27] which classified the structures into three distinct groups: vortices, upwelling, and downdraft. Signatures of these types of structures were also documented experimentally by Gupta et al. [28], but they did not effectively classify them. "Vortices" were defined as the region of cluster of surface-normal vorticity, where the ratio of surface-normal to the surface-parallel velocity is approximately zero. "Upwelling" was defined as the region of divergence of streamlines with large surface normal velocity. "Downdraft" is the region, which helps to nullify the effect of the upwelling and exhibit strong downward flow component from the free surface. These three types of structures were found to appear in a certain pattern: typical upwelling reaches the free surface and interacts to generate a donut-shaped vorticity pattern there. Upon interacting, these patterns generate a number of short-lived smaller vortices around the edge of the upwelling. A typical pattern concludes with the downdraft to wash away the excess fluid close to the free surface. The energy spectra from the locations close to the free surface show $k^{-(5/3)}$ and $k^{-(7/3)}$ region, which are characteristic of two dimensional turbulence and indicate occurrence of turbulent kinetic energy redistribution. However, at a distance from the free surface, usual characteristics of three-dimensional turbulence prevail.

Note that it is the unique nature of the interaction of these events originating from the bed and free surface that makes the shallow channel flow distinctive from the conventional open channel flow. As a result of this interaction, a clear separation of scales can be observed. Jirka and Uijttewaal [2] has aptly summarized the following: *"Shallow flows are largely unidirectional, turbulent shear flows driven by piezometric gradient and occurring in a confined layer of depth. This confinement leads to the separation of scale, three-dimensional turbulence $l_{3D} \leq H$, and large scale twodimensional turbulent motion, $l_{2D} \geq H$, with some mutual interaction."* Two plausible but much generalized solutions were provided by Nikora et al. [29] and Shen et al. [30]. According to Nikora et al. [29], due to smaller vertical extent of the flow, events from the bed reach the free surface and vice versa. When three cylinder-like fluid structure becomes sufficiently long to reach the free surface, they appear as smaller tornado-like vertical cylinders submerged in the channel flow. The complex interaction between these cylinders with the flow generates these two-dimensional structures. Shen et al. [30] indicated that the hairpin eddies from the bed may reach the free surface and, when the head part dissipates away due to the convection effect of the mean streamwise flow, and the remaining leg part of the hairpin vortex gives rise to these two-dimensional structures.

SHALLOW WAKE

Shallow wake has basically the same general characteristics of a deep wake. Upon accroaching a bluff body, a stagnation point along the upstream edge occurs with boundary layer along the wall. When the flow encounters sudden change in geometry, of an adverse pressure gradient, separation occurs from the sides of the body, giving rise to a von Karman vortex street. A detailed description of the deep, conventional wake can be found in Zdravkovich [31]. Whereas the overall broad flow picture remains the same, shallow flow imposes a number of different characteristics, which are crucial for the overall development of the flow features. First, the limited depth would restrict the onset of three-dimensional breakdown of the vortex street. As a result, the von Karman vortex street can be observed for a larger range of Reynolds number in a shallow wake. For example, the pattern of crude oil spillage from the ship Argo Merchant stranded at Nantauket shoals in Massachusetts shows a clear vortex street pattern at a very high Reynolds number $Re \approx 10^7$ [32]. Secondly, the bed friction tends to arrest the transverse growth of the disturbances associated with a typical vortex street pattern and stabilize the wake. This may occur either in the near-wake or in the far-wake region. For the case of a shallow wake, the stability number, as will be shown later, determines the effect of the bed friction and the subsequent wake stability. In the near wake region, if the stability number $S_{near} = C_f D/H$, where C_f is the skin friction coefficient, D and H are the characteristic diameter of the body and flow depth, respectively, exceeds a critical value, wake stabilization takes place. In the far-wake region, stability number is defined as $S_{far} = C_f(2b)/H$, where b is the local wake width. Similar wake stabilization takes place at the far wake region if S_{far} exceeds a certain critical value.

Stability Number

Crucial to the development of the shallow wake is the mutual interaction of the vertical shear due to the approaching flow with the horizontal (transverse) shear due to the variation of momentum deficit in the transverse direction. Therefore, the spatiotemporal growth of the shallow wake is controlled by their mutual interaction. The shallow wake is maintained in an equilibrium state as a balance of two different effects: on one hand, turbulent kinetic energy is extracted from the mean flow and is fed to the large-scale coherent structures by interaction with the lateral shear layer. On the other hand, in the bottom boundary layer, the turbulent kinetic energy is extracted from the large-scale structures and dissipated through the small-scale structures according to Kolmogorov theory of local isotropy. Thus, one can balance the loss of the turbulent energy due to the bottom friction by the production of the turbulent kinetic energy by the lateral shear layer. The energy fed into the large-scale coherent structures can be computed as the product of the coherent velocity fluctuations and the transverse mean stream wise velocity gradient as [33],

$$P = \langle u_c w_c \rangle \frac{\partial U}{\partial z}, \tag{1}$$

where the subscript "c" denotes the velocity associated with the coherent motions only. u and w indicate the instantaneous velocity in streamwise and transverse directions, respectively. Further, x- and z- denote the streamwise and transverse directions, respectively. The produced energy is dissipated through the small-scale fluctuations at the bottom boundary layer. Babarutsi and Chu [33] showed that if the velocity associated with the coherent structure is small compared to the mean flow, then the dissipation can be approximated as

$$F_b = \frac{C_f}{2H} U(2\langle u_c u_c \rangle + \langle w_c w_c \rangle), \tag{2}$$

where C_f is the skin friction coefficient and H is the depth of flow

Dividing the above two equations, the flux stability number can be approximated as

$$S_{\text{flux}} = \frac{F_b}{P} = \frac{C_f U}{2H(\partial U/\partial z)} \frac{2(\langle u_c u_c \rangle + \langle w_c w_c \rangle)}{\langle u_c w_c \rangle}. \tag{3}$$

The ratio on the right represents the ratio of the coherent Reynolds normal and shear stress, respectively. Chu et al. [34] proposed a gradient stability number for shallow flow, which has an inflexion point in the transverse distribution of the streamwise velocity, that is, $\partial^2 U/\partial z^2 = 0$. Without affecting the calculation, the term at the extreme right can be dropped safely and the velocity gradient can be expressed as $\Delta U = U_1 - U_2$, where U_1 and U_2 are characteristic velocities on each side of the shear layer of local thickness δ. Thus, the gradient stability number

$$S_{\text{grad}} = \frac{C_f}{H} \frac{U}{\partial U/\partial y}\bigg|_{\text{Inflexion}} \approx C_f \frac{\delta}{H} \frac{U_{\text{avg}}}{\Delta U} = \frac{C_f}{2R} \frac{\delta}{H}, \tag{4}$$

where

$$R = \frac{U_1 - U_2}{U_1 + U_2} = \frac{\Delta U}{2U_{\text{avg}}}. \tag{5}$$

This equation is applicable for any general shear layer. For the case of a shallow wake behind a bluff-body, it can be applied at the lee end of the recirculation bubble. Since $U_2 \approx 0$, $R = 1$. The width of the shear layer is approximated to be half of the characteristic diameter of the body (D/2). Using these, the wake gradient stability number can be found as

$$S_{\text{grad}} = \frac{1}{4} C_f \frac{D}{H}. \tag{6}$$

The wake stability number (S) is defined as four times the gradient stability number for sake of convenience,

$$S = 4S_{\text{grad}} = C_f \frac{D}{H}. \tag{7}$$

Unlike the gradient stability number, stability number (S) is a global parameter, and it does not depend on the choice of local coordinate system. Therefore, it is a more universally accepted number and mentioned extensively in shallow wake literature.

Effect of the Bed

As mentioned earlier, the uniqueness of a shallow wake depends on the combined effect of the bed and the free surface. Although these two effects are closely interrelated, it may be possible to view them independently. Wolonski et al. [35]

defined an island wake parameter, $P_{island} = U_a H^2/k_z D$, where U_a is the ambient flow velocity and k_z is the vertical eddy. The vertical eddy viscosity was approximated as $k_z = 0.15HU_\tau = 0.15HU(C_f/2)^{1/2}$, where U_τ is the friction velocity. They have suggested that a critical value of $P_{critical}$ of the order of unity exists, and a vortex street type of wake is not possible for $P_{island} < P_{critical}$. Their classification is somewhat incomplete and lags descriptive information, as they are based on satellite imaging and temperature mapping of the island wake and no direct measurements were conducted to bolster the classification.

An important investigation in shallow wake was performed by Ingram and Chu [36] by observing the oceanic shallow wake behind Rupert Bay, Ontario. Initially, they formulated an analytical description of the effect of the bed on the shallow flow. To complement the formulation, they performed flow visualization in the laboratory. A total of six laboratory observations and 26 events of oceanic shallow wakes at Reynolds numbers (Re = 4HU/ν) ranging from 4,700 to 11,000, and stability numbers ranging from 0.03 to 0.65 have been reported. They introduced the stability number as a measure of the bottom friction effect on the shallow wake. Based on a series of observations, they postulated a critical stability number of 0.48, above which the oscillating characteristic of the wake gets stabilized due to the bottom friction. Below the critical stability number, the effect of bed friction is not strong enough to stabilize the wake and vortex street type of wake can be observed. Figure 1 is adopted from the paper and depicts three different cases; (a) Re = 11, 000, $S_w = 0.031$, (b) Re = 11, 000, $S_w = 0.013$ and (c) Re = 4, 700, $S_w = 0.054$. Clearly seen in this figure is the suppression of the wake oscillation with change in the Reynolds number and stability number. The frictional effect of bed is also shown to affect the entrainment coefficient of the wake. However, the range of Reynolds number and the stability number for the investigation is not large enough to draw firm conclusions about the state of the wake.

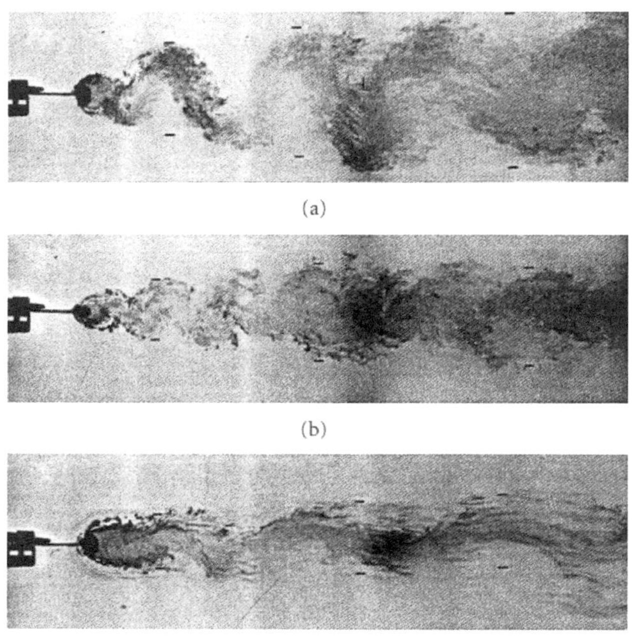

(a)

(b)

(c)

Figure 1: Qualitative demonstration of the effect of the bed on the development of shallow wake (Reproduced with the permission of American Geophysical Union).

Along the same lines, extensive laboratory experiments (57 with circular cylinder, 54 with solid plate, and 34 with porous plate) and visualization of shallow wake were performed by Chen and Jirka [37] covering a wide range of Reynolds number and stability number (Re_h = 570−7, 300, Re_d = 3, 000−4, 14, 000, S = 0.01−0.94). Based on visual observations, they divided the shallow wake into three different categories: vortex street type (VS), unsteady bubble wake (UB), and steady bubble wake (SB). If the perturbation due to the separating shear layer is dominant, the shallow wake resembles the well-known von Karman vortex street as seen in a conventional deep wake. Although qualitatively similar, the vortex street of shallow wake and the conventional von Karman vortex street have differences too. The vortex street (VS) shows a streaky and fuzzy appearance due to the presence of the small-scale structures arising from the bed frictional effect. This fuzziness sometimes causes blurring out of the well-observed roll up of the vortices. This type of wake is visible for $S \leq 0.2$. If the stability number is increased ($0.2 \leq S \leq 0.5$), the bed friction effect increases and the flow separates from both sides of the body simultaneously, not alternatively. A nearly steady bubble is formed immediately downstream of the body. But the end of the bubble oscillates alternatively and this leads to periodic breakdown of the vortices. This is called the unsteady bubble (UB) wake. The steady bubble (SB) mode can be observed when the bed friction effect decreases the momentum to such an extent that the recirculation bubble stays almost stable at the lee of the body and no shedding is observed. By analytical approximation of the wake momentum deficit, Chen and Jirka [37] defined a length scale L beyond which the wake momentum deficit disappears

$$L = \frac{2H}{C_f} = \frac{2D}{S}. \tag{8}$$

Unstable shallow wakes get stabilized after a streamwise distance of the order of L, mainly due to the increasing momentum deficit, and subsequent increase in the stability parameter. They also observed that the width of the shallow wake greatly exceeds the depth of the flow, and as a result the flow structures are mainly two-dimensional in nature. They suggested that the majority of the effect of the bed friction is agglomerated at the bottom boundary layer and may not contribute towards the three-dimensionality of shallow flow. The bed friction effect is averaged over the full depth of flow and is manifested in the form of the wake stability number.

Negretti et al. [38] mounted artificial roughness mats on the bed of a shallow wake flow to investigate the effect of the increased roughness. Two series of experiments were conducted; for the first series, roughness elements were placed in the lateral shear layer region and for the second series, the roughness elements were placed at the end of the recirculating bubble. For the first case, roughness was found to decrease the magnitude of the Reynolds stress in addition to narrowing the wake width. Also, the wake behaved very similar to a stable bubble wake, since the introduced roughness reduces the momentum exchange with the ambient flow. When the roughness element was placed at the back of the re-circulating bubble, the usual wake behaviour was observed till the flow reached the roughness and the fluid structures were destroyed while flowing over the roughness. This study explicitly indicates that the bed roughness has the ability to alter the near-wake as well as the far-wake structure. Mainly these three studies investigated the effect of bed friction only, and explicitly, although a number of investigators examined the effect of bed friction as a part of a study intended to look into some other aspect of the shallow wake.

Effect of Free Surface

Quantification of the effect of the free surface is always difficult both computationally and experimentally. Computationally speaking, until recently there was no proper model available for simulating the free surface other than imposing the rigid-lid assumption on it. Experimentally speaking, using a laser is very difficult because of the wavy characteristics of the free surface and the appearance of the bubbles. Despite these difficulties, a number of studies were performed in this direction and were mainly concerned about exploring the relationship between the development of free surface deformation and turbulent anisotropy [17, 19, 39–41].

Logory et al. [41] used LDV and PIV to investigate the effect of free surface on an otherwise deep wake. Due to the large water depth, the effect of bed is expected to be confined close to the bed and not penetrate beyond a certain vertical distance from the bed. As a result, close to the free surface, the sole effect of free surface will be realized with reduced contamination arising from bed friction. In fact, LDV measurement of the wake flow behind the bluff body, measured at half-depth, matches well with the published data for the deep wake. But close to the free surface (depth 12~13 mm below), the wake width was found to be doubled in size, while the centerline velocity remained almost constant. The contour of the mean streamwise velocity in the streamwise plane downstream of the body, adopted from Logory et al. [41] is shown in Figure 2. The self-similar behavior of the mean and root-mean-square velocity is also lost at the free-surface location. PIV measurements at a depth ~20 mm below the free surface showed evidence of "upwelling" and "downdraft" as seen in a shallow flow, in addition to the large-scale structures originating due to the presence of the body. At the near surface layer, opposite signed streamwise vorticity was found to exist in pairs at both sides of the wake centerline, which induces flow from the centerline towards the ambient flow region. In the deep wake, vortex structures are predominantly two-dimensional, but close to the surface, the vortex structures tend to reorient themselves in either surface-normal or surface-parallel direction. The magnitude of the surface normal vorticity was found to be smaller compared to the deep wake region. The origin of these structures is believed to be the quasi-streamwise rollers originating from the turbulent boundary layer on the bluff body. The existence of these surface-normal structures was also confirmed by the LIF (laser induced fluorescence) flow visualization. A histogram of the vorticity of the surface-parallel structures revealed that they are inclined at 45–50° angles from the wake centerline and induce a velocity away from the wake centerline directed to the upstream direction. The resulting motion from this complex vorticity distribution causes the wake widening at the near-surface region as the surface-parallel structures carry low momentum fluid towards the ambient flow region.

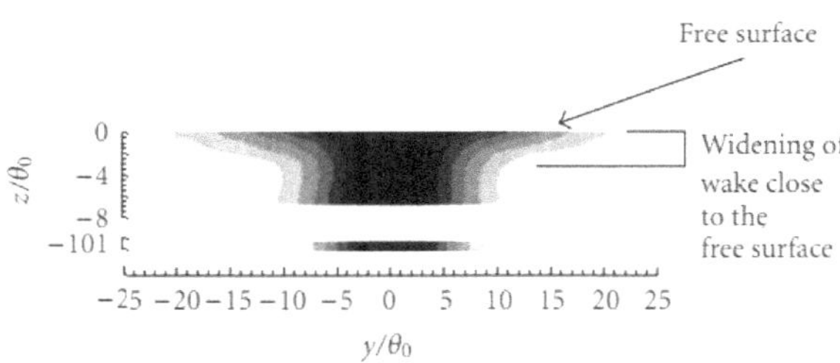

Figure 2: Widening of the wake close to the free-surface (Adopted from Logory et al., [41]). The contour color represents mean velocity; white to dark shade represents higher to lower range of streamwise velocity. θ_0 (= 3.1 mm) is the half-momentum thickness of the wake (Reproduced with the permission of American Institute of Physics).

Based on the available literature and his own analysis, Maheo [17] summarized the near surface dynamics of a typical shallow flow, for the case of turbulent jet and wake. He found that the free surface only affects the velocity components in a thin layer close to the surface. The thickness of this surface layer was approximately either one-half of the local width or the vorticity thickness of the wake or jet, respectively. Inside this layer, the fluctuation in the vertical direction gets retarded and the turbulent kinetic energy in the vertical direction was redistributed to the horizontal velocity components in the streamwise and transverse directions. The streamwise vorticity tubes buried inside the surface layer were found to generate surface currents which affect the dynamics of flow. Figure 3 shows a typical signature of the surface currents close to the free surface. The motions of these structures contribute towards widening of the surface shear flows, compared to the deep flow case. He found that the widening of the turbulent shear layer at the near surface can be as much as 20% for a turbulent wake and 25% for a turbulent jet.

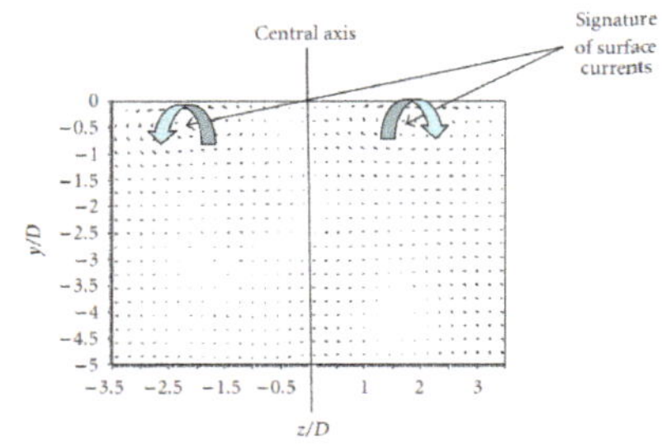

Figure 3: Signature of the surface currents as seen in a plane normal to the streamwise direction downstream of a bluff body. (Reproduced by permission from Maheo).

Effect of Approaching Velocity

The incoming open channel flow approaches and senses the presence of the body. When the approaching flow meets the body, a vertical stagnation line is generated and the total flow divides along the upstream sides of the body. The resultant stagnation pressure at the body at a certain elevation is higher than at an elevation closer to the bed due to the nonuniform incoming velocity profile. Consequently, a vortex tube is generated and elongated along the sides of the body, due to the accelerating flow region (Region III, [31]) at the sides of the body. This particular system of vortex tube is called the horseshoe vortex, and it plays a crucial role in the development of shallow wake. Figure 4 illustrates a typical flow model of shallow wake. The sign of vorticity of the approaching flow as well as the horseshoe vortex is shown. Due to the complex interaction process of the bed and free surface with the flow structures, a shallow wake has a considerable three-dimensionality associated with it.

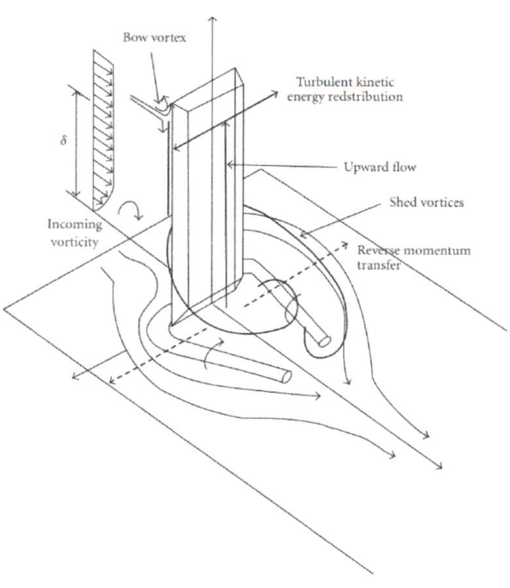

Figure 4: A simple model of the shallow wake, as constructed from the available literature.

Rao et al. [42] investigated the effect of the approaching velocity on the bed-structure interaction for a typical shallow wake by a qualitative dye visualization study for flow depths of 150, 200, and 250 mm, respectively. The maximum stream wise velocity and Reynolds number ranges were U = 0.08–0.21 m/s and Re_d = 1500–4400, respectively. Dyes of different colors were injected into the flow through ports situated at different elevations on the cylinder. Figure 5 shows the side view of a circular cylinder, immersed in a shallow channel flow of 100 mm depth. The red dye from the lower port gets entrapped into the horseshoe vortex and followed the vortex pattern closely. The blue dye follows the vortex shedding from the side of the body. A distinguishing feature of the horseshoe vortex is the upward three-dimensional rollup of the horseshoe vortex and the subsequent interaction with the vortex street emanating from the sides of the body. This interaction leads to the three-dimensionality and vertical variability of the shallow wake. Amongst different conclusions, they found a vertical oscillation of the shear layer at the sides of the body, the source of which was tracked back to the horseshoe vortex wrapping around the body. As mentioned earlier, they connected the dynamics of the horseshoe vortex to be the possible source of the three-dimensionality of the shallow wake. However, the interactions needed to be quantified.

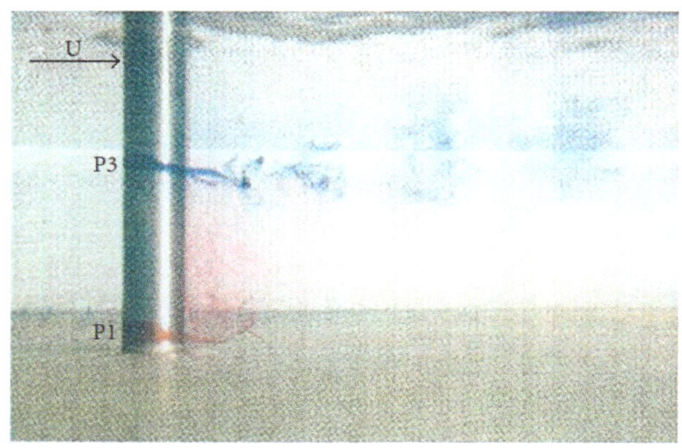

Figure 5: Qualitative demonstration of the dynamics of horseshoe vortex around a circular cylinder as the source of three-dimensionality of shallow wake (Reproduced by permission from Visualization Society of Japan).

Wang et al. [43] investigated the effect of variable momentum thickness of the approaching boundary layer on different aspects of the downstream wake flow. The momentum thickness of the approaching boundary layer was adjusted to be 0.07D, 0.13D, and 0.245D, where D (=20 mm) is the characteristic width of the body. The Reynolds number of the flow was Re_D = 11, 500. In this experiment, the flow was allowed to move over the top of the body, that is, the body does not pierce the free surface. This does not really comply with the standard definition of a shallow wake, but some of the conclusions may be worth mentioning. With increasing momentum thickness of the approaching boundary layer, an upward flow originating from the bed and towards the free surface was noticed. This upward flow contributed towards the reduction of the reverse flow region close to the bed. At the same time, the magnitude of the Reynolds stress decreased close to the bed. Also, the variable thickness of the approaching flow was found to alter the nature of the vortex shedding from the body (symmetric/anti-symmetric). These studies indicate that the approaching flow has a significant influence on the shallow wake flow.

Two-Dimensional Coherent Structures of the Shallow Wake

The turbulent channel flow can be disturbed due to several reasons which can result in oscillation and instability of the flow. If the strength of these instabilities sometimes exceeds a certain threshold, this can grow in space extracting energy from the mean flow and grow in size. During the process of growth, sometime, the transverse size of these eddies may exceed the vertical flow depth. Due to purely kinematic reason, the vorticity vector of these eddies, termed as two-dimensional coherent structures (2DCS), must be perpendicular to the bed or the free surface. Of course, smaller coherent structure whose size is less than the depth of flow contains no preferential direction of orientation and exhibits three-dimensional behaviour. The appearance of the 2DCS has been consistently reported by a number of researchers [37, 44–46].

Holmes et al. [47] and Bonnet et al. [48] reviewed the recent literature on the importance and the characteristics of coherent structures in different kinds of turbulent flow. Although there is no general consensus about the definition of a coherent structure, Hussain [49] formed a working definition as follows: "A coherent structure is a connected (large-scale) turbulent fluid mass with an instantaneously phase-correlated vorticity over its spatial extent". Carmer and Jirka [50] showed that the transport capacity of these large-scale structures is about two-orders of magnitude higher than the small-scale structures. Therefore, characterization of the large-scale coherent structures will lead to a better modeling of discharge and dispersion of pollutants released in shallow rivers or lakes.

Jirka [51] has indicated three different types of forcing mechanisms for the generation of the large-scale coherent structures:

Topological Forcing (Type A)

Presence of a body (e.g., island, headlands, jetties, etc.) in a shallow flow represents the strongest mechanism for generation of 2DCS. With development of these eddies, the interaction between them and the bed increases. As a result, more energy is extracted from the mean flow and is fed to the rotational energy of the eddy. In this case, turbulent kinetic energy flows from the smaller to the larger scale and therefore exhibit inverse energy cascade.

Internal Shear Flow Instability (Type B)

Varying momentum flux (deficit or excess) due to the velocity gradient in the lateral direction may trigger this type of instability. Such triggering action can be caused by the presence of a source flow like shallow jet, wake, or mixing layer.

Secondary Instability of the Base Flow (Type C)

This is the weakest type of instability and experimental evidence is sparse (except for [29]). Consider the case of a shallow channel flow containing a vertically sheared boundary layer and 3D turbulent coherent structures. This flow will be maintained in equilibrium by the bed friction. However, a slight imbalance in the flow will lead to the redistribution of the turbulent kinetic energy in the entire range of wave numbers. Supercritical flow (inertia dominated gravity forces) gives a 2D turbulence inverse cascade, while subcritical flow (gravity dominates inertial force) gives direct cascade of energy. The distortion of the vortex line caused by the secondary instability will lead to the formation of 2DCS provided the magnitude of the disturbance exceeds a certain critical level.

Following the work of Dracos et al. [44] and Uijttewaal and Tukker [45], the generation and development of the 2DCS can be divided into a number of stages in time and space. The "near-field" of any shallow flow is characterized mainly by the two-dimensional transverse shear layers and contains small-scale three-dimensional structures. The "middle-field" has strong three-dimensionality because of the interaction of the bed and free surface. At the "far-field", the structures generated earlier grow in size greater than the depth of flow. To preserve the kinematic condition, these eddies must have two-dimensional character and will have dominant vorticity vector perpendicular to the free surface. Also, the mean flow, whose transverse scale is of the same order as the large turbulent eddies, became approximately two-dimensional.

Following the generation, the 2DCS grows and develops by entraining outside ambient fluid. In addition to this, two or more 2DCS may merge together to generate a larger 2DCS [45]. It must be mentioned here that the limited depth restricts the mechanism of vortex stretching that is present in three-dimensional turbulent flow. Merging of 2DCS to form larger 2DCS leads to the flow of turbulent kinetic energy from the high to the small wavenumber range and therefore exhibits inverse energy cascade [52]. The possible existence of the two-dimensional turbulence in shallow flow was demonstrated by Dracos et al. [44] and Chen and Jirka [37] by computing the velocity spectra. The 2DCS lose their energy owing to the effect of the bed friction. Dimensionally, it can be shown that an eddy of size $2H/C_f$ will loose all its energy in one rotation due to bed friction [50].

Other Related Literature and Vertical Variability of Shallow Wake

There are other studies in the literature related to the shallow wake. These include, for example, searching for suitable scale to absorb the effect of bounding surfaces, and investigation of the variation of the flow characteristics as a function of the vertical elevation from the bed. Balachandar et al., [53] classified the shallow wake into shallow-shallow and deep-shallow wake based on visual observation and video-imaging of dye concentration injected in shallow wake. The depth of flow, mean free stream velocity, and Reynolds number were varied between 10–40 mm, 76–157 mm/s, and 1570–4810, respectively. When the stability number was lower than a critical value ($S_w < 0.008$) then the von Karman type vortex street was observed. For stability numbers greater than the critical value, either intermittent vortex street was observed or was completely annihilated. This classification is more like that of Wolansky [54] for island wake as discussed earlier. Figure 6, adapted from Tachie [55], shows an example of the deep-shallow and shallow-shallow wake, respectively. The left figure corresponds to higher depth of flow (H/D = 2.00), and the right figure corresponds to lower depth of flow (H/D = 0.25). The vortex street can be seen for the case of higher depth of flow, but it is completely annihilated for lower flow depth due to increased bed friction effect. In a later study to characterize turbulent shallow wake, Balachandar et al., [56] performed velocity measurements downstream of a flat plate immersed in shallow channel flow. Measurements were carried out at three different vertical locations, H/4, H/2, and 3H/4. Several normalizing parameters were attempted in an effort to collapse the mean streamwise velocity defect at different streamwise locations and depth of flow. $C_d\delta_v$ was identified as one of the proper normalizing length scales, where C_d and δ_v are the drag coefficient and the half-width of the wake, respectively. At a given distance from the bed, the approaching velocity upstream was found out to be one of the normalizing scales for streamwise velocity deficit, for which the data collapse fairly well onto a single profile for all streamwise and vertical measurement locations. It should be mentioned here that for the deep wake, the approaching velocity is constant with vertical distance from the bed, whereas for shallow flow, it is a function of the vertical distance. A friction parameter β was proposed to represent the relative effect of the bed friction and transverse shear. Above a critical value, $\beta_c = 0.2$, the organized motion of the vortices began to disappear

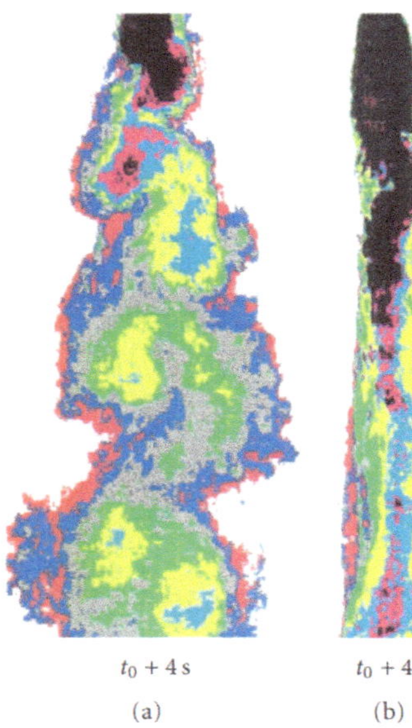

Figure 6: Example of deep-shallow and shallow-shallow wake (Reproduced by permission from Tachie).

To quantify the effect of the bed roughness, Tachie and Balachandar [57] measured velocity distribution of a shallow wake generated on a smooth and a rough wall. The roughness was found to affect both the mean and turbulence parameters of the of the wake flow. For approximately the same value of Reynolds number, the width of the wake was found to be much smaller for the rough bed compared to the smooth bed. Also, the ratio of the transverse shear to the bed friction was found to be smaller for the case of the rough bed. Shallowness of the flow was also found to enhance the turbulent fluctuations in the flow.

In a series of papers, the general features, stability, and control of shallow wake have been further addressed [58–60] using PIV. The experimental procedure for all these investigations is more-or-less similar; a circular cylinder is placed in a shallow channel flow. In the first paper, the focus was on the spatiotemporal evolution of the vortex formation process. In the second and third investigations, they addressed the effect of Reynolds number on the flow and explored different possible ways to stabilize the wake, respectively. The shallow wake was found to be characterized by large-scale vertical structures (or 2DCS) emanating from the body downstream of the cylinder. However, their investigation discovered an axial flow from the bed towards the free surface through the core of these structures, supporting the finding of Cohn and Koochesfahani [61]. This flow was believed to contribute significantly to the three-dimensionality of the shallow wake structure. The instantaneous streamline topology and vorticity contours at different elevations from the bed compare well, in a general sense, with a conventional deep-wake, but a signature of the three-dimensionality was also observed. The location of the saddle point in the instantaneous streamline topology was found to be a function of the distance from the bed. With increasing distance from the bed, the spanwise extent of the instantaneous vorticity contour as well as the streamline topology grows in the transverse direction, probably due to the diminishing effect of the bed. Close to the bed, a unique structure of the instantaneous streamline was observed, called as owl face of the first kind [62], which has a very well-defined pair of foci and saddle points. With increasing distance from the bed, the organization of the foci as well as the saddle points is lost. At the near-bed location, the vorticity contour was found to be tilted toward the central region, but with increasing distance from the bed, the tendency of tilt decreases. Additionally, the peak magnitude of vorticity was also found to be dissipated rapidly close to the bed, than at some locations far from the bed. The contour of the time-mean Reynolds stress was found to be very well defined in accordance with the shear layer separation, except close to the bed. In the near-bed region, additional cluster of Reynolds stress was noticed which explains the interaction of the horseshoe vortex with the shear layer at the sides of the cylinder. Unlike the vorticity, the peak magnitude of the Reynolds stress remains almost the same and does not change with the distance from the bed. The stability number of the shallow wake was found to play an important role in the resulting mode of the wake. For example, increasing the stability number leads to the attenuation of the appearance of the large-scale coherent structures in the wake. Despite this, a well-defined varicose instability was observed. The reason of this particular instability can be attributed to the dynamics of horseshoe vortex wrapping of the body. At a even higher stability number, the wake was completely stabilized, but smaller amplitude rotational motions were noticed to set off now and then in the near-wake region. In the third investigation of the aforementioned series, Cohn and Koochesfahani [61] introduced a variable width streamwise slot cut through the cylinder. For smaller slot size, the cluster of vorticity preserved their shape, but appeared downstream compared to the case of no-slot. Also the frequency of undulation of the emanating jet from the slot may affect the shedding frequency of the vertical structures.

At this point, it is obvious that the flow variables of a shallow wake are different at different distances from the bed, and this can be termed as vertical variability of shallow wake. Singha et al. [46] explored the vertical variability of a typical shallow wake in an open channel flow. The stability number of the wake was small (0.0015), and velocity measurements were performed at near-bed, mid-depth, and near-surface vertical locations to reveal the vertical variability. Figure 7 is adopted from the paper and clearly illustrates the difference of the wake structure at different vertical locations. They also identified some key features of the flow: effect of the bed friction to dampen the transverse growth of the wake, existence of sweep and ejection type of events at the free surface contributing towards redistribution of the turbulent kinetic energy at the near-surface location, the effect of the structures at near-bed and free-surface region towards enhanced momentum transfer at these locations. In a continued study, Singha [63] applied POD and closed-streamline based coherent structure identification scheme to detect the location, size, and strength of the coherent structures at near-bed, mid-depth, and near-surface. The finding clearly indicates that the bed as well as the free surface tends to reduce the size and strength of the coherent structures, but the effect is more drastic for the bed rather than the free-surface. The largest and strongest structures can be found at the mid-depth location only. Also, the horseshoe vortex as well as free surface events was found to affect the location of appearance of the coherent structures at near-bed and near-surface vertical locations.

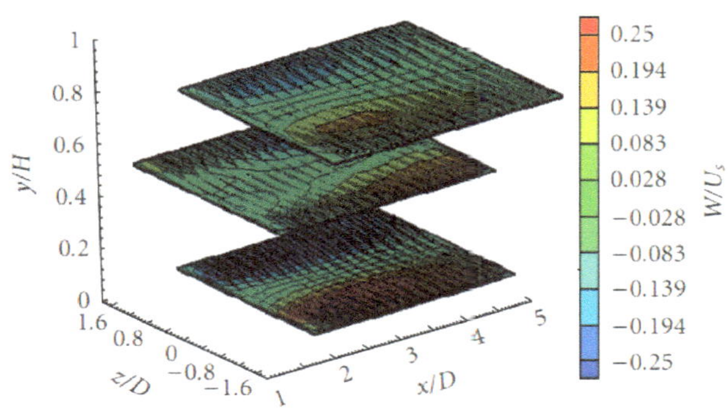

Figure 7: A vector plot representing the mean-flow field in the three horizontal planes (y/H = 0.10, 0.51 and 0.76) superimposed by the contour of transverse velocity component. This figure clearly proves the vertical variability of the shallow wake. In this figure W represents velocity in z-direction.

SUMMARY AND FUTURE PROSPECTS

The related literature on shallow wakes in an open channel flow has been reviewed to highlight the effect of three important parameters, namely, the effect of the bed, the effect of the free surface, and the effect of the nonuniform approaching flow. The possible cause and effect of these three parameters on the uniqueness of shallow wake have been discussed. The origin and development of 2DCS have been reviewed.

Based on the review of the previous literature, a probable model of the flow geometry downstream of a bluff body immersed in a shallow open channel flow can be hypothesized as follows: the vertically sheared approaching flow, incident on the body, encounters the horizontal shear due to the presence of the body. This will generate a typical horseshoe vortex which wraps around the body (effect of bed) and extends in the downstream direction. A typical signature of the horseshoe vortex can be found in Fu and Rockwell [60] based on the Reynolds stress contour downstream of the body close to the bed. Due to the presence of the horseshoe vortex, one would expect a pair of vortices (clockwise for negative z and counterclockwise for positive z) close to the bed in the streamwise plane. Above these pair of vortices, the presence of two large vortices (counterclockwise for negative z, and vice versa) in the streamwise vertical plane (y-z plane behind the body) has been demonstrated by Eibeck [64] which induces an upward flow along the center of the body. Close to the body, within a few diameters downstream, the flow field in the near-wake region can be very dissipative due to intense interaction between the two above-mentioned vortices. However, the legs of the horseshoe vortices will dissipate faster than the other vortices, and then only the second pair of vortices will remain dominant through the entire flow depth. Close to the bed, along the centerline downstream of the body, the flow travels towards the saddle point and then upwards towards the free-surface. This particular axial flow from the nearbed to free surface has been experimentally documented by Koochesfahani [65]. When this flow reaches the vicinity of the free surface, it interacts with the turbulent events that originate near the free surface [14] and leads to energy redistribution [46] between the streamwise and transverse velocity components.

Based on the status of the present literature, the above flow picture starts to evolve, but there still exist missing links in the flow pattern. For example, the flow properties of the wake are different at different elevations from the bed and these properties can be expressed as a function of the distance from the bed, or some dimensionless numbers, which may take into account the damping effect of the bed, or the wall-effect of the free-surface. The unique nature of the vortex shedding for shallow wake, including the effect of the near-bed horseshoe vortex and the axial flow from the bed towards the free surface, has not been addressed in sufficient detail. The examination of the instantaneous streamline topology as a function of the distance of the bed may lead to the possibility of defining a common feature of the topology with variable distance from the bed. Most notable is the appearance of the 2DCS in the shallow wake flow, and the investigation of their characteristics is limited. A robust method of identifying and extracting these 2DCS from the flow is lacking. Once, successfully identified, the effect of the dynamics of these 2DCS on the shallow wake can be addressed in a clearer way.

REFERENCES

1. H. J. D. Vriend, "Complex flows in civil and environmental engineering," in Shallow Flows, G. H. Jirka and W. S. J. Uijttewaal, Eds., Taylor and Francis, London, UK, 2004.
2. G. H. Jirka and W. S. J. Uijttewaal, Shallow Flows, Taylor & Francis, London, UK, 2004.
3. R. S. Scorer, Experimental Aerodynamics, Ellis Horwood Limited, Chichester, UK, 1978.
4. R. E. Falco, "Coherent motions in the outer region of turbulent boundary layers," Physics of Fluids, vol. 20, no. 10, pp. S124–S132, 1977.
5. M. R. Head and P. Bandyopadhyay, "New aspects of turbulent boundary-layer structures," Journal of Fluid Mechanics, vol. 107, pp. 297–338, 1981.
6. S. K. Robinson, "Coherent motions in the turbulent boundary layer," Annual Review of Fluid Mechanics, vol. 23, no. 1, pp. 601–639, 1991.

7. R. J. Adrian, "Hairpin vortex organization in wall turbulencea," Physics of Fluids, vol. 19, no. 4, Article ID 041301, 2007. ·

8. R. F. Blackwelder and L. S. G. Kovasznay, "Time scales and correlations in a turbulent boundary layer,"Physics of Fluids, vol. 15, no. 9, pp. 1545–1554, 1972.

9. T. Theodorsen, "The structures of turbulence," in 50th Jahre Grenzschictforschubg, H. Gortier and W. Tolmien, Eds., Friedrich Vieweg & Sohn, Braunschweig, Germany, 1995.

10. S. J. Kline and S. K. Robinson, "Quasi-coherent structures in the turbulent boundary layer—part I: status report on a community wide summery of the data," in Near Wall Turbulence, S. J. Kline and S. H. Afgan, Eds., Hemishpere, New York, NY, USA, 1989.

11. Y. Wu and K. T. Christensen, "Spatial structure of a turbulent boundary layer with irregular surface roughness," Journal of Fluid Mechanics, vol. 655, pp. 380–418, 2010. ·

12. R. J. Volino, M. P. Schultz, and K. A. Flack, "Turbulence structure in rough- and smooth-wall boundary layers," Journal of Fluid Mechanics, vol. 592, pp. 263–293, 2007. ·

13. J. C. Lin, M. Ozgoren, and D. Rockwell, "Space-time development of the onset of a shallow-water vortex,"Journal of Fluid Mechanics, vol. 485, pp. 33–66, 2003. ·

14. V. Roussinova, A. M. Shinneeb, and R. Balachandar, "Investigation offluid structures in a smooth open-channel flow using proper orthogonal decomposition," Journal of Hydraulic Engineering, vol. 136, no. 3, pp. 143–154, 2010. ·

15. K. Satoru, U. Hiromasa, O. Fumimaru, and M. Tokuro, "Turbulence structure and transport mechanism at the free surface in an open channel flow," International Journal of Heat and Mass Transfer, vol. 25, no. 4, pp. 513–521, 1982.

16. C. E. Willert and M. Gharib, "The interaction of modulated vortex pairs with free surface," in Proceedings of the ASME Symposium of Free-Surface Turbulence, FED, vol. 101, pp. 25–36, ASME, June 1994.

17. P. Maheo, Free surface turbulent shear flow, Ph.D. dissertation, California Institute of Technology, 1999.

18. J. C. R. Hunt and J. M. R. Graham, "Free stream turbulence in near plane boundaries," Journal of Fluid Mechanics, vol. 84, no. 2, pp. 209–235, 1978.

19. D. T. Walker, R. I. Leighton, and L. O. Garza-Rios, "Shear-free turbulence near a flat free surface," Journal of Fluid Mechanics, vol. 320, pp. 19–51, 1996.

20. R. Balachandar and V. C. Patel, "Velocity measurements in a developed open channel flow in the presence of an upstream perturbation," Journal of Hydraulic Research, vol. 43, no. 3, pp. 258–266, 2005.

21. D. G. Dommermuth, "Initialization of vortical free-surface flows," Journal of Fluids Engineering, vol. 116, no. 1, pp. 95–102, 1994.

22. C. Zhang, L. Shen, and D. K. P. Yue, "The mechanism of vortex connection at a free surface," Journal of Fluid Mechanics, vol. 384, pp. 207–241, 1999.

23. A. Weigard and M. Gharib, "A plane mixing layer with cross-shear," Journal of Fluids Engineering, vol. 117, p. 374, 1995.

24. K. Lam and S. Banerjee, "On the condition of streak formation in a bounded turbulent flow," Physics of Fluids A, vol. 4, no. 2, pp. 306–320, 1992.

25. S. Komori, R. Nagaosa, Y. Murakami, S. Chiba, K. Ishii, and K. Kuwahara, "Direct numerical simulation of three-dimensional open-channel flow with zero-shear gas-liquid interface," Physics of Fluids A, vol. 5, no. 1, pp. 115–125, 1993.

26. R. A. Handler, T. F. Swean Jr., R. I. Leighton, and J. D. Swearingen, "Length scales and the energy balance for turbulence near a free surface," AIAA Journal, vol. 31, no. 11, pp. 1998–2007, 1993.

27. Y. Pan and S. Banerjee, "A numerical study of free-surface turbulence in channel flow," Physics of Fluids, vol. 7, no. 7, pp. 1649–1664, 1995.

28. R. Gupta, S. Kumar, and S. Banerjee, "Characteristics of attached vortices in free surface channel flows," in Proceedings of the 12th US National Congress of Applied Mechanics, 1994.

29. V. Nikora, R. Nokes, W. Veale, M. Davidson, and G. H. Jirka, "Large-scale turbulent structure of uniform shallow free-surface flows," Environmental Fluid Mechanics, vol. 7, no. 2, pp. 159–172, 2007. ·

30. L. Shen, X. Zhang, D. K. P. Yue, and G. S. Triantafyllou, "The surface layer for free-surface turbulent flows," Journal of Fluid Mechanics, vol. 386, pp. 167–212, 1999.

31. M. Zdravkovich, Flow around Circular Cylinder, vol. 1, Oxford University Press, New York, NY, USA, 1997.

32. M. Van Dyke, An Album of Fluid Motions, Parabolic Press, Stanford, Calif, USA, 1982.

33. S. Barbarutsi and V. H. Chu, "Experimental study of turbulent mixing layers in shallow open channel flow," Tech. Rep. 85-1, McGill University, 1985.

34. V. H. Chu, J. H. Wu, and R. E. Khayat, "Stability of turbulent shear flows in shallow channel," inProceedings of the 20th IAHR Congress, vol. 3, Moscow, Russia, 1983.

35. E. Wolonski, J. Imberger, and M. L. Heron, "Island wakes in shallow coastal waters," Journal of Geophysical Research, vol. 89, pp. 10553–10569, 1984.

36. R. G. Ingram and V. H. Chu, "Flow around islands in Rupert Bay: an investigation of the bottom friction effect," Journal of Geophysical Research, vol. 92, no. 13, pp. 14521–14533, 1987. ·

37. D. Chen and G. H. Jirka, "Experimental study of plane turbulent wakes in a shallow water layer," Fluid Dynamics Research, vol. 16, no. 1, pp. 11–41, 1995.

38. M. E. Negretti, S. A. Socolofsky, A. C. Rummel, and G. H. Jirka, "Stabilization of cylinder wakes in shallow water flows by means of roughness elements: an experimental study," Experiments in Fluids, vol. 38, no. 4, pp. 403–414, 2005. ·

39. D. T. Walker and V. G. Johnston, "Observation of turbulence near the free surface in the wake of a model ship," in Dynamics of Bubbles and Vortices Near a Surface, I. Sahin and G. Tryggvason, Eds., AMD, vol. 119, ASME, New York, NY, USA, 1991.
40. D. G. Anthony and W. W. Willmarth, "Turbulence measurements in a round jet beneath a free surface,"Journal of Fluid Mechanics, vol. 243, pp. 699–720, 1992.
41. L. M. Logory, A. Hirsa, and D. G. Anthony, "Interaction of wake turbulence with a free surface," Physics of Fluids, vol. 8, no. 3, pp. 805–815, 1996.
42. S. K. Rao, D. Sumner, and R. Balachandar, "A visualization study of fluid-structure interaction between a circular cylinder and a channel bed," Journal of Visualization, vol. 7, no. 3, pp. 187–199, 2004.
43. H. F. Wang, Y. Zhou, C. K. Chan, and K. S. Lam, "Effect of initial conditions on interaction between a boundary layer and a wall-mounted finite-length-cylinder wake," Physics of Fluids, vol. 18, no. 6, Article ID 065106, 2006.
44. T. Dracos, M. Giger, and G. H. Jirka, "Plane turbulent jets in a bounded fluid layer," Journal of Fluid Mechanics, vol. 241, pp. 587–614, 1992.
45. W. S. J. Uijttewaal and J. Tukker, "Development of quasi two-dimensional structures in a shallow free-surface mixing layer," Experiments in Fluids, vol. 24, no. 3, pp. 192–200, 1998.
46. A. Singha, A.-M. Shinneeb, and R. Balachandar, "PIV-POD investigation of the wake of a sharp-edged flat bluff body immersed in a shallow channel flow," Journal of Fluids Engineering, vol. 131, no. 2, Article ID 021202, 2009.
47. P. Holmes, J. L. Lumley, and G. Berkooz, Turbulence, Coherent Structures, Dynamical Systems and Stability, Cambridge University Press, 1996.
48. J. P. Bonnet, J. Delville, M. N. Glauser et al., "Collaborative testing of eddy structure identification methods in free turbulent shear flows," Experiments in Fluids, vol. 25, no. 3, pp. 197–225, 1998.
49. A. K. M. F. Hussain, "Coherent structures—reality and myth," Physics of Fluids, vol. 26, no. 10, pp. 2816–2850, 1983.
50. C. F. V. Carmer and G. H. Jirka, "On turbulent and transport in shallow wake flows," in Proceedings of the IAHR Conference, Beijing, China, 2001.
51. G. H. Jirka, "Large scale flow structures and mixing processes in shallow flows," Journal of Hydraulic Research, vol. 39, no. 6, pp. 567–573, 2001.
52. R. H. Kraichnan, "Inertial ranges in two-dimensional turbulence," Physics of Fluids, vol. 10, no. 7, pp. 1417–1423, 1967.
53. R. Balachandar, M. F. Tachie, and V. H. Chu, "Concentration profiles in shallow turbulent wakes," Journal of Fluids Engineering, vol. 121, no. 1, pp. 34–43, 1999.
54. E. Wolansky, "Water circulation in a topographically complex environment," in Proceedings of the International Symposium on the Physics of Shallow Estuaries and Bays, University of Miami, 1984.
55. M. F. Tachie, Scaler transport in Intermediate two-dimensional turbulent wakes, M.S. thesis, University of Saskatchewan, 1997.
56. R. Balachandar, S. Ramachandran, and M. F. Tachie, "Characteristics of shallow turbulent near wakes at low Reynolds numbers," Journal of Fluids Engineering, vol. 122, no. 2, pp. 302–308, 2000.
57. M. F. Tachie and R. Balachandar, "Shallow wakes generated on smooth and rough surfaces," Experiments in Fluids, vol. 30, no. 4, pp. 467–474, 2001.
58. H. Akilli and D. Rockwell, "Vortex formation from a cylinder in shallow water," Physics of Fluids, vol. 14, no. 9, pp. 2957–2967, 2002.
59. H. Fu and D. Rockwell, "Shallow flow past a cylinder: transition phenomena at low Reynolds number,"Journal of Fluid Mechanics, vol. 540, pp. 75–97, 2005.
60. H. Fu and D. Rockwell, "Shallow flow past a cylinder: control of the near wake," Journal of Fluid Mechanics, vol. 539, pp. 1–24, 2005.
61. R. K. Cohn and M. M. Koochesfahani, "Effect of boundary conditions on axial flow in a concentrated vortex core," Physics of Fluids A, vol. 5, no. 1, pp. 280–282, 1993.
62. A. E. Perry and T. R. Steiner, "Large scale vortex structures in turbulent wakes behind bluff bodies: 1. vortex formation process," Journal of Fluid Mechanics, vol. 174, pp. 233–270, 1987.
63. A. Singha, Shallow wake in open channel flow—a look into the vertical variability, Ph.D. dissertation, University of Windsor, Windsor, Canada, 2009, http://winspace.uwindsor.ca/wintheses/singha2.pdf.
64. P. A. Eibeck, "An experimental study of the flow downstream of a circular and tapered cylinder," Journal of Fluids Engineering, vol. 112, no. 4, pp. 393–401, 1990.
65. M. M. Koochesfahani, "Vortical patterns in the wake of an oscillating airfoil," AIAA Journal, vol. 27, no. 9, pp. 1200–1205, 1989.

CHAPTER 2

SECONDARY CURRENTS AND TURBULENCE OVER A NON-UNIFORMLY ROUGHENED OPEN-CHANNEL BED

Thorsten Stoesser, Richard McSherry and Bruno Fraga

Keywords: Turbulence, bed roughness, large-eddy simulation, secondary currents

INTRODUCTION

Secondary currents are considered an important aspect in river engineering because they affect the primary mean flow field, the spanwise variation of bed shear stresses (and consequently the bed and banks erosion), the transport of bed-load and suspended sediments as well as the conveyance and mixing of dissolved matters. In straight open channel flow, secondary currents are a result of turbulence anisotropy (Prandtl's second kind [1]). Near the water surface, secondary currents transport low momentum fluid from the side walls towards the center of the channel, where high-momentum fluid is suppressed below the free-surface. This triggers a downflow of near surface fluid towards the bed, which results in the velocity dip observed in natural channels (a good summary of early work on this topic is found in Nezu and Nakagawa's 1993 textbook [2]). The distribution of wall shear stresses along the wetted perimeter in an open-channel is affected by these secondary currents, resulting in local bed shear maxima wherever there is downflow of high momentum fluid and in local bed shear minima wherever there is upward movement (Figure 1). The non-uniform distribution of the wall shear stress in the transverse direction of open channels is important in explaining the longitudinal striping of sediments that has been observed in channels and rivers by Vanoni [3], Culbertson [4], Karcz [5], Colombini [6], and others. On the basis of field observations, Nezu and Nakagawa [7] inferred multi-cellular secondary currents and associated sand ribbons. With Laser Doppler Anemometry (LDA) and an ultrasonic bedform instrument, Nezu et al. [8] and Onitsuka and Nezu [9] found that the organized fluid motions and the associated sediment transport occurred intermittently on a movable plane sand bed and, after the sand ridges were formed, the secondary-current cells appeared stable in the entire channel cross section. Such secondary currents are also generated on beds with smooth and rough striping [10,11,12,13,14]. Gyr and Kinzelbach [15] argue that secondary currents lead to the demixing of sediments in a channel, by transporting fine material laterally towards the smooth strips while maintaining the rough strips through bed amouring. This feedback process further stabilizes the smooth-rough-strip patterns.

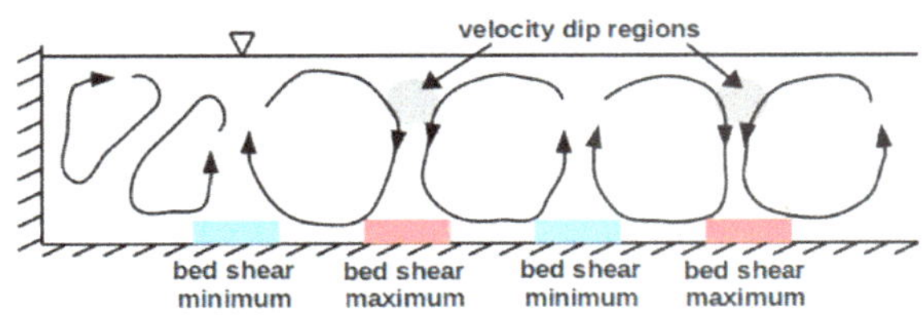

Figure 1: Organization of secondary currents in open channel flows, with associated velocity dip regions and bed shear maxima and minima.

Most of the numerical work to date focused on the calculation of secondary currents in closed channels. The first calculation of secondary flow in straight ducts was carried out by Launder and Ying [16], who derived a method that is based on approximating the Reynolds stresses with algebraic expressions by simplifying the corresponding transport equations proposed by Hanjalic and Launder [17]. This model has subsequently been employed, sometimes with modifications, in numerous studies (e.g., [18,19]) to calculate secondary flows in ducts. Recently, new explicit algebraic stress models (EASM) have become popular as an alternative to full Reynolds Stress Models (RSM) to reduce the computational effort [20,21]. A few Direct Numerical Simulation (DNS) studies of flow in ducts exist (e.g., [22,23]).

Citation: Thorsten Stoesser, Richard McSherry and Bruno Fraga, Secondary Currents and Turbulence over a Non-Uniformly Roughened Open-Channel Bed, Water 2015, 7, 4896-4913; doi:10.3390/w7094896.

The direct computation of all quantities and the high resolution in space and time allowed these researchers to gain insight into the details of the flow and to investigate the details of the Reynolds stress budgets. Among large-eddy simulations of the flow in square ducts some were performed at similar Reynolds numbers to the DNSs described above (e.g., [24]) and others at much higher Reynolds numbers [25]. A detailed study of flows in narrow ducts with a free surface is provided by Brogliaet al. [26]. They investigated the mean flow field and a variety of turbulence statistics and highlighted the complex structure of the flow near the corners formed by the free surface and the side wall. Broglia et al. [26] were able to explain the existence and origination of secondary motions in the context of stress balance equations. Shi et al. [27] performed an LES in a smooth rectangular channel at a Reynolds number of approximately 90,000 and showed that LES allows very accurate predictions of turbulence statistics and boundary shear stress distributions are feasible. Some recent works show the potential of LES to describe the secondary flow patterns and turbulence statistics of curved open-channels (e.g., [28,29,30]). However, flow in bends exhibits different mechanics as the centripetal force contributes to the generation of spanwise motion (Prandtl's first kind [1]) together with the aforementioned turbulence anisotropy.

Though recent progress has been made to understand turbulence-driven secondary currents in rivers, there are still many knowledge gaps that have to be addressed [31]. In this paper results of an LES of open-channel flow over non-uniformly distributed roughness are reported. LES resolves the energetic, large-scale motion of the flow [32], thus allows to simulate directly flow features created by turbulence, such as the turbulence driven secondary currents to be observed herein. The objectives of this study are to provide a detailed description of turbulence statistics of the flow over a non-uniformly roughened bed, shed light on the generation and virtue of secondary currents and the effect of non-uniform roughness on the bed shear stress, quantify the apparent stress and analyze the streamwise momentum balance. The study intends to further the present level of understanding of the interdependence of secondary currents, turbulence production and anisotrphy, and the distribution of shear stresses momentum in open channel flows.

NUMERICAL FRAMEWORK

The large-eddy simulation code (Hydro3D) used in this study is based on a finite volume discretization on a Cartesian grid with collocated variable arrangement [33] and has been validated thoroughly for many different flows [34,35,36,37,38,39,40,41]. The Hydro3D code solves the filtered Navier-Stokes equations for incompressible fluid flow (e.g., [42]):

$$\frac{\partial \hat{u}_i}{\partial x_i} = 0 \qquad (1)$$

$$\frac{\partial \hat{u}_i}{\partial t} + \frac{\partial \hat{u}_i \hat{u}_j}{\partial x_j} = -\frac{\partial \hat{p}}{\partial x_j} + \nu \frac{\partial (2S_{ij})}{\partial x_j} - \frac{\partial \tau_{ij}}{\partial x_j} \qquad (2)$$

where $\hat{u}_i$ and $\hat{u}_j$ (i or j = 1, 2, or 3) are the resolved velocity vectors (i.e., $\hat{u}_1 = \hat{u}$, $\hat{u}_2 = \hat{v}$, and $\hat{u}_3 = \hat{w}$ denoting the velocity components in x, y and z axis direction, respectively) and $\hat{p}$ is the resolved pressure divided by the density. These quantities are filtered in space. Similarly, x_i and x_j represent the spatial location vectors in x, y, and z axis direction, respectively, υ is the kinematic viscosity and S_{ij} is the filtered strain-rate tensor, defined as $S_{ij} = \frac{1}{2}\left(\frac{\partial \hat{u}_i}{\partial x_j} + \frac{\partial \hat{u}_j}{\partial x_i}\right)$. The term τ_{ij} results from the unresolved sub-grid scale fluctuations and needs to be modeled by a sub-grid scale (SGS) model. The dynamic version [43] of the original Smagorinsky [44] sub-grid scale (SGS) model is employed to approximate the anisotropic part of τ_{ij}, *i.e.*, τ_{ij}^a, as $\tau_{ij}^a = \tau_{ij} - \frac{2}{3}\delta_{ij}k_r \; (= -2\nu_t S_{ij})$, where δ_{ij} is the Kronecker delta and k_r is the residual kinetic energy. An anisotropic filter ($\Delta = (\Delta x \Delta y \Delta z)^{1/3}$) together with the characteristic filtered rate of strain ($|S| = (2S_{ij}S_{ij})^{1/2}$) are used to compute the sub-grid scale eddy viscosity ν_t as $\nu_t = (C_s\Delta)^2|S|$. The Smagorinsky constant C_s is calculated locally by making use of the flow information available from the smallest resolved scales. A double filtering procedure leads to a closed expression, commonly referred to as Germano's identity, relating the smallest resolved stresses to the largest unresolved stresses at the two different filter levels. This additional information is then used to determine the model parameter C_s through local averaging. The filtered equations are integrated over finite control volumes and convective and diffusive fluxes are computed using central differences, which ensure second-order accuracy in space. An explicit three-step Runge-Kutta scheme is used to integrate the equations in time providing third-order accuracy.

BOUNDARY CONDITIONS AND SETUP

The computational setup of the LES is selected in analogy to the laboratory experiments of Wang and Chen [1], the data of which are used to validate the LES results. The channel cross-section is divided into alternating rough and smooth strips of width λ, which equals the water depth, h. The channel width to depth ratio is B/h = 8.0 (Figure 2). The roughness in the rough strips is composed of uniform gravel and the relative submergence of the roughness is h/d_m = 31, with d_m being the mean grain diameter. The Reynolds number based on the hydraulic radius (h_r) and the bulk velocity u (bulk) is Re = 34,000. The flow is driven by a pressure gradient dp/dx that unambiguously provides the global shear velocity u* from $\bar{\tau} = dp/dx \cdot h_r$ and $u_* = \sqrt{\bar{\tau}/\rho}$, with $\bar{\tau}$ being the global shear stress.

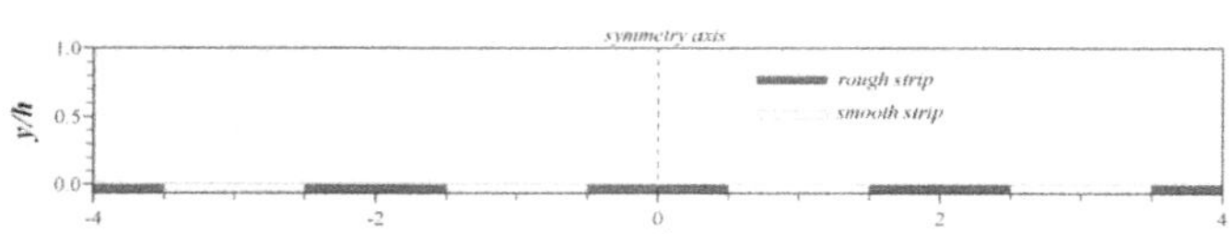

Figure 2: Cross-sectional view of the setup of the large-eddy simulations (LES).

The computational domain of the simulation spans 16 h in streamwise, 8 h in spanwise and h in the vertical directions, respectively. The streamwise extent is chosen to be big enough to include all relevant turbulence structures, which is verified by performing two-point correlations of the streamwise velocity fluctuation over the smooth strips. This justifies the use of periodic boundary conditions to be applied in the streamwise direction mimicking a fully developed turbulent channel flow. The free surface is set as a frictionless rigid lid and is treated as a plane of symmetry. This treatment is reasonably accurate considering the low Froude number (Fr ≈ 0.35) of the simulation; it is generally accepted that for open channel flows with Froude numbers below approximately 0.5, deformations at the free surface can be considered to be negligibly small and the free surface pressure can be considered to be hydrostatic [45,46,47]. On the smooth strips and the smooth side wall the no-slip condition is applied. The roughness closure model needed to simulate the flow over the roughness employs the method proposed and validated by Stoesser [48]. The bed roughness is incorporated by using a forcing term that is added to the right-hand side of the normalized momentum equations. For the U-equation this force reads in time-discrete form:

$$f = \begin{cases} C - D + \Delta P + \dfrac{\Delta V}{\Delta t}(U_t - U^n) & \text{inside the roughness} \\ 0 & \text{elsewhere} \end{cases} \quad (3)$$

in which C = discrete convection term; D = discrete diffusion term; P = resolved pressure divided by the density; U^n = resolved velocity in streamwise direction in the previous time step; and U_t = target velocity.

Two computational grids were used and only the results of the fine grid simulation are presented and discussed. The fine grid consists of approximately 41 million grid points. The grid is uniform in the horizontal plane using 600, 576 cells in the x and z directions, respectively. The grid spacing in wall units is $\Delta x^+ \approx 48$ and $\Delta z^+ \approx 32$, based on the global shear velocity. In the vertical, the grid varies such that it is lowest near the bed and then stretched towards the water surface, where the grid spacing is $\Delta y^+ \approx 18$. Over the rough strips the vertical spacing is approximately $\Delta y^+ \approx 16.0$ and over the smooth strips $\Delta y^+ \approx 4.0$, which is due to the non-uniform distribution of the wall shear stress across the channel. Figure 3 shows the computational grid in a cross-section (for clarity only every third grid line is plotted). The simulation was initially run for a period of 80 eddy turn-over times, ($t_e = h/u_*$) to develop the flow and was then continued for another 260 t_e to acquire first and second order statistics.

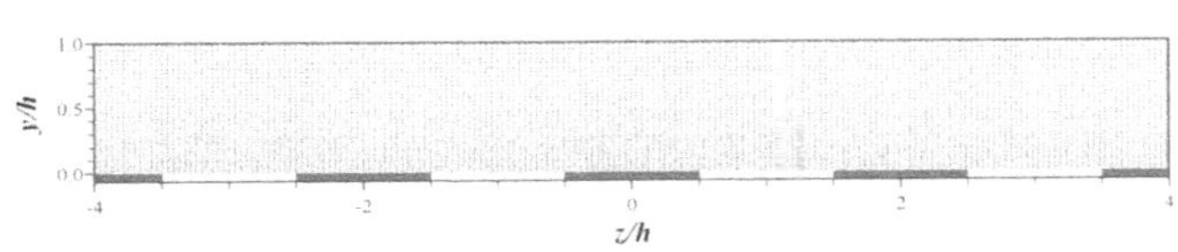

Figure 3: Cross-sectional view of the fine grid (only every 3rd grid line is plotted).

RESULTS AND DISCUSSION

Time-Averaged Flow

The validation of the LES method employed herein is carried out first by comparing different quantities between the experiment of Wang and Chen [1,2] and the large-eddy simulation. Figure 4 shows measured (a) and calculated (b) secondary currents in a part of the cross-section. The LES predicts accurately the pair of secondary vortices that forms as a result of channel geometry and non-uniform roughness in the cross section. The downward movement of fluid over the rough strips and the upward movement over the smooth strips is matched quite accurately. Some deviations are found in the location of the vortex core and the size of the vortices.

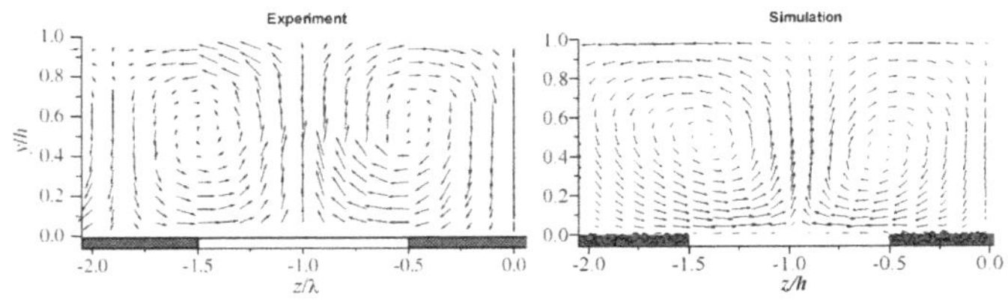

Figure 4: Streamwise-averaged velocity vectors in the vertical-spanwise plane, showing secondary currents: (**a**) experiment; (**b**) simulation.

The measured vortex core is at approximately y/h = 0.5, whereas the simulated core is at approximately y/h = 0.4. The measured vortices are almost symmetric whereas the LES predicted vortices differ slightly in size and shape. A possible reason for these slight discrepancies is the rather poor grid resolution in the wall normal direction at the channel side walls. A grid spacing of $\Delta z^+ \approx 32$ is not considered sufficient to properly resolve the turbulent boundary layer in these regions; wall functions and/or increased near-wall resolution will therefore be implemented in the continuation of the study. However, the overall agreement is quite satisfying. Figure 5 presents measured (left part) and calculated (right part) contours of the time-averaged streamwise velocity. Overall the agreement is quite good: simulations and experiments describe quite accurately the location of the high streamwise momentum regions located above the rough strips. The primary flow is influenced strongly by the prevailing secondary currents: The downward movement of high-momentum fluid over the rough strips creates pockets of high streamwise velocity there and the well-known velocity dip. The flow bulges over the smooth strips, which is due to low-momentum near-bed fluid being transported away from the wall. The corner vortex pair causes the flow to bulge towards the corners by transporting high-momentum surface fluid towards the bed-sidewall corner. The differences in this figure and also the ones mentioned above are most likely a result of the modeling inherent in LES, such as the SGS model, the rigid lid assumption, the treatment of roughness or the periodic boundary conditions. This validation of the LES method allows further detailed analysis.

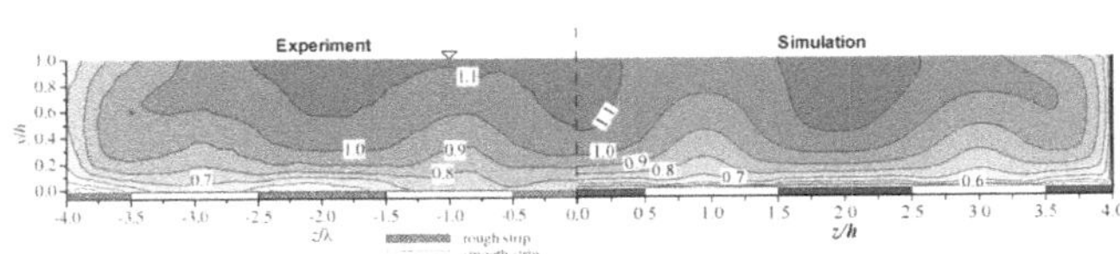

Figure 5: Contours of time-averaged streamwise velocity, normalized on the bulk velocity, for experiments ((**left**) half of the cross section) and LES (**right**). The contours provided by the simulation were averaged in the streamwise axis.

Figure 6 presents velocity vectors and contours of the stream function φ. The flow is symmetric about z/h = 0.0, so for better visibility from now on only the left hand side is shown. Five secondary vortices form in each half of the channel, the shapes of which differ quite significantly from one another. The vortex that is located closest to the centerline rotates clockwise and spans one water depth in both spanwise and wall-normal directions. At the smooth sidewall a corner vortex pair forms with a counter-clockwise rotating vortex near the bed and a clockwise rotating vortex near the free-surface. This vortex pair is not symmetric about the corner-bisector, due to the difference in roughness between the lower (bed-sidewall) and upper (sidewall-free surface) corners. The remaining two vortices have also opposite directions of rotation and differ in size and intensity.

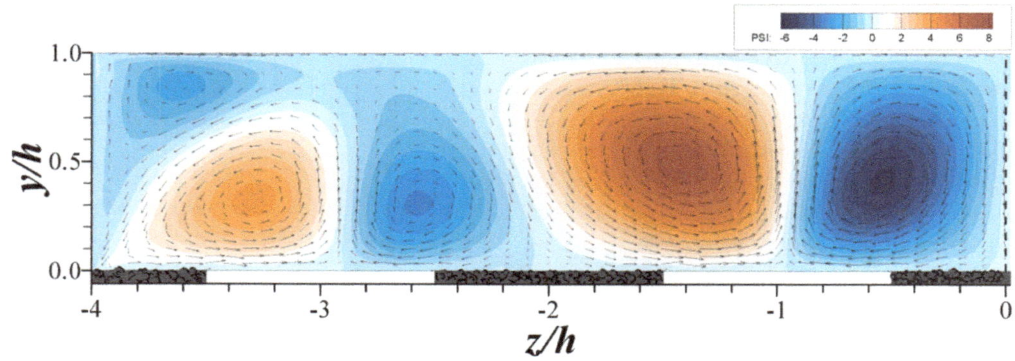

Figure 6: Contours of the streamfunction together with the vectors of the secondary flow in the left half of the cross-section.

Second Order Turbulence Statistics

Figure 7 compares measured (left part) normalized primary shear stresses, $-\overline{u'v'}/u_*^2$, with the one obtained from the simulation (right part). The measurements seem to suffer from insufficient averaging time, however, overall the match is reasonable. The simulation seems to underpredict the positive values of the shear stress found at the free surface in the middle of the channel, which might be a result of the artificial slip boundary condition used in the LES. There is a considerable area of positive shear stress in the side-wall-free-surface-corner, which is predicted correctly by the LES, a result of negative gradients of the streamwise velocity, or the fact that streamwise momentum is transported away from the free surface during inward motion (i.e., u' < 0 and w' < 0), respectively. Near the bed the shear stress is considerably higher above the rough strips than over the smooth strips. The magnitudes of the normalized shear stress over the rough (z/h = 0.0) and smooth strips (z/h = −1.0) in comparison with the theoretical straight line for two-dimensional flows is provided inFigure 7a. The peak of the shear stress over the rough strip is more than 50% greater than the squared mean shear velocity, whilst the shear stress over the smooth strips is suppressed to approximately half of the squared shear velocity. This has two reasons, the first, most obvious one, is that shear stresses are generally higher over rough walls than over smooth walls. The second reason is a result of the prevailing secondary flow. The transport of high momentum towards the bed over the rough strips leads to steeper streamwise velocity gradients there and hence to enhanced shear. The opposite occurs over the smooth strips, where low momentum fluid is convected away from the wall, thereby reducing streamwise velocity gradients. The streamwise and wall-normal turbulence intensity profiles are presented in Figure 8b. They also show that over the rough strip significantly higher values are observed than over the smooth strip. Over the smooth strip the streamwise turbulence intensity profile features a distinct peak (with $u'/u_* = 2.2$) near the bed. Over the rough strip the peak intensity is about 25% higher (with $u'/u_* = 2.6$) than over the smooth strip and there is considerable streamwise turbulence until approximately 30% of the water depth (i.e., y/h = 0.3). Interestingly, further away from the bed (i.e., y/h > 0.3) the streamwise turbulence over the smooth strips is slightly higher than over the rough strips. A similar trend is found for the wall normal turbulence intensities.

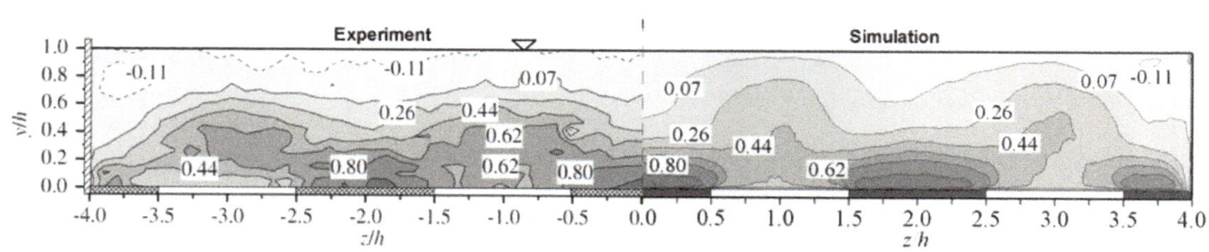

Figure 7: Contours of measured (**left**) and simulated (**right**) shear stress, normalized on the squared global shear velocity, in the cross-section.

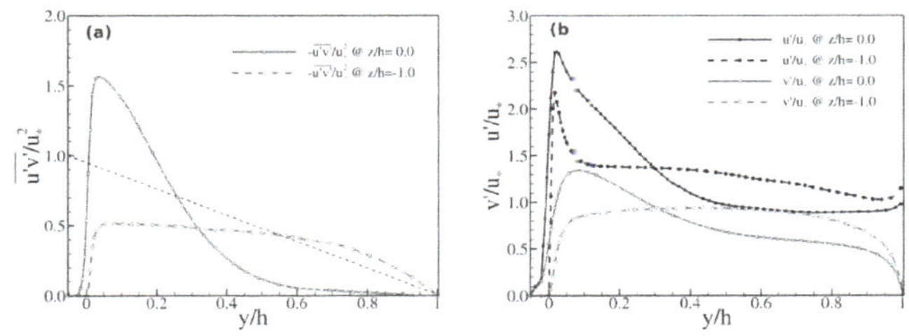

Figure 8: Profiles of the primary shear stress (**a**) and streamwise and wall normal turbulence intensities (**b**) over smooth (z/h = −1.0) and rough (z/h = 0.0) strips. In both plots the global shear velocity has been used for normalisation.

Figure 9a presents contours of the normalized kinetic energy in the left half of the domain, providing a better view of the spatial distribution of turbulence in the cross section. There are pockets of high turbulent kinetic energy over the rough bed, a result of bed roughness but also of the steeper velocity gradients. Turbulence levels in terms of kinetic energy, over the rough bed are about 70% higher than over the smooth strips and approximately 50% higher than adjacent to the smooth side walls. On the other hand the turbulent kinetic energy decreases much quicker over the rough strips than over the smooth strips, which is owed to the fact that over the smooth strips the wall-normal velocity gradients (i.e., dU/dy) remain throughout the water depth and that the wall-normal Reynolds stress attains non-zero values only very close to the water surface (see Figure 3 and Figure 7). On the other hand over the rough strips dU/dy and $\overline{u'v'}$ are negligibly small already at approximately half the water depth and dU/dy even becomes negative close to the water surface. Contours of the dominant turbulence production term $P_{12} = -\overline{u'v'}\frac{\partial U}{\partial y}$ are plotted in Figure 9b and visualize the statements made above quite well. Clearly, turbulence production occurs mainly over the rough strips, and P_{12} is relatively small close to the smooth wall. The region of very large P_{12} values (i.e., P_{12} larger than about 0.07) extends for between 4 and 6 grid points in the vertical direction above the rough strips, which corresponds to a distance of approximately 16 to 24 wall units.

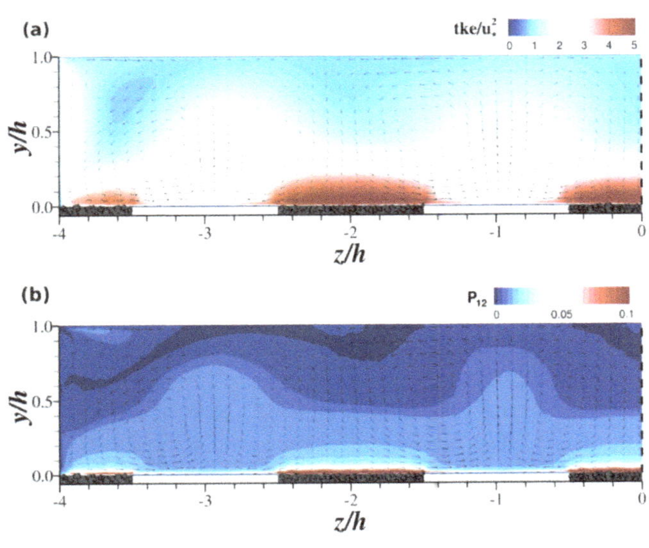

Figure 9: Contours of the normalized turbulent kinetic energy (**a**) and normalized dominant turbulence production term (**b**) in the left half of the cross-section.

Figure 10 presents the distribution of the three shear stress components together with secondary flow vectors in one half of the cross-section. The primary shear stress, Figure 10a, has been discussed above and its distribution is shown again to highlight interesting features in conjunction with the other two components. Positive $\overline{u'v'}$ values are found mainly in the free-surface corner but also close to the free surface over the rough strips, which are highlighted by the isoline of $\overline{u'v'} = 0$ (black line). Except very close to the rough wall (where $\overline{u'v'}$ attain their peaks) spanwise shear stresses, $\overline{u'w'}$, depicted in Figure 10b, are of comparable magnitude to the primary shear stress and maximum values are observed near the side wall and in the cores of the secondary circulation vortices. This suggests that streamwise momentum transfer in the spanwise direction is quite significant. High values of $\overline{u'w'}$ are also present at the free surface and near the smooth wall, where W is zero and |V| attains local maxima, indicating streamwise momentum transfer in the spanwise direction by both turbulence and convection. Another interesting feature of the spanwise shear stress is that $\partial\overline{u'w'}/dz < 0$ for upflow, i.e., V > 0, and $\partial\overline{u'w'}/dz > 0$ for downflow, i.e., V < 0 (which also applies to flow in smooth bed channels, [4]) and that in areas of dominant or plain up- and downflow $\overline{u'w'} \approx 0$. Figure 10c presents cross-plane, or secondary, shear stresses. $\overline{v'w'}$ reaches local maxima of approximately 10% of the squared shear velocity, which is quite considerable, and are not only found in the channel corners but also at the interface of rough and smooth strips. Significant levels of the secondary shear stress, $\overline{v'w'}$, and the spanwise shear stress, $\overline{u'w'}$, are a result of the prevailing secondary currents and, away from the side wall, exhibit a high level of symmetry with respect to the center of the strips (e.g., z/h = 0, z/h = −1.0, z/h = −2.0). This clearly supports the hypothesis that the interplay between secondary currents and cross-stream sediment demixing, i.e., rough-smooth strip formation process as described in [15], leads to a stabilization of such longitudinal bedforms.

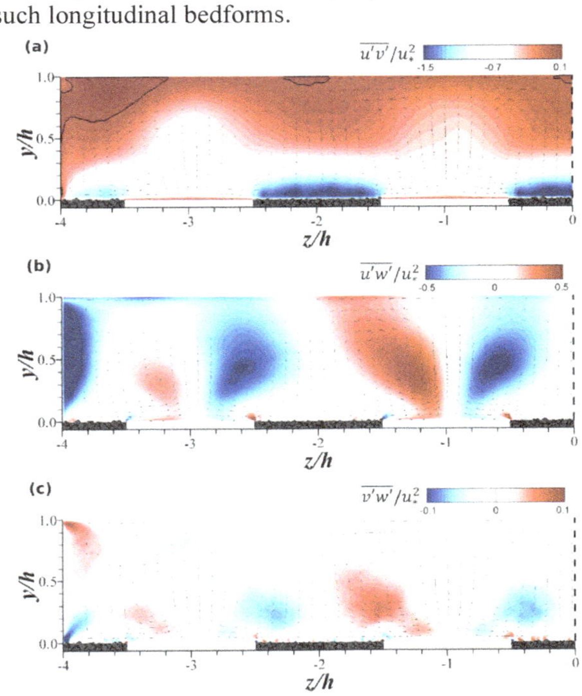

Figure 10: Contours of the normalized primary shear stress (**a**); nor-

malized spanwise shear stress (**b**) and secondary shear stress (**c**) in the left half of the cross-section.

The origin of secondary currents was first explained in 1926, when Prandtl [1] suggested that the secondary motion in non-circular ducts was caused by turbulence. He was the first to distinguish between secondary currents that are caused by vortex stretching (first kind) and secondary currents that are generated by the anisotropy of turbulence (second kind). The streamwise vorticity equation is used to describe the origin of secondary currents and can be obtained by eliminating the pressure term in the streamwise component of the momentum equation [49]. For a steady, incompressible, uniform turbulent flow in a straight channel this equation has the exact form [4]:

$$\underbrace{V\frac{\partial\Omega}{\partial y}+W\frac{\partial\Omega}{\partial z}}_{A}=\underbrace{\frac{\partial^2}{\partial y\partial z}\left(\overline{w'w'}-\overline{v'v'}\right)}_{B}-\underbrace{\left(\frac{\partial^2}{\partial z^2}-\frac{\partial^2}{\partial y^2}\right)\overline{v'w'}}_{C}+\underbrace{\nu\left(\frac{\partial^2\Omega}{\partial y^2}+\frac{\partial^2\Omega_1}{\partial z^2}\right)}_{D} \quad (4)$$

in which $\Omega_1 \equiv \frac{\partial V}{\partial z}-\frac{\partial W}{\partial y}$ is the streamwise vorticity. Term A represents the convection of streamwise vorticity by the mean flow and D represents the viscous diffusion of Ω. Terms B and C are generation terms of streamwise vorticity and both involve turbulent Reynolds stresses. The origin of secondary currents has been argued about amongst researchers [4,50,51], however, Broglia et al. [26] suggest that in open channels the turbulence anisotropy of the normal stresses (i.e., $\overline{w'w'}-\overline{v'v'}$) term dominates the secondary shear stress term in the free surface corner in particular. Figure 11 presents the distribution of normalized turbulence anisotropy in one half of the domain and supports previous findings which suggest that the secondary currents have their origin in the channel corners, where both $\overline{w'w'}-\overline{v'v'}$ and $\overline{v'w'}$ have local maxima or minima, respectively. The normal stress anisotropy is an order of magnitude greater than the shear stress, and in the corner regions it is of opposite sign. This indicates significant contribution of $\overline{w'w'}-\overline{v'v'}$ to streamwise vorticity generation. Substantial gradients of $\overline{w'w'}-\overline{v'v'}$ are also found above the rough strips and near the bed at the interface of rough and smooth strips, suggesting that this kind of bedform enhances the generation of secondary currents. In summary, the above discussions showed that secondary currents and the arrangement of non-uniform bed roughness, chosen here, form a closely interrelated system which appears to form stable smooth-rough-strip patterns.

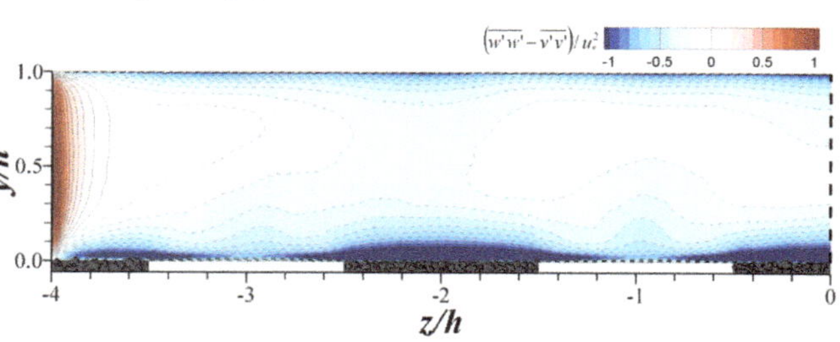

Figure 11: Contours of the normalized normal stress anisotropy in the left half of the domain.

Bed Shear Stress and Apparent Shear Stress

The presence of secondary currents markedly affects the distribution of the wall shear stress along the wetted perimeter of the channel [52,53]: Bed shear stresses are magnified in regions of downflow and are reduced in regions of upflow. The presence of non-uniform bed roughness enhances this pattern, due to the fact that downflow occurs over rough strips where the bed shear stress is naturally greater. Figure 12a presents the spanwise distribution of the normalized bed shear stress, τ_b/μ_*^2, in the center section of the channel (i.e., −1 ≤ z/h ≤ 1). The bed-shear stress over the rough bed is obtained from the vertical distribution of $\overline{u'v'}$, while the bed shear stress over the smooth strips is accurately calculated from the velocity gradient at the first grid point off the wall. Additionally plotted are experimental data of Studerus [13]. Simulation and experimental data agree quite well, though the bed roughness and experimental conditions are not exactly the same. However, the ratio of bedform width to water depth, λ/h, is the same, i.e., λ/h = 1.0, which seems to be the dominant parameter and another indicator of the aforementioned bedform stabilization process. The bed shear stress over the rough strips is about twice as big as the global shear stress, i.e., $\rho\mu_*^2$, and approximately four times greater than over the smooth strips. Bed shear stress non-uniformity, as a result of sediment demixing, leads to streamwise momentum loss near the rough bed, which has to be compensated for by lateral transport of streamwise momentum via convection and/or turbulence. The presence of secondary currents and non-zero spanwise shear stresses result in the so called apparent stress. The contributions of secondary currents and transverse shear stresses to the squared bulk shear velocity, u_*^2, can be analyzed with the Reynolds- and depth-averaged streamwise momentum equation, which, for a uniform flow in a straight channel, reads:

$$\frac{\partial}{\partial z}\underbrace{\int UW dy}_{\int SC}+\frac{\partial}{\partial z}\underbrace{\int \overline{u'w'}dy}_{\int RS}+\frac{\tau_b}{\rho}=u_*^2 \quad (5)$$

here the viscous stress contribution is neglected because it is very small except very close to the sidewalls. This equation expresses that the spanwise gradient of the depth integrated secondary current (SC) and Reynolds stress (RS) terms together with the bed shear stress sum to the integral shear velocity, however both ∫SC and ∫RS can have opposite signs, so that they may compensate each other. Figure 12b plots the two components of the apparent stress, which are normalized with the squared bulk shear velocity, u_*^2, in the center section of the channel (i.e., −1 ≤ z/h ≤ 1). The two components have opposite signs over the entire section plotted and the flow is perfectly symmetric about z/h = 0.0, where the two components are zero. Also plotted is the sum of the two, which indicates that the apparent shear stress is mainly carried by the secondary currents except very close to the sidewall, where the viscous stress dominates and ∫SC approaches zero. The peaks of ∫SC and ∫RS do not occur at the same location, and the largest secondary current contribution is approximately half the squared bulk shear velocity and occurs at the interface of rough and smooth channel. The peaks of the Reynolds stress contributions are found a bit further away from the interface and are approximately 40% of the squared bulk shear velocity. The sum of ∫SC and ∫RS form a straight line which changes sign at the center of smooth and rough strips and the gradient of the sum changes sign at the interface.

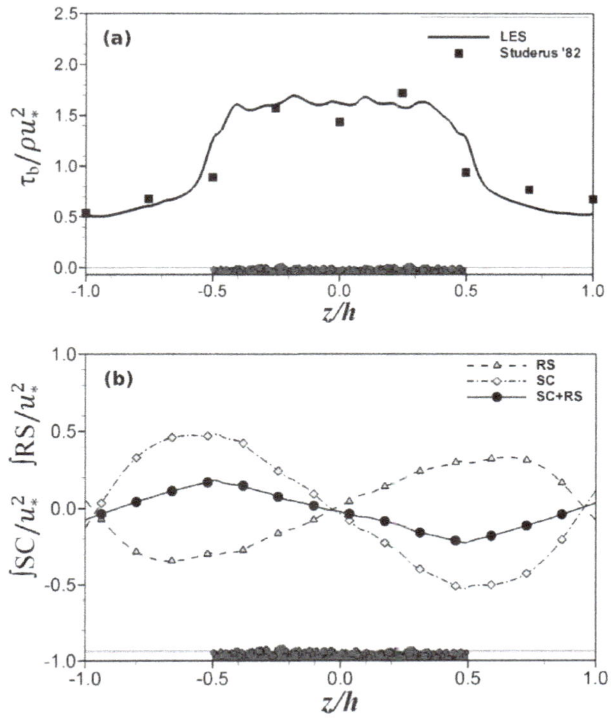

Figure 12: Spanwise distribution of bed-shear stress (**a**) and apparent shear stress (**b**) over smooth and rough strips.

Momentum Balance

The transport of streamwise momentum can be analyzed with the Reynolds averaged form of the x-momentum equation, which, for a uniform flow in a straight channel, reads:

$$0 = \underbrace{-\frac{1}{\rho}\frac{\partial p}{\partial x}}_{P} + \underbrace{\nu\left(\frac{\partial^2 U}{\partial y^2} + \frac{\partial^2 U}{\partial z^2}\right)}_{V} - \underbrace{\left(V\frac{\partial U}{\partial y} + W\frac{\partial U}{\partial Z}\right)}_{SC_1 + SC_2} - \underbrace{\left(\frac{\partial \overline{u'v'}}{\partial y} + \frac{\partial \overline{u'w'}}{\partial z}\right)}_{RS_1 + RS_2} \qquad (6)$$

The two terms denoted SC represent the convective transport of streamwise momentum, term P is the pressure gradient that drives the flow, term RS is transport due to turbulent stresses and term V is transport due to viscous stresses. The latter is negligibly small except very close to the wall. While in a 2D flow, i.e., in the absence of secondary currents only P, V and RS_1 are non-zero, terms SC_1, SC_2 and RS_1 are non-zero in a 3D flow (in the presence of secondary currents). The contributions of the four major contributors (i.e., SC_1, SC_2, RS_1, RS_2) to the transport of streamwise momentum are provided in Figure 13. The four components are normalized by P and are all plotted with their respective sign to make obvious which component compensates the other. The contour colors are selected such that small contributions are not visible. First of all, and for obvious reasons, very close to the bed and very close to the sidewall RS_1 and RS_2 dominate the momentum transfer in the flow. Secondly, the above-discussed importance of secondary currents for the transport of streamwise momentum is found in Figure 13, i.e., removal of streamwise momentum over the smooth strips through upflow and addition of streamwise momentum towards the wall through downflow over rough strips (Figure 13a); very small contributions of streamwise momentum through the spanwise shear stress except in the channel corners (Figure 13b); the supression of turbulence away from the rough bed due to the downflow (Figure 13c); and strong lateral gradients of the transverse shear stress in upflow regions over the smooth bed (Figure 13d). Figure 13 also demonstrates which of the four components balances which, and where. For instance transport of streamwise momentum by the secondary currents is compensated by wall normal turbulent transport over the rough strips, while it is compensated by spanwise turbulent transport over the smooth walls. Significant differences are observed in the respective variations of SC_1 and SC_2: these result from the fact that the vertical gradient of streamwise velocity, dU/dy, is generally much larger than the spanwise gradient, dU/dz. It should be noted, however, that near the strip interface there are also strong gradients in the spanwise direction; these are very localized but produce magnitudes of SC_2 that are similar to those of SC_1.

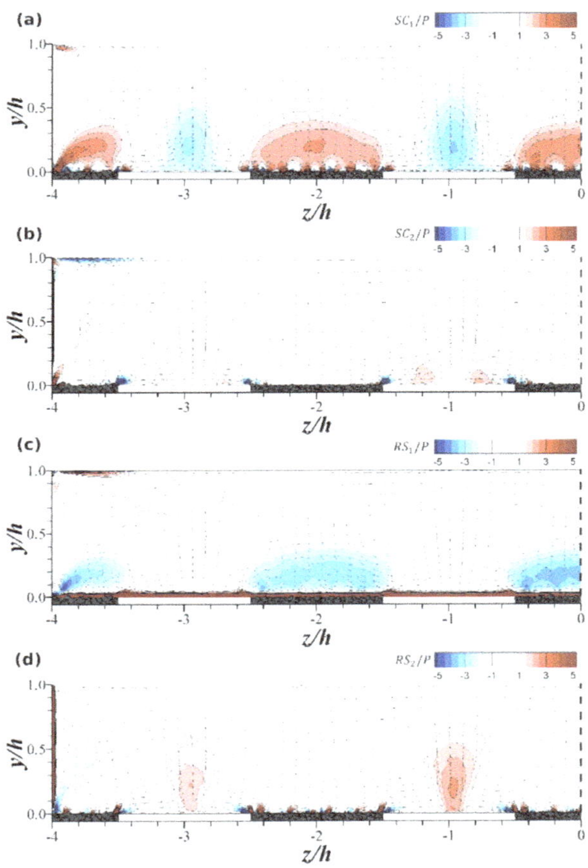

Figure 13: The four dominating components of streamwise momentum transport.

CONCLUSIONS

Large-eddy simulations of flow in a rectangular channel with non-uniform bed roughness were performed. Setup and boundary conditions were chosen in analogy to a laboratory experiment to ensure code validation and confirmation of the LES' capability to accurately reproduce such flows. Good agreement between LES and experimental data was found and additional turbulence statistics were extracted from the LES, presented and discussed. The secondary flow together with the non-uniform roughness on the channel bed significantly affects the time-averaged streamwise flow and the second order turbulence statistics. Near the bed, turbulence is enhanced over rough strips and suppressed over smooth strips. The secondary flow and bed roughness affect the streamwise velocity and second order turbulence statistics considerably. Significant lateral momentum transfer takes place due to both advection and turbulence. The bed shear stresses over the smooth strips are approximately four times less than over the rough strips, a result of near bed low momentum fluid being transported from the rough strips to the smooth strips and high momentum fluid being convected from the surface towards the bed. The presence of secondary currents yields nonzero spanwise and cross-plane shear stresses that result in an apparent stress in the spanwise direction and affect the streamwise momentum balance greatly. The naturally high bed shear stress over rough beds is enhanced further by downflow of high-momentum fluid from the surface towards the bed. The generation terms in the streamwise vorticity equation, i.e., normal stress anisotropy and secondary shear stress, are quantified and it was confirmed that turbulence anisotropy gradients dominate over secondary shear stress gradients. Statistical evidence of the feedback process between the secondary flow and bed roughness formation was provided.

ACKNOWLEDGMENTS

This research was jointly supported by the BP/The Gulf of Mexico Research Initiative via the Gulf Integrated Spill Response Consortium and the UK Engineering and Physical Sciences Research Council.

REFERENCES

1. Prandtl, L. Ueber die ausgebildete Turbulenz. In Proceedings of the 2nd International Congress for Applied Mechanics, Zurich, Switzerland, 12–17 September 1926; pp. 62–74.
2. Nezu, I.; Nakagawa, H. Turbulence in Open-Channel Flows. IAHR-Monograph; CRC Press: Rotterdam, The Netherlands, 1993.
3. Vanoni, V.A. Transportation of suspended sediment by water. Trans. ASCE **1946**, 111, 67–133.
4. Culbertson, J.K. Evidence of secondary circulation in an alluvial channel. USGS Prof. Pap. **1967**, 575, 214–216.
5. Karcz, I. Reflection of the origin of scour: Small-scale longitudinal streambed scours. In Fluvial Geomorphology; Morisawa, M., Ed.; Binghamton State University of New Youk Publications: New York, NY, USA, 1973; pp. 149–173.
6. Colombini, M. Turbulence driven secondary flows and the formation of sand ridges. J. Fluid Mech. **1993**, 254, 701–719.
7. Nezu, I.; Nakagawa, H. Self-forming mechanism of longitudinal sand ridges and troughs. In Proceedings of the 23rd IAHR Congress, Delft, The Netherlands, 21–25 August 1989; pp. 65–72.
8. Nezu, I.; Nakagawa, H.; Kawashima, N. Cellular secondary currents and sand ribbons in fluvial channel flows. In Proceedings of the 6th Asia and Pacific Division of International Association for Hydraulic Research (APD-IAHR) Congress, Delft, The Netherlands, 26–29 August 1988; pp. 51–58.
9. Onitsuka, K.; Nezu, I. Generation mechanism of turbulence driven secondary currents in open-channel flows. InIUTAM Symposium on Geometry and Statistics of Turbulence; Kluwer Academic: Boston, MA, USA, 2001.
10. Wang, Z.-Q.; Chen, N.S. Secondary flows over artificial bed strips. Adv. Water Resour. **2005**, 28, 441–450.
11. Wang, Z.-Q.; Chen, N.-S. Time-mean structure of secondary flows in open channel with longitudinal bedforms. Adv. Water Resour. **2006**, 29, 1634–1649.
12. McLean, S.R. The role of non-uniform roughness in the formation of sand ribbons. Mar. Geol. **1981**, 42, 49–74.
13. Studerus, X. Sekundärströmungen im Offenen Gerinne Über Rauhen Längsstreifen. Ph.D. Thesis, Swiss Federal Institute of Technology, Zurich, Switzerland, March 1982. (In German).
14. McLelland, S.J.; Ashworth, P.J.; Best, J.L.; Livesey, J.R. Turbulence and secondary flow over sediment stripes in weakly bimodal bed material. J. Hydraul. Eng. **1999**, 125, 463–473.
15. Gyr, A.; Kinzelbach, W. Bed forms in turbulent channel flow. Appl. Mech. Rev. **2004**, 57, 77–93.
16. Launder, B.E.; Ying, W.M. Secondary flows in ducts of square cross-section. J. Fluid Mech. **1972**, 54, 289–295.
17. Hanjalic, K.; Launder, B.E. A Reynolds stress model of turbulence and its application to thin shear flows. J. Fluid Mech.**1972**, 52, 609–638.
18. Naot, D.; Rodi, W. Calculation of secondary currents in channel flow. ASCE J. Hydraul. Div. **1982**, 108, 948–968.
19. Demuren, A.O.; Rodi, W. Calculation of turbulence driven secondary motion in noncircular ducts. J. Fluid Mech. **1984**,140, 189–222.
20. Shih, T.H.; Zhu, J.; Lumley, J.L. A new Reynolds stress algebraic equation model. Comput. Methods Appl. Mech. Eng.**1995**, 125, 287–302.
21. Wallin, S.; Johansson, A.V. An explicit algebraic Reynolds stress model for incompressible and compressible turbulent flows. J. Fluid Mech. **2000**, 403, 89–132.
22. Gavrilakis, S. Numerical simulation of low-Reynolds-number turbulent flow through a straight square duct. J. Fluid Mech. **1992**, 244, 101–129.
23. Huser, A.; Biringen, S. Direct numerical simulation of turbulent flow in a square duct. J. Fluid Mech. **1993**, 257, 65–95.
24. Madabhushi, R.K.; Vanka, S.P. Large eddy simulation of turbulence-driven secondary flow in a square duct. Phys. Fluids A **1991**, 3, 2734–2745.
25. Su, M.D.; Friedrich, R. Investigation of fully developed turbulent flow in a straight duct with large eddy simulation. J. Fluids Eng. **1994**, 116, 677–684.
26. Broglia, R.; Pascarelli, A.; Piomelli, U. Large-Eddy simulations of ducts with a free surface. J. Fluid Mech. **2003**, 484, 223–253.
27. Shi, J.; Thomas, T.G.; Williams, J.J.R. Large-eddy

simulation of flow in a rectangular open channel. J. Hydraul. Res. **1999**, 37, 345–361.

28. Van Balen, W.; Uijttewaal, W.S.; Blanckaert, K. Large-eddy simulation of a mildly curved open-channel flow. J. Fluid Mech. **2009**, 630, 413–442.
29. Stoesser, T.; Ruether, N.; Olsen N.R.B. Calculation of primary and secondary flow and boundary shear stresses in a meandering channel. Adv. Water Resour. **2010**, 33, 158–170.
30. Song, C.G.; Seo, I.W.; Kim, Y.D. Analysis of secondary current effect in the modelling of shallow flow in open channels. Adv. Water Resour. **2012**, 41, 29–48.
31. Nikora, V.; Roy, A.G. Secondary flows in rivers: Theoretical framework, recent advances and current challenges. In Proceedings of the 7th Gravel-Bed Rivers Conference, Tadoussac, QC, Canada, 6–10 September 2010.
32. Stoesser, T. Large-eddy simulation in hydraulics: Quo Vadis? J. Hydraul. Res. **2014**, 52 441–452.
33. Stoesser, T.; Nikora, V. Flow structure over square bars at intermediate submergence: Large eddy simulation study of bar spacing effect. Acta Geophys. **2008**, 56, 876–893.
34. Bomminayuni, S.; Stoesser, T. Turbulence statistics in an open-channel flow over a rough bed. J. Hydraul. Eng. **2011**,137, 1347–1358.
35. Bai, J.; Fang, C.D.; Stoesser, T. Transport and deposition of fine sediment in open channels with different aspect ratios.Earth Surf. Process. Landf. **2013**, 38, 591–600.
36. Kara, S.; Stoesser, T.; Sturm, T.W. Turbulence statistics of flow in compound open channels. J. Hydraul. Res. **2012**, 50, 482–494.
37. Kara, S.; Stoesser, T.; Sturm, T.W.; Muluhasan, S. Flow dynamics through a submerged bridge opening with overtopping. J. Hydraul. Res. **2015**, 53, 282–294.
38. Kara, M.C.; Stoesser, T.; McSherry, R. Calculation of Fluid Structure Interaction: Methods, Refinements, Applications.ICE Proc.: Eng. Comput. Mech. **2015**, 168, 59–78.
39. Kim, S.; Stoesser, T. Closure modelling and direct simulation of vegetation drag in flow through emergent vegetation.Water Resour. Res. **2011**, 47.
40. Kim, D.; Kim, J.H.; Stoesser, T. Hydrodynamics, turbulence and solute transport in ozone contact chambers. J. Hydraul. Res. **2013**, 51, 558–568.
41. Fraga, B.; Stoesser, T.; Lai, C.C.K.; Socolofsky, S.A. A large-eddy simulation-based Eulerian-Lagrangian approach to predict bubble plume dynamics. Ocean Model. **2015**. submitted for publication.
42. Rodi, W.; Constantinescu, G.; Stoesser, T. Large-Eddy Simulations in Hydraulics; CRC Press: Rotterdam, The Netherlands, 2013.
43. Germano, M.; Piomelli, U.; Moin, P.; Cabot, W.H. A dynamic subgrid-scale eddy viscosity model. Phys. Fluids **1991**, 3, 1760–1765.
44. Smagorinsky, J. General circulation experiments with the primitive equations—Part I: The basic experiment. Mon. Weather Rev. **1963**, 91, 99–152.
45. Kirkhill, G.; Constantinescu, G. Flow and turbulence structure around an in-stream rectangular cylinder with scour hole. Water Resour. Res. **2010**, 46.
46. Koken, M.; Constantinescu, G. An investigation of the dynamics of coherent structures in a turbulent channel flow with a vertical side-wall obstruction. Phys. Fluids **2009**, 21, 085104.
47. Paik, J.; Sotiropoulos, F. Coherent structure dynamics upstream of a long rectangular block at the side of a large aspect ratio channel. Phys. Fluids **2005**, 17.
48. Stoesser, T. Physically realistic roughness closure scheme to simulate turbulent channel flow over rough beds within the framework of LES. J. Hydraul. Eng. **2010**, 136, 812–819.
49. Einstein, H.A.; Li, H. Secondary currents in straight channels. Trans. Am. Geophys. Union **1958**, 39, 1085–1088.
50. Perkins, H.J. The formation of streamwise vorticity in turbulent flow. J. Fluid Mech. **1970**, 44, 721–740.
51. Gessner, F.B. The origin of secondary flow in turbulent flow in a corner. J. Fluid Mech. **1973**, 58, 1–25.
52. Knight, D.W.; Demetriou, J.D.; Hamed, M.E. Boundary shear in smooth rectangular channels. J. Hydraul. Eng. **1984**,110, 405–422.
53. Nezu, I.; Rodi, W. Experimental study on secondary currents in open channel flow. In Proceedings of the 21st IAHR Congress, Delft, The Netherlands, 13–18 August 1985; pp. 114–119.

CHAPTER 3

FLOW PATTERNS IN AN OPEN CHANNEL CONFLUENCE WITH INCREASINGLY DOMINANT TRIBUTARY INFLOW

Laurent Schindfessel, Stéphan Creëlle and Tom De Mulder

Keywords: Open channel confluence, flow pattern, hydrodynamics, discharge ratio, large-eddy simulation, impinging

INTRODUCTION

Confluences of open channels are important elements in hydraulic networks of rivers and man-made canals. The associated flow patterns govern the transport of solutes and sediments in the network and influence the water levels of the incoming channels. The flow features that appear in an open channel confluence can be conceptualized as follows (Figure 1after Best [1]): At the point where the two incoming flows meet, a stagnation zone develops, *i.e.*, a zone of reduced flow velocity. From the stagnation zone, a mixing layer departs, which delineates the merging streams. At the downstream junction corner, the tributary flow may detach, causing the formation of a separation zone. Next to the separation zone, the merging flows passing through a narrowed cross-section are contracted, leading to increased velocities.

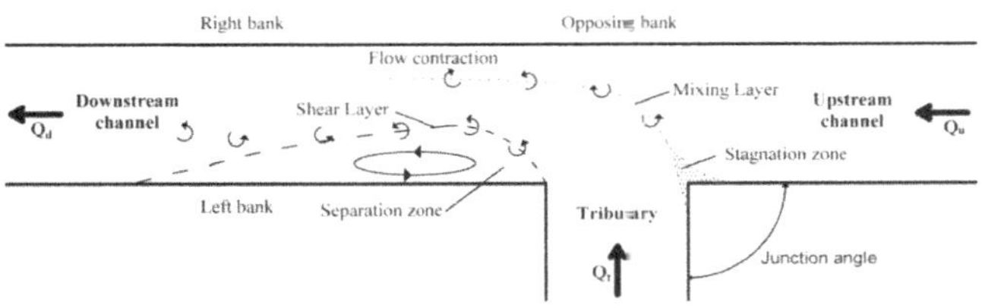

Figure 1: Conceptual model of an open channel confluence (after Best [1]).

Citation: Laurent Schindfessel, Stéphan Creëlle, and Tom De Mulder, Flow Patterns in an Open Channel Confluence with Increasingly Dominant Tributary Inflow, Water 2015, 7, 4724-4751; doi:10.3390/w7094724.

From the very beginning of laboratory research on confluences, starting with Taylor [2], the ratio of incoming discharges has been recognized as a key parameter [1,2,3]. However, laboratory research was mainly performed with discharge ratios in the range of $0.1 \lesssim q \lesssim 0.9$, at least for the case of asymmetrical confluences with a fixed concordant bed and subcritical flow throughout [12]; the type of confluence that will be further considered in this paper. In the above-mentioned equation, the discharge ratio q is defined as the ratio of upstream to downstream channel discharge (see Figure 1):

$$q = \frac{Q_u}{Q_d} = \frac{Q_u}{Q_u + Q_t} \tag{1}$$

Although it is known that the three-dimensionality of the flow increases with a decreasing discharge ratio, *i.e.*, when the tributary becomes dominant, indications of the flow features at $q < 0.1$ are scarce. For example, Webber & Greated [3] observed a decreasing accuracy of their theoretical model at lower q. Gurram *et al.* [6] were the first to provide a qualitative view on the case where $q = 0$ (*i.e.*, all flow coming from the tributary) by means of photographs. They noted the significant impact of the tributary flow on the opposing wall, and alluded to flow in a mitre bend. The uniqueness of $q = 0$ has also been demonstrated for transcritical flow [4], where a local maximum in the water height at the opposing wall is found when $q = 0$, a phenomenon that is absent at the other investigated discharge ratios. Moreover, the flow field indicates inflow from the confluence area into the upstream channel when $q = 0$. For subcritical flows, Shumate [8] reported that the case $q = 0.083$ (*i.e.*, the case with the smallest discharge ratio amongst his experiments) did not follow the trends deduced from the cases with higher discharge ratios. He reported the separation zone was shorter than expected due to the reflection of the lateral flow in the upstream channel. All the foregoing elements offer evidence that distinct flow features are present at sufficiently small discharge ratios. However, no systematic study of these patterns has been undertaken.

As preliminary research, Schindfessel *et al.* [12] experimentally compared two discharge ratios, $q = 0.25$ and $q = 0.05$, in a 90° angle confluence flume and found some particularities at the small discharge ratio. The impinging tributary flow led to a recirculating eddy in the upstream channel, as well as to changes in the helicoidal cells.

Field-based research also acknowledged the importance of the discharge ratio (e.g., [14,15]), not only on time-averaged flow, but also on intermittent features [16,17]. According to the literature overview provided by Leite Ribeiro *et al.* [18], the only morphological study including $q < 0.1$ was provided by Rhoads *et al.* [15]. These authors found that low discharge ratios are associated with a distinct bed topography, whereby the scour hole is located close to the opposing bank (seen from the tributary), leading to erosion of that bank in the long term. This is caused by the tributary flow protruding far into the confluence zone. Though this confirms the importance of small discharge ratios, values of $q < 0.1$ occurred only a limited number of times within the time period covered by the measurements. Hence, the aforementioned results cannot be linked to a specific discharge ratio. For the sake of completeness, Best [1] monitored a flood event in which q decreased to less than 0.1, resulting in an increased penetration of the tributary avalanche into the confluence, and an erosion of the upstream channel avalanche.

In summary, the knowledge of flow patterns when the tributary discharge is dominant, say $q < 0.1$, is not that well developed, neither in laboratory nor in field experiments. Nevertheless, such discharge ratios can be present in the case of confluent rivers with a distinct hydrological regime [1,15]. Other examples of confluences with extreme discharge ratios pertain to more engineered situations, e.g., the junction of a river and a channel bypassing a weir or a navigation lock in the river (meant for spilling the river discharge or for fish migration purposes), or networks in a sewer or drainage system.

Recognizing the knowledge gap, this paper offers a systematic study of the evolution of the flow patterns at a (specific) open channel confluence when the tributary discharge becomes increasingly dominant. Questions that arise are:

1. What happens at extremely low discharge ratios when the tributary flow impinges on the opposing wall? How does this influence the small inflow coming from the upstream channel?
2. Are known trends regarding flow patterns in the function of the discharge ratio confirmed at limiting small discharge ratios?
3. Does the small discharge ratio influence the position of the mixing interface and possible helicoidal cells?
4. To what extent does the small discharge ratio influence possible intermittent flow features?

In order to elucidate these issues, a 90° confluence with a fixed concordant bed was studied by means of numerical simulations that are (partially) validated by laboratory experiments. Though the focus is on a detailed study of velocity fields, both engineering and earth sciences may benefit from an improved understanding of flow features at confluences in which the tributary discharge is dominant.

LABORATORY EXPERIMENTS FOR NUMERICAL MODEL VALIDATION

Experiments are carried out in an existing 90° angle confluence flume with concrete walls. All the channels of the confluence are horizontal, concordant and have a chamfered rectangular section with a width W of 0.98 m, as depicted inFigure 2. Note that the chamfered cross-section is typical for rectangular sections made with *in situ* cast concrete, where the chamfering is meant to simplify the removal of the formwork. While to the best of our knowledge no other laboratory studies used a chamfered section, movable bed studies and field sites exhibit a wide range of cross-sections, which may show some similitude to a chamfered rectangular section. At the upstream boundaries, flow straightening screens are installed, so that a relatively uniform and stabilized flow is retrieved. The downstream water level ($h = 0.415$m) is controlled by means of a weir.

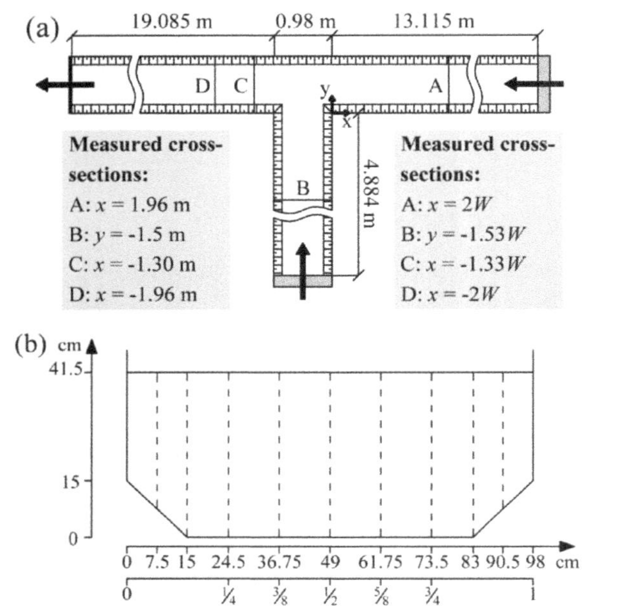

Figure 2: (**a**) Top view of the laboratory experiment; (**b**) Cross-section of channels.

Of the two discharge ratios that are investigated experimentally, one ($q = 0.25$) is inside the range covered in literature ($0.1 \lesssim q \lesssim 0.9$), and one ($q = 0.05$) is deliberately chosen outside that range. In both experiments, the total downstream discharge Q_d is kept constant at a value of 40 L/s. In addition, the downstream water level is constant, and hence the downstream Froude number Fr_d is constant as well, equalling 0.05. Although the Froude number is small compared to other laboratory experiments (see overview in [12]), the current value is in the typical range for lowland rivers [19]. As expected for such low subcritical flows, measurements by means of ultrasonic water level sensors show that the variation in the water level over the confluence is negligible. The Reynolds number based on the hydraulic radius equals 9.8×10^4.

Flow velocity at the water surface is measured by means of surface particle tracking velocimetry (PTV). The PTV technique is conducted by seeding the water surface with floating polypropylene tracers, which are coated and have a maximum size of about 5 mm. A 1920 × 1080 camera takes images of the tracer at a frequency of 30 Hz. In processing, these images are rectified by means of the software FIJI, and the individual particle paths are tracked at a rate of 10 Hz. Eventually, the useful image recordings have a length of about 3 min. By using the PTV technique, it is possible to measure the surface velocities with a very high spatial resolution, providing better results than the particle image velocimetry technique adopted in preliminary experiments reported in [12].

Complementary to the measurements of surface velocity, Acoustic Doppler Velocimetry (ADV) measurements are performed in a selection of cross-sections, as indicated in Figure 2. The ADV equipment, namely a Nortek Vectrino II, is a profiling instrument able to measure profiles of 3 cm, but the sweet spot offers the largest accuracy over the profile. The vertical height of the measurement cell equals 2 mm, and only measurements with a sufficiently high signal-to-noise ratio and correlation larger than 80% are retained. The selected despiking algorithm is that of Goring & Nikora [20], as discussed by Wahl [21]. Following experience from preliminary measurements, the measuring time of 2 min is sufficient to capture the mean velocity.

Additionally, the total measurement error of ADV is estimated by performing different measurements at the same point, while resetting the incoming discharges in the confluence flume and repositioning the instrument at the given measurement point. This verification is done in the upstream channel, and results in a standard deviation of the mean of about 0.001 m/s in the longitudinal direction and 0.002 m/s in the lateral and vertical directions. It should be mentioned that the aforementioned uncertainties are lower bounds for the ones in the confluence area, where complex 3D flows are present. Especially near strong local gradients, a small positioning error will result in a relatively large velocity error, *i.e.*, measurement error is inseparable from measurement location. As a check on both measurements techniques, PTV measurements are compared to ADV results close to the water surface, showing a good agreement.

The adopted coordinate system is presented in Figure 2. Since the *z*-axis, originating at the bed, points upwards, a right-handed system is established. In this work, the instantaneous velocity components in the longitudinal (x), lateral (y) and vertical (z) direction are denoted as u, v and w, respectively, whereas the corresponding time-averaged velocity components are denoted as $\overline{u}$, $\overline{v}$ and $\overline{w}$, respectively. As a length scale and velocity scale, the width of the channels W (= 0.98 m) and the bulk velocity in the downstream channel U_d (= 0.104 m·s^{-1}), respectively, are utilized. The (downstream) water depth h (= 0.415 m) is also used as a length scale in the vertical direction, when appropriate.

NUMERICAL MODEL SIMULATIONS

Model Description

In order to simulate the 3D complex flow patterns in a confluence, several numerical modeling strategies are available, based on different assumptions; hence governed by different equations, and requiring different amounts of computational effort. Since the research objectives include the study of dynamic flow features like coherent structures, intermittent flow patterns and upwelling of the flow, use will be made of 3D large-eddy simulations (LESs), which require more computer resources than RANS models but are more appropriate to resolve the aforementioned (time-dependent) features. The theory behind this type of simulation has been explained extensively in literature [22,23,24]. Briefly, the Navier-Stokes equations are spatially filtered, such that smallest scales (eddies) are removed. Their influence is modeled by a so-called sub-grid scale (SGS) model. The large eddies are retained and are resolved on a mesh in space and time, by solving a discretized version of the filtered equations. LESs have already been applied successfully to study confluence hydrodynamics (e.g., [13,16,17,25]).

The present simulations are run in the OpenFOAM suite (version 2.2.2, OpenFOAM foundation, London, UK), which is an open-source code for computational fluid dynamics. As an SGS model, the standard Smagorinsky model is selected, since this has successfully been applied to laboratory confluences [25], and sensitivity to modeling dissipation in the smallest scales is small compared with the influence of boundary conditions and mesh resolution [26,27].

In order to reduce the required computational efforts, and to simulate rough walls more easily, the near-wall boundary layer is not fully resolved (as in [13]), neither is a near-wall RANS model applied (as in [16,17]) but it is simply modeled by means of a wall function (as in [25]). The wall function that was implemented by the authors in OpenFOAM discerns three sublayers of the turbulent boundary layer:

$$\begin{cases} v^+ = \dfrac{1}{\kappa}\ln(z^+) + 5.5 + B & \text{if} \quad z^+ \geq 30 \\ v^+ = 2\dfrac{1}{\kappa}\ln(z^+) - 3.05 + B & \text{if} \quad 30 > z^+ > 5 \\ v^+ = z^+ & \text{if} \quad z^+ \leq 5 \end{cases} \tag{2}$$

where v^+ and z^+ are the wall coordinates; k is the von Karman constant and the parameter B accounts for the wall roughness [28]:

$$\begin{cases} B = 0 & \text{if} \quad k_s^+ < 2.25 \\ B = \dfrac{1}{\kappa}\ln\left(\dfrac{k_s^+ - 2.25}{87.75} + C\,k_s^+\right)\sin[0.4258\,(\ln(k_s^+) - 0.811)] & \text{if} \quad 2.25 \leq k_s^+ < 90 \\ B = \dfrac{1}{\kappa}\ln(1 + C\,k_s^+) & \text{if} \quad k_s^+ \geq 90 \end{cases} \tag{3}$$

where k_s^+ is the dimensionless roughness height and C is a tuning parameter (equal to 0.5). Although the assumptions underlying the wall function may not be valid in complex flows, the wall function, *i.e.*, Equation (2) above, has been used successfully by van Balen *et al.* [26,29] in a mildly curved bend, (though with a more simple formulation for the wall roughness parameter B than Equation (3)). In the present work, the wall roughness height used in Equation (3) is set equal to 1 mm, which provides the best correspondence with the experiments in the confluence flume with the concrete walls. Roughness has indeed a significant impact on flow in a confluence [30].

For an LES, it is important that the incoming flow contains sufficient and correct turbulence scales [22,23,24]. Therefore, the inlet velocity field is copied from a section downstream of the inlet, onto which random fluctuations are imposed. This implementation resembles a precursor iteration, only with the precursor domain attached to the domain of interest.

Such a situation is advantageous, since no additional memory storage is required. At the downstream boundary, convective boundary conditions are applied. In the present simulations, the free surface is modeled by means of a so-called rigid lid. This approach has been applied to many types of open channel flows, see e.g., [16,17,23,24,26,27,29], although more complex surface tracking methods are also in use [9,13,25,31]. However, because of the small Froude number in the present experiments, the rigid lid is certainly a valid approach, and preferable because of reduced computational efforts.

The mesh contains 4.2 million cells, with 106 × 37 cells in the lateral and vertical direction. All channels in the computational domain of the confluence are 5 m long, which might seem long for the inlet channels. However, recalling that this length includes the precursor simulation, it ensures fully developed inlet flow. The mesh is slightly coarsened towards the inlet and outlet boundaries, where the largest cell aspect ratios of the domain can be found, equaling 3. Discretization in space (finite volume) and time are second order accurate. The time-step in the LES equals 0.025 s (= 2.65×10^{-3} W/U_d). After about 7.5 flow through times (= 760 s = 80 W/U_d), collection of flow data started ($t = 0$), over an additional 16.5 flow through times (= 1640 s = 175 W/U_d).

Two simulations were performed corresponding to the two discharge ratios (q = 0.25 and q = 0.05) in the experiments. Additionally, the limiting case $q = 0$, *i.e.*, all discharge coming from the tributary, was simulated in a third LES. The mesh remains the same for all the simulations, which aids the intercomparison of the results.

Verification of the Simulations

As a first step in the verification process, it is checked whether the dimensionless wall distance z^+ is in an acceptable range for the use of a wall function. It is found that z^+ is generally sufficiently high (20 to 45) to allow the use of a wall function (requiring $20 \leq z^+ \leq 500$ [22,23,24]). However, in regions of very low velocity, e.g., still flow in the upstream channel in the case where $q = 0$, the use of a wall function introduces some inaccuracies [27], due to z+ being lower than 20. However, to keep inaccuracies as small as possible, the adopted wall function includes the buffer layer and the laminar sublayer.

In a second step, the percentage of resolved turbulent kinetic energy is assessed, as advocated by Pope [32]. Practically, the unresolved sub-grid energy is estimated by employing the formulations of Coussement *et al.* [33]. In contrast to an LES where the full boundary layer is resolved, the main part of the turbulent kinetic energy in the cells close to the wall is modeled by the wall function. Accordingly, these cells will by definition match the criterion to a lesser extent. This explains why not all cells exhibit a resolution of minimum 80%, the threshold value advocated by Pope [32] (Figure 3a). Notably in the upstream channel at $q = 0.25$, the relatively low velocity causes the first cell to be in the buffer layer. Hence, much turbulent kinetic energy is modeled here, leading locally to a less resolved LES. At even lower velocity in the upstream channel (*i.e.*, at decreasing q), the cells closest to the wall are located in the viscous sublayer (recall that all discharge ratios are simulated using the same mesh), so that the LES is more resolved (Figure 3a). Of the cells that resolve less than 80% turbulent energy in q = 0.25, 72% is located in the upstream channel ($x > 0$). This confirms that the region of interest, *i.e.*, the confluence area ($x < 0$), is well resolved.

Comparison of the simulation results with a finer mesh (consisting of 4 times the aforementioned number of cells) did not result in significant differences; hence, it is concluded that the results from the adopted mesh are sufficiently mesh-independent.

Further verification includes determining the Kolmogorov time scale, and linking this to spectral density. For the point with coordinates (x = −1.33 W, y = 0.75 W, z = 0.95 h), the Kolmogorov scale equals about 1 s, which corresponds to the end of the −5/3 law in the spectrum, as is to be expected (Figure 3b). The upper axis of Figure 3b provides the number of cells associated with an eddy of a certain frequency (adopting a typical mesh width Δ of 0.01 m), indicating that the structures associated with the Kolmogorov time scale are about 10 cells large. This corresponds to rules of thumb regarding the minimum number of cells necessary to resolve an eddy (5 to 20 [22]; ≥ 6 [23]). Additionally, the −5/3 region is present over about an order of magnitude, affirming the LESs are sufficiently well resolved.

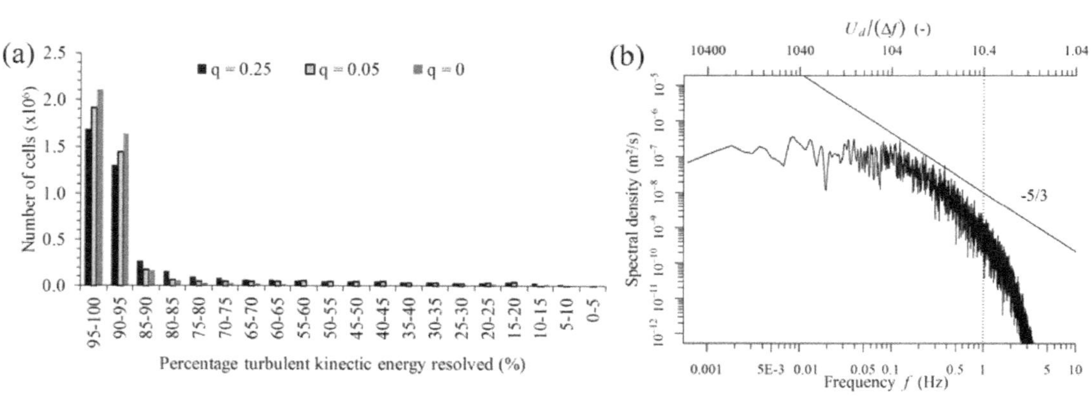

Figure 3: (**a**) Percentage of turbulent kinetic energy resolved in the simulations; (**b**) Spectral density of the turbulent kinetic energy at the point with coordinates ($x = -1.33W$, $y = 0.75$ W, $z = 0.95$ h) at $q = 0.25$. The frequency associated with the Kolmogorov scale is indicated by a dotted line.

Validation of the Simulations

A comparison of the LES velocity results with the experimental ADV data (in the cross-sections $x = -1.33\ W$ and $x = -2\ W$) generally exhibits a close correspondence (Figure 4). The velocity gradient over the shear layer of the separation zone might appear sharper in the numerical simulations, but this can be attributed to the linear interpolation of $\overline{u}$ between the measured verticals. Although the lateral and vertical velocity also show a good correspondence, a zone of reduced $\overline{v}$ and $\overline{w}$ is present near the top right in the case where $q = 0.25$ (Figure 4a,c), while the LESs predict a cell rotating in the clockwise direction. Nevertheless, the LESs correctly simulate that this cell changes direction in the $q = 0.05$ case (to be discussed in Section 4.2.4), see Figure 4e,f. This indicates that trends are correctly simulated. Finally, a comparison is also made between the PTV (surface) data and the LES (at a height of $z = 0.9\ h$ to avoid the influence of the rigid lid) in Figure 5. It should be mentioned that in zones of upwelling flow (see Figure 6 in the results section for a presentation of flow features), the PTV might be biased to record a reduced upwelling, as these zones are rapidly depleted of tracers. This explains some differences between the PTV and LES at the end of the separation zone, and in the zone of upwelling flow near the right bank in Figure 5d. However, the correspondence, both in the longitudinal and lateral direction, demonstrates the appropriateness of the rigid lid approximation, and confirms the reliability of the LES results over a wide range of the confluence area.

Compared with earlier simulations [34], the adopted wall function and inlet condition in this paper prove superior in modeling the experimental set-up.

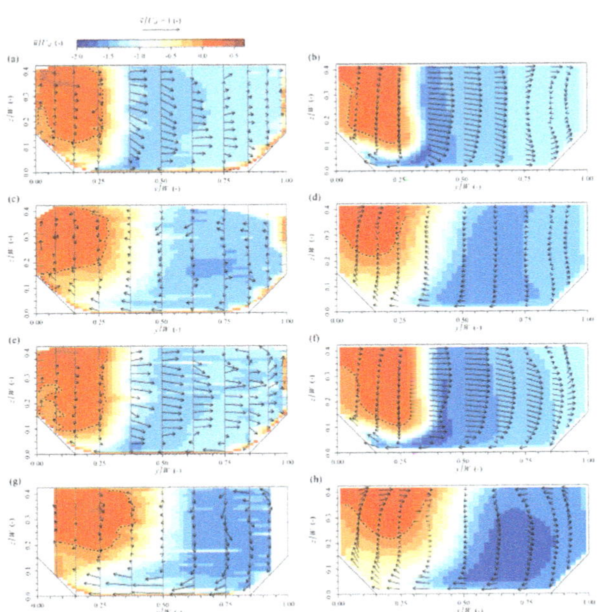

Figure 4: Comparison of experimental measurements on the left and numerical results on the right, for the measured cross-sections. The color scale indicates the time-averaged longitudinal velocity, which is interpolated between the measured verticals for the ADV data, and arrows represent lateral and vertical velocity. The dotted line indicates $\overline{u} = 0$, *i.e.*, separates upstream and downstream flow. In the left figures, the measured locations are indicated with grey lines, and the arrows are located in the sweet spots. (**a**,**b**) $q = 0.25$, $x = -1.33\ W$; (**c**,**d**) $q = 0.25$, $x = -2\ W$; (**e**,**f**) $q = 0.05$, $x = -1.33\ W$; (**g**,**h**) $q = 0.05$, $x = -2\ W$.

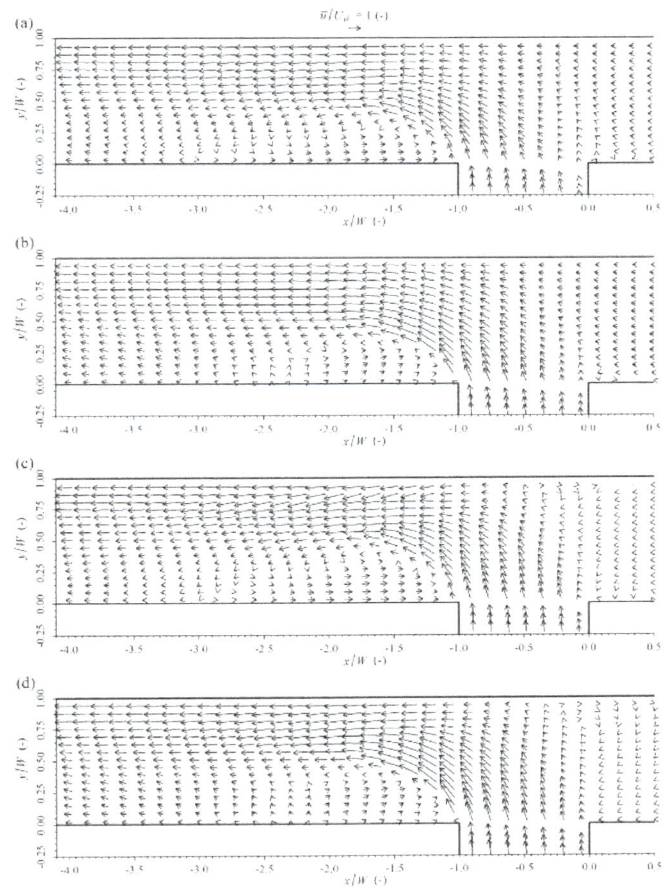

Figure 5: Comparison of experimental measurements of the surface velocity (PTV) and the numerical predictions. (**a**) $q = 0.25$, experimental; (**b**) $q = 0.25$, LES; (**c**) $q = 0.05$, experimental; (**d**) $q = 0.05$, LES.

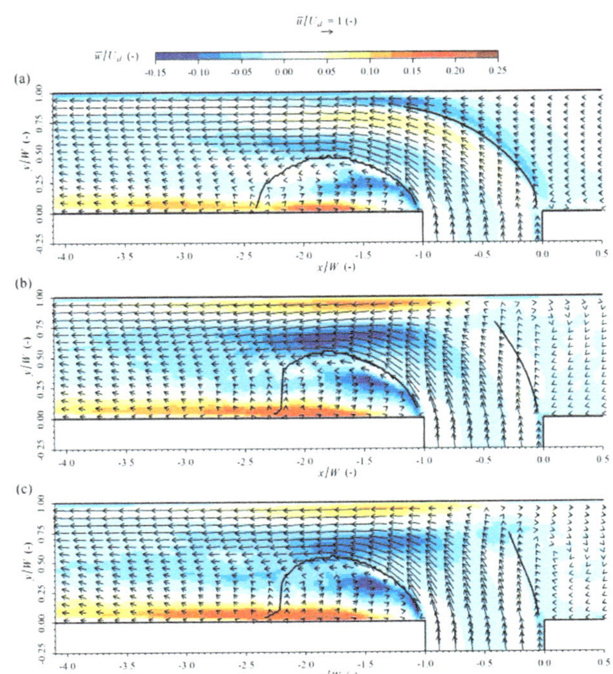

Figure 6: Time-averaged flow patterns near the surface ($z = 0.9\ h$).

Arrows indicate the longitudinal and lateral velocity component, while the color represents the vertical velocity component. The location of the mixing interface and the extent of the separation zone are indicated by a line. (**a**) $q = 0.25$; (**b**) $q = 0.05$; (**c**) $q = 0$.

RESULTS

In Section 4.1, the flow features will be presented for the reference case characterized by a moderate discharge ratio $q = 0.25$, *i.e.*, the case in which the tributary incoming flow rate Q_t is (only) 3 times larger than the upstream channel incoming flow rate Q_m (or $U_tQ_t/U_mQ_m = 9$). The changes in flow features at small discharge ratios will be presented in Section 4.2 for the cases with $q = 0.05$ (or $Q_t/Q_m = 19$ and $U_tQ_t/U_mQ_m = 361$) and $q = 0$ (or $Q_t/Q_m \to \infty$ and $U_tQ_t/U_mQ_m \to \infty$).

Reference Case with Moderate Discharge Ratio (q = 0.25)

To illustrate the flow patterns, Figure 6a presents a plot of the (time-averaged) velocity at a height of $1/10h$ below the surface ($z = 0.9h$). The arrows represent the velocity vector in the horizontal plane, while the color scale represents the vertical velocity component $\overline{w}$. For clarity, Figure 6a also presents the location of the mixing layer and the extent of the separation zone. These locations are determined as follows.

The mixing layer is determined by means of the streamline departing from the confluence apex, in the horizontal plane $z = 0.9h$ (*i.e.*, neglecting the vertical velocity component), since Mignot *et al.* [11] also used this starting point for a 2D streamline marking the mixing layer. The separation zone, which is present in the schematized sections when the tributary flow is sufficiently important, has been delineated in literature by different methodologies. For example, Best & Reid [35] used surface dye tracers and neutrally buoyant particles. Gurram *et al.* [6] measured the separation zone by means of dye injections. Finally, Shumate [8] utilized his three-dimensional velocity measurements. In this paper, the (time-averaged) velocity field is also utilized to determine the extent of the separation zone. To this end, the y-coordinate Y at which there is a net zero discharge is determined for a given cross-section (with coordinate x) and at a given elevation above the bed (with coordinate z):

$$\int_0^Y \overline{u}(x, y, z)\, \mathrm{d}y = 0 \tag{4}$$

The time-averaged flow in the reference case $q = 0.25$ will now be presented. Briefly, the flow has all the well-known features of a confluence flow (Figure 1). At the upstream confluence corner, a stagnation zone is present. The streamlines, which start there, separate the tributary inflow from the upstream channel inflow. The merging of the two flows causes the upstream channel flow to deflect towards the right bank, as can be derived from the path of the streamline between the incoming flows. These streamlines actually delineate a shear layer (also called the mixing interface), *i.e.*, a zone of strong momentum gradients in the direction normal to the flow [11]. The vertical velocity shows that the flow from the upstream channel dives near the mixing layer, while the tributary flow surfaces (due to a helicoidal cell, see Section 4.2.3). At the downstream corner of the confluence, the tributary flow detaches from the wall, and a separation zone is formed. In the separation zone, the flow near the wall has typically an upstream oriented velocity component, *i.e.*, recirculating flow is present. In the separation zone, the flow generally dives, because the entrainment by the tributary causes a decrease in water level [6,7,35]. However, near the left bank in the separation zone, upwelling flow is present. This is caused by flow intruding the separation zone, starting at the bed. As a consequence, the separation zone is smaller near the bed than at the top of the water column [8,11,12,13]. Preliminary numerical simulations have indicated that this upwelling is enhanced by the presence of the chamfers [34]. Next to the separation zone, the joining flow from the two merging channels is contracted. The disappearance of gradients, which is enforced by the flow contraction, fades the shear layer, though the two flows can still be discerned downstream by means of the vertical velocity. Eventually, the flow takes the full width of the downstream channel, though fully uniform flow is not yet achieved in the numerical model.

Thus, the time-averaged flow corresponds to the known flow patterns (Figure 1) according to the conceptual model proposed by [1]. Experiments in the well-known range $q \gtrsim 0.1$ have shown that the three-dimensionality of these features intensifies at decreasing q [6,7,8,13].

Cases with a Small Discharge Ratio (q = 0.05 and q = 0)

Confluence Area and Upstream Channel

When the discharge ratio decreases, *i.e.*, less flow is coming from the upstream branch, the tributary flow is less deflected by the upstream channel. This can be illustrated by means of the average inflow angle α at the cross-section $y = 0$ with which the tributary flow enters the confluence:

$$\alpha = \int_A \operatorname{atan}(\, \overline{v} \,/\, -\overline{u}\,)\, \mathrm{d}A \tag{5}$$

The results show that the average inflow angle increases from $\alpha = 65°$ at $q = 0.25$ to $\alpha = 71°$ at $q = 0$. Note that this angle is less than the geometrical confluence angle $\theta = 90°$, as was also observed in other laboratory experiments [4,7]. As a consequence of both the increased tributary discharge and the increased inflow angle, the tributary flow protrudes further towards the opposing bank. This can be seen by observing the location of the mixing layer in Figure 6: the mixing layer shifts towards the upstream channel, as the inflow from the upstream channel decreases. When comparing the cases $q = 0.25$ and $q = 0.05$, and observing the position of the mixing layer, one can see that the tributary flow stretches to the opposing bank in the case $q = 0.05$. As is observed when looking at the vertical velocity, it is more correct to say that the tributary flow impinges on the opposing bank: near the right bank in the downstream channel, a zone of upwelling flow is present, which already starts in the confluence area. This is consistent with the preliminary findings of Schindfessel *et al.* [12], and is also found lower in the water column in the present simulations. For the $q = 0$ case, the velocity vectors on the $x = -0.30\ W$ line are even more orthogonal to the wall. Thus, the impinging is a new flow feature, absent at $q = 0.25$, but present at $q =$

0.05 and $q = 0$. While the main part of the impinging flow is directed to the downstream channel, a part is pushed into the upstream channel. There, it drives a horizontal recirculating eddy (or gyre), which blocks the upstream channel incoming flow (Figure 6). Looking at the section $x/W = 0$, presented in Figure 7, clarifies the flow patterns at the interface between upstream channel and the confluence area. At $q = 0.25$, nearly all velocity is directed in the downstream direction, apart from some zones near the bottom. The highest longitudinal velocities are present near the right bank, while closer to the left bank (*i.e.*, to the upstream confluence corner), the longitudinal velocity magnitude is lower. This corresponds to the known flow patterns (Figure 1 after the conceptual model proposed by Best [1]): near the upstream confluence angle, a zone of stagnating flow develops, while the flow from the upstream channel is directed to the right bank by the momentum coming from the tributary inflow.

Due to the appearance of a horizontal recirculating eddy at small discharge ratio, however, flow from the confluence area protrudes into the upstream channel (upstream oriented velocity $\overline{u}$ in Figure 6b,c). Where the highest downstream velocity was noted in the $q = 0.25$ case, the highest upstream velocity is now found. Therefore, the position of the fastest downstream flow is shifted to the other bank as the discharge ratio approaches zero. This rupture from the trend at more moderate discharge ratios is attributed to the appearance of the recirculating eddy, induced by the impinging flow. Note that lateral and vertical velocity components are negligible to the longitudinal component, *i.e.*, the recirculating eddy is a horizontal flow feature. The eddy also reduces the stagnation zone, in the sense that at $q \leq 0.05$, no region of substantially lower flow is present near the confluence apex (left bank in Figure 7).

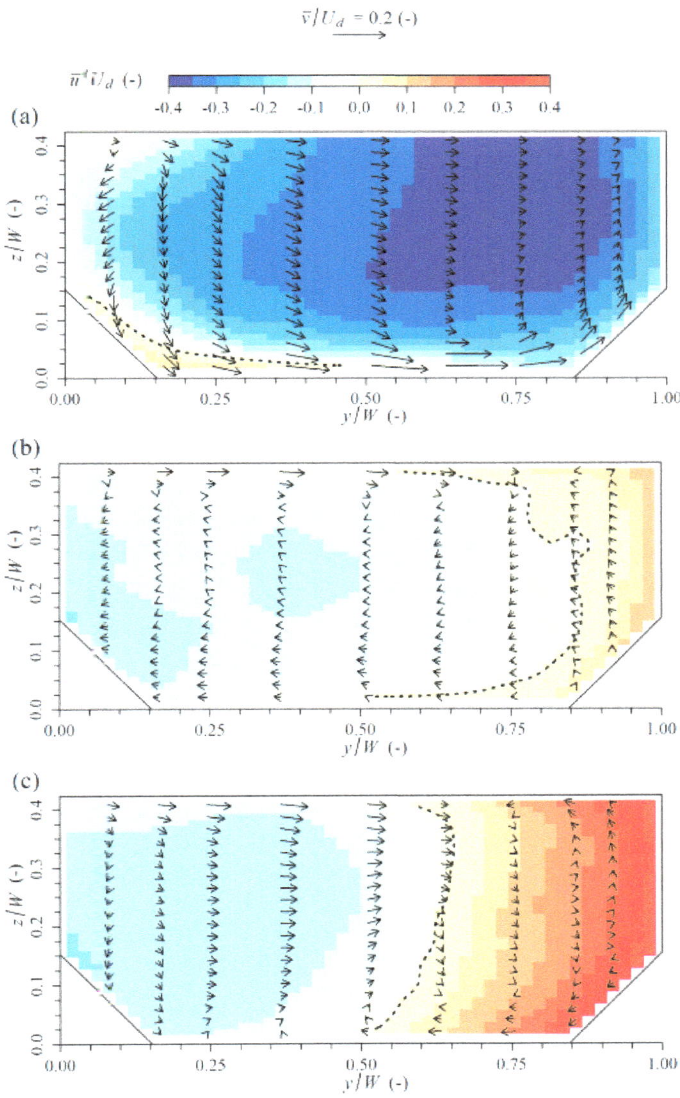

Figure 7: Cross section at $x/W = 0$, looking downstream. Color scale represents the time-averaged downstream velocity $\overline{u}$. Arrows represent lateral and vertical velocity. The dotted line indicates $\overline{u} = 0$, *i.e.*, separates upstream and downstream flow; (**a**) $q = 0.25$; (**b**) $q = 0.05$; (**c**) $q = 0$.

To quantify the importance of the recirculating eddy, and its growth when the discharge ratio decreases, the total discharge Q_u through the section $x/W = 0$ (*i.e.*, the last section of the upstream channel) is split into two parts: $Q_u = Q_{u,out} + Q_{u,in}$. The "outflow" discharge $Q_{u,out}$ represents the surface integral of all flow velocities oriented from the upstream channel to the confluence, whereas $Q_{u,in}$ denotes the surface integral of all flow velocities oriented from the confluence to the upstream channel.

Although the recirculating eddy appears first at $q = 0.05$, its influence cannot be seen in the discharge parameters (Table 1). At $q = 0$, however, when there is no discharge in the upstream channel, the recirculating eddy pumps a discharge in the upstream channel that is of similar strength as the flow of $q = 0.05$. Obviously, the outflow discharge equals the inflow discharge, as there is no net discharge in the upstream channel. Table 1 thus quantifies that the exchange of mass from the confluence area into the upstream channel starts to be significant when $q < 0.05$.

Table 1: Composition of the total discharge Q_u through the section $x/W = 0$, for different discharge ratios q.

q (–)	Q_u ($m^3 \cdot s^{-1}$)	$Q_{u,in}$ ($m^3 \cdot s^{-1}$)	$Q_{u,out}$ ($m^3 \cdot s^{-1}$)
0.25	0.010	0.000	0.010
0.05	0.002	0.000	0.002
0	0.000	0.002	0.002

Mixing Layer

The mixing layer, *i.e.*, the shear surface between the incoming flows, is delineated (pragmatically) by the streamline starting at the confluence apex (see above). Compared with mixing layers in natural bathymetries [16,17], the mixing layer of the $q = 0.25$ case is relatively curved, as is typical for asymmetrical 90° confluences [11]. Due to the momentum inflow from the tributary, the mixing layer extends nearly to the opposing bank when entering the downstream channel (Figure 6). Only few vortices form in the mixing layer, and their strength is considerably smaller than the Kelvin-Helmholtz vortices that can be found in the separation zone (not shown). This may be attributed to the curvature of the mixing layer [36]. In the cases with dominant tributary flow, the increased flow from the tributary pushes the mixing layer towards the upstream channel, leading to a less curved mixing layer. When calculating the velocity gradient over the mixing layer, it is observed that the initial gradient is significantly higher at $q = 0$ compared with $q = 0.25$ (not shown). This is not only attributed to the larger velocity difference between the incoming channels, which is an obvious consequence of a smaller discharge ratio, but also to a decrease in the extent of the stagnation zone. The stagnation zone is not only reduced on the tributary side by the tributary's increased discharge, but also on the upstream channel side, by the recirculating eddy.

Separation Zone

The maximum width and length of the separation zone, determined as explained previously, are shown in Table 2together with the predictions based upon Best & Reid [35]. The widths compare reasonably well with the regression line, yet the length of the separation zone is consistently smaller. This can be explained by the chamfered rectangular section in this research, in contrast to the rectangular section of Best & Reid [35], as the chamfers aid the vertical flow of high momentum fluid from the bottom to the top of the separation zone, effectively decreasing the length [34].

As the tributary discharge increases, the velocity difference between tributary and separation zone increases as well. Since the velocity difference is the driving force in a shear layer, the shear in the separation zone strengthens with smaller q, resulting in a higher turbulent kinetic energy (TKE) (Figure 8). Not only is the absolute value of the TKE in the shear layer larger at smaller q, the region over which high TKE values are found is also larger (notably in the downstream direction). Note that the maximum TKE is not localized on the limit of the separation zone as defined in this work, but is slightly shifted to the side of the separation zone.

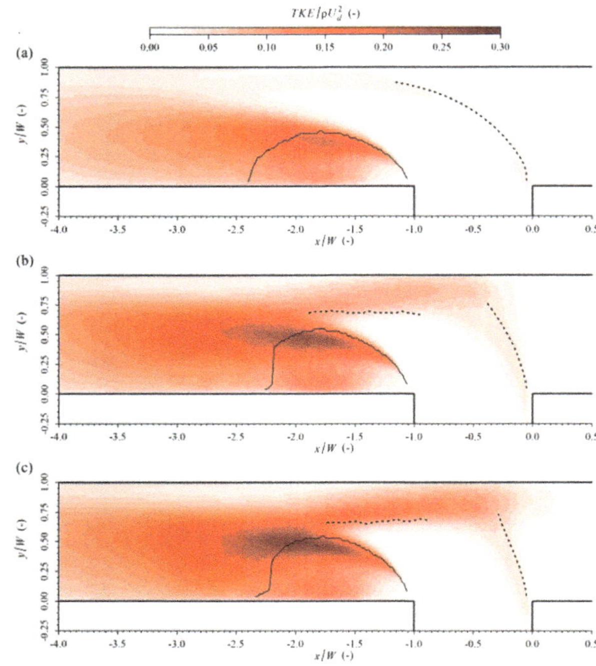

Figure 8: Turbulent kinetic energy (TKE) near the surface ($z = 0.9\ h$). The location of the mixing interface, the velocity gradient between impinging and non-impinging flow and the extent of the separation zone are indicated by a line. (**a**) $q = 0.25$; (**b**) $q = 0.05$; (**c**)$q = 0$.

Table 2: Dimensions of the separation zone ($z = 0.9\ h$).

	width/W (–)		length/W (-)	
q (-)	Present Work	Best & Reid [34]	Present Work	Best & Reid [34]
0.25	0.50	0.45	1.40	2.29
0.05	0.55	0.50	1.25	2.53
0	0.55	0.51	1.35	2.59

The largest TKE values can be found at about mid-depth, and are located closer to the wall, corresponding to the three-dimensional nature of the separation zone. Further, at lower depths, the TKE values in the separation zone and the region of high TKE values enlarge when the discharge ratio diminishes.

Contracted Flow and Flow Recovery Areas

The impinging flow that was indicated in Section 4.2.1 also has consequences for the flow patterns in the downstream channel. More specifically, the impinging causes the flow to surface near the right bank, as is well illustrated by the vertical velocity (Figure 6). In the $q = 0.05$ case, a large region of upwelling flow is noticed near the right bank ($y/W \approx 1$). Additionally, this zone is flanked by downward velocities in the contracting section. Such flow behavior was absent in the $q = 0.25$ case, where upwelling and diving were positioned differently, and were much weaker. The upwelling caused by impinging near the right bank in the $q = 0$ case is weaker than in the $q = 0.25$ case. Although this might be unexpected, decreased vertical velocities compared with $q = 0.05$ are already present lower in the water column at a more upstream position (near the opposing wall, around $x = -0.5\ W$ and $z = 0.50\ h$, not shown). This decreased vertical velocity is associated with increased water levels (pressure on the rigid lid) in the case $q = 0$ *versus* $q = 0.05$: the increased water pressure at $q = 0$ counters the upward vertical motion, thus explaining the decreased upwelling noticed in Figure 5.

In order to clarify the secondary currents in the downstream channel, the streamwise-oriented vorticity (SOV) is employed. This parameter is calculated from the time-averaged velocity field and is defined as $(\nabla \times \bar{u}) \cdot \bar{u} / \|\bar{u}\|$, whence it follows that it differs by a factor of $2\|\bar{u}\|$ from the helicity [37]. Using this parameter, Figure 9 shows the helicoidal cells in the section $x = -1.33\ W$, revealing that there are three regions of augmented SOV. First, the shear layer between the separation zone and the contracted flow exhibits elevated values of SOV. However, the velocity gradients present in this layer create vorticity, but are not indicative of a helicoidal cell. Second, near the bottom ($z < 0.05\ W$), a small clockwise-rotating cell is present (positive value of SOV), having its center around $y = 0.40\ W$. It transports fluid from the contracted flow, having high velocity, to the separation zone, corresponding to the upwelling flow discerned before (Figure 6). Note that this intrusion of high velocity fluid dissolves the separation zone (indicated by a dotted line in Figure 9) from below. At a smaller discharge ratio, this bottom cell increases in strength, causing more upwelling in the separation zone. Both the presence and the increased strength at smaller q are in correspondence with the experiments (Figure 4). Third, near the right bank, one or more helicoidal cells are present. At $q = 0.25$, one clockwise cell is clearly present near the surface, with a counterclockwise cell beneath it. At the smaller q, only a single counter-clockwise cell remains, located near the surface. Consequently, lateral surface velocities near the right bank have a changed sign, being the result of the impinging flow. This is a very noticeable change of small discharge ratios. Although the third cell in $q = 0.25$ is not clearly visible in the experimental results, the cell of $q = 0.05$ is definitely present in the experiments (Figure 4), confirming that the LES predicts the effect of diminishing q correctly.

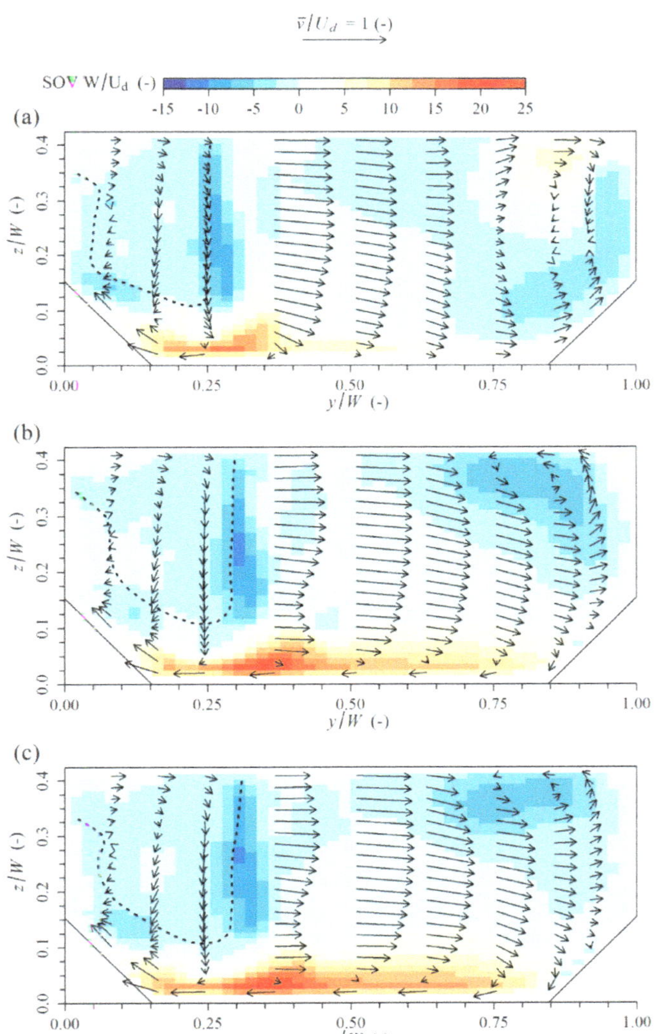

Figure 9: Cross section at $x = -1.33\ W$, looking downstream. Color scale represents the streamwise-oriented vorticity of the mean flow. Arrows represent time-averaged lateral and vertical velocity. The dotted line indicates $\bar{u} = 0$, *i.e.*, separates upstream and downstream flow; (**a**) $q = 0.25$; (**b**) $q = 0.05$; (**c**) $q = 0$.

In order to characterize the recovery towards uniform flow in the downstream channel, a momentum coefficient β is defined:

$$\beta = \frac{\int_A \bar{u}^2 \, dA}{Q_d U_d} \tag{6}$$

For a fully uniform flow β = 1, while the more non-uniform the velocity distribution is over the cross-section, the more β exceeds the unit value. Figure 10 presents the stated momentum coefficient, based on a selection of cross-sections. From comparing $q = 0.25$ and $q = 0.05$, it follows that at all longitudinal positions, a decreasing discharge ratio corresponds to an increasing non-uniformity. This is consistent with the increasing width of the separation zone (see above), which is the primary source of non-uniformity. However, at a discharge ratio equal to 0, the momentum coefficient levels out, *i.e.*, at some longitudinal positions it is marginally higher than in case $q = 0.05$, while at other positions it is slightly lower.

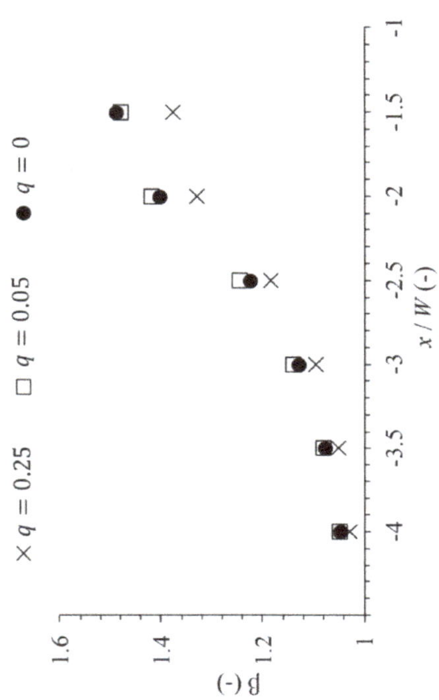

Figure 10: Momentum coefficient β of mean flow, characterizing the non-uniformity of the velocity distribution at different sections, as a function of the discharge ratio.

Additionally, the results of section $x = -4\ W$ show that the recovery to uniform flow takes a longer distance, if the discharge ratio is smaller, notwithstanding the larger cross-sectional velocities when q diminishes. This finding can also be noted in Figure 6.

Turbulent Kinetic Energy

The increased initial velocity gradient increases the TKE in the mixing layer (Figure 8), and causes vortices in the mixing layer to be slightly stronger when tributary flow is dominant and to be less dissipated than in the reference case. In all cases, a zone of elevated TKE originates in the mixing later, as was observed in literature [11,16,17], but in the cases $q = 0.05$ and $q = 0$, the TKE significantly increases in value when it reaches the opposing bank. At mid-depth, the highest TKE levels are found in the mixing layer and near the opposing bank, while the contracted flow has less elevated levels of TKE (not shown). Closer to the surface, however, the larger TKE values are found in the flow contraction (Figure 8). Additionally, a velocity difference in the direction normal to the streamwise velocity is present in the time-averaged velocity field, indicated in Figure 8 as a distinct shear interface. This velocity difference can, to some extent, also be seen in the velocity plots of Figure 6. From Figure 9, on the other hand, it can be discerned that this shear layer is localized where impinging flow and non-impinging flow meet, *i.e.*, large gradients in lateral velocity are present there. Although flow contraction is known to decrease velocity gradients, the velocity difference remains relatively constant when moving downstream, corresponding to the still important upwelling flow. The (three-dimensional) interface, where impinged flow meets with non-impinged flow, will lead to stronger fluctuations of the local velocity, and can thus explain the elevated levels of TKE in that region.

Intermittent Flow Patterns and Implications for Time-Averaged Results

In the laboratory experiments, was noticed that flow near the right bank exhibits a larger standard deviation in the case where $q = 0.05$ *versus* $q = 0.25$ [12]. In addition, it was visually observed that lateral flow is sometimes directed towards the bank, while at other instances in time, it is directed away from the bank. The change in sign of the lateral velocity seems to occur with a certain intermittency, as indicated by long ADV measurements at the point with coordinates ($x = -1.33\ W$, $y = 0.75\ W$, $z = 0.95\ h$) indicated in Figure 11.

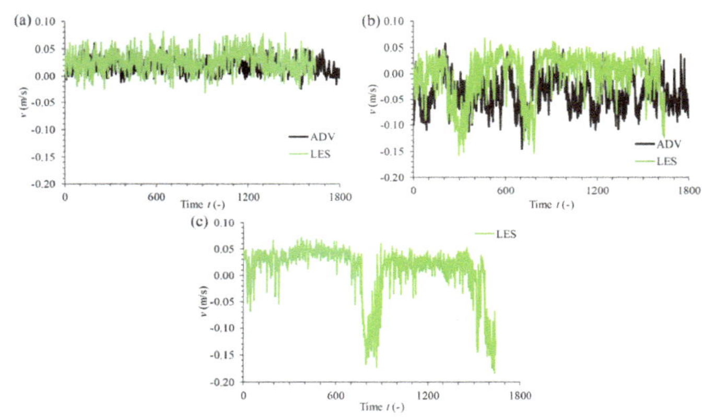

Figure 11: Time series of lateral velocity v at the point with coordinates ($x = -1.33\ W$, $y = 0.75\ W$, $z = 0.95\ h$), in the experiment (ADV) and in the LES. **(a)** $q = 0.25$; **(b)** $q = 0.05$; **(c)** $q = 0$.

At this point, the numerical simulations of case $q = 0.25$ seem to match closely to the experimental values, as is visible both from the time-series as well as from a probability density function (Figure 11a and Figure 12). For the case $q = 0.05$, the experimental and numerical probability density functions match less well, though similar intermittent patterns in velocity direction are observed, having similar amplitudes (Figure 11b). Although there is an acceptable agreement between the measured and simulated mean velocity at this point, the period of these events seems to be much longer in the simulations than in the experiment. However, as the simulation exhibits intermittency having a similar amplitude, the numerical results can be used to further elucidate this phenomenon.

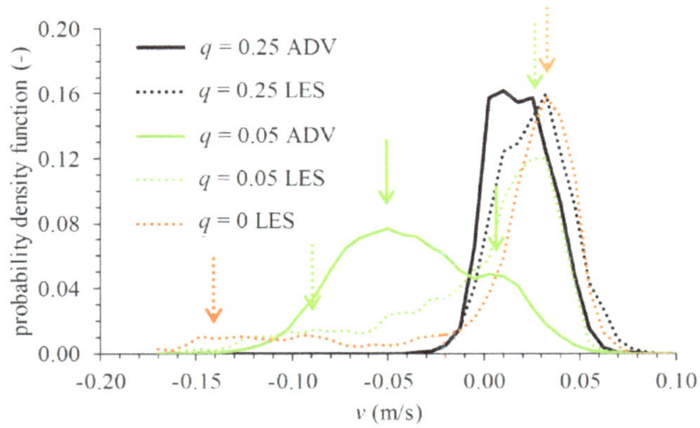

Figure 12: Probability density functions of the time series of lateral velocity v shown inFigure 11, both for the experiment (ADV) and for the LES.

Figure 13 presents instantaneous streamlines at two instances in time, representing the extreme patterns within the range for the case q = 0.05, and Figure 14 shows the cross-section $x = -1.33$ W at these instances. In the flow patterns that are present most of the time, the tributary flow impinging on the opposing bank is located near the top of the water column, and is reflected towards the bottom of the downstream channel (Figure 13a and Figure 14a). The flow patterns associated with the upwelling event, however, are substantially different. The streamlines illustrate that the impinged flow moves upwards,*i.e.*, strong upwelling is present, forming a strong helicoidal cell. The helicoidal cell is also discernible in the downstream channel (Figure 14b), explaining the negative v component in the time series (Figure 11b). Also note that the tributary flow reaching the opposing bank, *i.e.*, the flow that impinges, is located rather low in the water column compared with the flow patterns that are present most of the time in the LES.

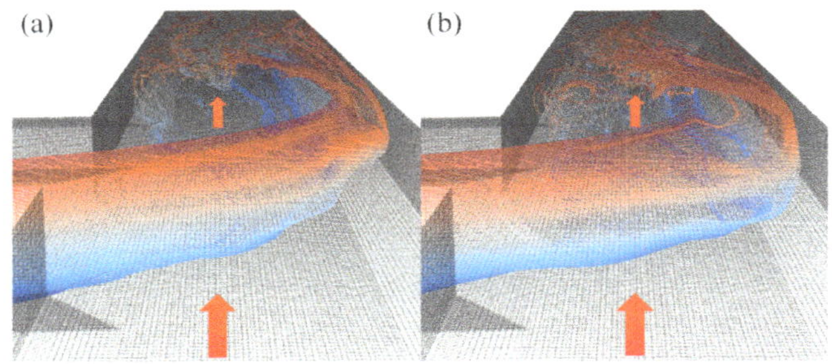

Figure 13: Instantaneous streamlines when q = 0.05, starting at x = $-0.20W$ in the tributary channel, at (**a**) t = 540 s; (**b**) t = 740 s. Color scale indicates the vertical height at which the streamline is located. Red arrows indicate flow direction.

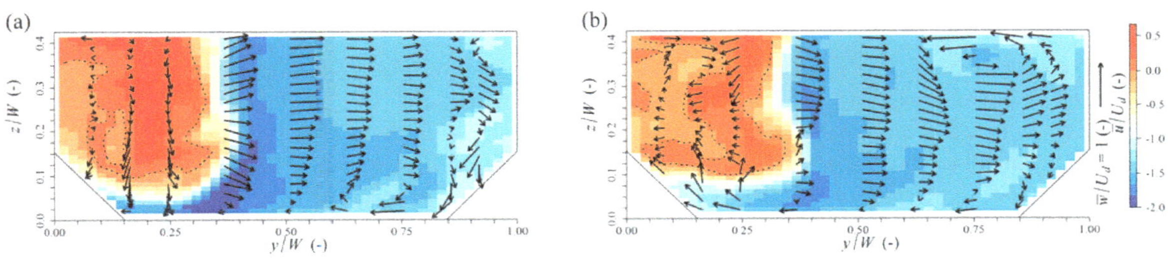

Figure 14: Instantaneous velocity in the section $x = -1.33$ W when q = 0.05, looking downstream, at (**a**) t = 540 s; (**b**) t = 740 s. Color scale indicates longitudinal velocity u, while arrows show the lateral and vertical velocity.

It is emphasized that such time dependent changes are also noted in the experiment, discernible by the two peaks in the histogram of Figure 12, indicated by the full green arrows. Changing surface velocity near the right bank was also observed visually by means of the PTV tracers. Notwithstanding that the LESs show larger periods, some similar events of high negative v are present (dotted green arrows in Figure 12). Although the probability density function accentuates the difference between the experiment and LES, the limb of the numerical histogram in the negative v is clearly non-zero.

For q = 0, similar intermittent shifts in velocities are noted (orange arrows in Figure 12), yet with seemingly larger intervals. In addition, the velocity amplitude is slightly larger. The instantaneous flow patterns of the modes resemble those of Figure 14, indicating a similar origin. Again, the position where the tributary impinges on the opposing bank seems to be associated with the intermittency.

The intermittent flow features that are observed trigger the question whether the aforementioned time-averaged flow was recorded over a sufficiently long period to be representative, even when the number of flow-through times seems acceptable. To assess this question, the time evolution of the time-averaged velocity is investigated. A good example is the q= 0 case: the difference in time-average velocity between 0 to 600 s and 0 to 1000 s is representative for the effect of the short but strong changes in flow patterns. Comparing these values in the horizontal plane, $z = 0.90h$ indicates that that the time-averages change only 0.006 m/s on average. Similar conclusions can be drawn when comparing time-averaged values at 900 s and 1800 s in the case where q = 0.05. Consequently, these results underline that the time-averaged flow fields presented before are representative, despite the fact that they contain a limited number of upwelling events. Although this shows that the simulation time was sufficiently long, it should be noted that the presence of the helicoidal cell near the right bank (Figure 9) in the time-averaged velocity field can mainly be attributed to the upwelling event.

DISCUSSION

Multiple authors have argued that the three-dimensional nature of the flow tends to intensify when the discharge ratio decreases (e.g., [6,7,8,13]). This trend is certainly confirmed by the present results, but moreover, some characteristic features of small discharge ratios are observed.

The most important novel feature of confluence flow at a small discharge ratio (*i.e.*, when the contribution of the tributary is sufficiently dominant), is that the tributary flow impinges on the opposing bank. Although it is obvious to focus on the flow in the downstream channel, this research indicates that the presence of a recirculating eddy in the upstream channel, induced by the impinging flow, cannot be neglected. From Figure 6 it follows that the recirculating eddy gains a lot of strength between $q = 0.05$ and $q = 0$, indicating the aforementioned effects will be weak when $q > 0.05$. This might explain why a recirculating eddy was not discerned in Shumate's [8] experiments at $q = 0.083$. One of the consequences of the recirculating eddy is that it effectively reduces the stagnation zone on the upstream channel side. Additionally, the stagnation zone is reduced on the tributary side when q diminishes, simply because more momentum is present. This was noticed by means of the average inflow angle α, which decreases with decreasing q, *i.e.*, the tributary inflow at $y = 0$ is less deflected. The observed inflow angle in the present simulations compares well with other studies, certainly when accounting for the much higher spatial resolution in LES compared with experiments: Hsu *et al.* [7] found an average angle of $\alpha = 75°$ at $q = 0.1$ and Hager [4] found $\alpha = 73°$ at $q = 0$. As the stagnation zone is thus reduced on both sides of the confluence apex, the stagnation point (*i.e.*, where the streamline dividing the incoming flows starts [38]) will not shift as much to the confluence side as expected from analytical considerations [38]. A possible consequence of the recirculating eddy is that the exchange of solutes between the confluence area and the upstream channel may drastically change for discharge ratios smaller than 0.05. In these situations, sedimentation in the center of the upstream channel may be expected. Bergeron & Roy [39] for example, noted the accumulation of gravel at the entrance of the upstream channel after the break of an ice jam in the tributary. Although the gravel bar probably originates from the period during which the tributary channel was jammed (causing less transport capacity at the confluence), the accumulation could have been reinforced during the period of small discharge ratio when the ice jam broke [39].

In the downstream channel, the separation zone is highly three-dimensional; hence, it is difficult to characterize its width or length in one number [6,8,9,10]. Nevertheless, the separation zone gradually widens at the surface when the discharge ratio decreases, closely corresponding to the logarithmic regression provided by Best & Reid [35], even when the discharge ratio is outside their experimental range. The determination of the length of the separation zone suffers from the presence of upwelling flow near the end of the separation zone [8]. The upwelling flow prevents the separated flow from reattaching to the bank, an issue that is enhanced by the chamfered section [34]. This might explain why the length of the separation zone does not match with the results of Best & Reid [35]. Moreover, in field-based research, it was found that flow separation is sometimes absent. The flow separation might be suppressed by the gentle curvature of natural confluence corners, or by improved flow alignment introduced by the bathymetry [40,41,42]. Additionally, sediment can easily accumulate in the separation zone, inhibiting reverse flow [1].

The secondary currents in the case where $q = 0.25$ show dissimilarities in comparison to those in the case of rectangular sections with similar width-to-depth ratio yet different Froude numbers, where one strong clockwise rotating cell is present [8,9,10,31]. The current simulations do show a clockwise cell, but it is located close to the bottom. Since impinging flow at a small discharge ratio causes important changes in secondary currents, similar changes can be expected in rectangular sections, though this statement is not verified in this work.

By means of a momentum coefficient β, it is also shown that the cross-sectional non-uniformity of the velocity distribution in the downstream channel increases when q decreases. However, only a slight change of β is noted when comparing $q = 0.05$ and $q = 0$. A comparison between the present values and literature is not easy, as Hsu *et al.* [7] defined a momentum coefficient β in the function of the flow in the contracted section (excluding the separation zone), and the smallest discharge ratio measured by Ramamurthy *et al.* [5] equaled $q = 0.40$. Despite this, it can be concluded based on the current finding that the curve fit of Hsu *et al.* [7], stating that the momentum coefficient (of the contracted section) is inversely proportional to q, when $0.1 \lesssim q \lesssim 0.9$, is not valid for very small q; the momentum coefficient does not become infinitely large when $q$becomes extremely small—In contrast, it only changes marginally between $q = 0.05$ and $q = 0$. The curve proposed by Ramamurthy *et al.* [5] for high Froude numbers, *i.e.*, a quadratic curve stabilizing at small q, seems to be more qualitatively correct. The flow patterns observed in the downstream channel can be tied to field-based research. For instance Rhoads *et al.*[15] noted that in the case of a dominant tributary flow, the scour hole migrates to the right bank of the downstream channel, and depositions take place near the inner bank. These findings are in line with the conceptual model of Best [1], and can be linked to the increased flow contraction at smaller q. Multiple studies also associate erosion of the outer bank with a small discharge ratio [14,15,39]. When making abstraction of differences in junction angle, bed geometry, *etc.*, this corresponds to the tributary flow impinging on the opposing bank.

The velocity difference over the shear layer in the separation zone occurs at a moderate discharge ratio that is larger than that in the mixing layer. As a consequence, the largest TKE values are present in the separation zone. When q decreases, TKE increases in these shear layers. An important difference when tributary flow becomes dominant, is the appearance of a velocity gradient in the contracted section between the flow being impinged and reflected by the opposing wall, and the non-impinged flow. Furthermore, levels of TKE increase in the downstream channel as the discharge ratio decreases, which is consistent with the increased non-uniformity in this section. The former may introduce new paths of sediment transport, while the latter may induce stronger scour, in the case of movable bed confluences where bed-forming flows are associated with small q values.

The intermittent flow patterns that were observed both in the experiment and in the LES have quite large intervals in time between two consecutive events. Instantaneous flow patterns show that these patterns are related to the position where the tributary flow impinges on the opposing wall, relating the intermittency to the discharge ratio. It is not yet clear whether the different vertical position of the impinging flow is a trigger for or a consequence of the intermittent flow patterns. Further research is needed to discover the cause of the intermittency. Knowledge of the cause might aid in improving the LESs, which simulate the amplitude correctly, but not the interval. Moreover, it seems that these intermittent flow patterns are dissimilar from those reported by Constantinescu *et al.* [16,17], where bimodal oscillations originate in helicoidal cells around the mixing layer. Further, in case of discordant beds, intermittent upwelling events are attributed to shear layer instabilities via streamwise eddies [25]. In the present research, however, the intermittent flow patterns are associated with the small discharge ratio and the impinging on the opposing bank.

CONCLUSIONS

This paper investigated the flow patterns in a subcritical junction using LESs, focusing on the case where the tributary provides the dominant part of the incoming discharge. Recalling the research questions stated in the introduction, one can conclude that increasingly dominant tributary flows, *i.e.*, small discharge ratios, lead to:

1. New features in the flow patterns induced by the impinging of the tributary flow on the opposing bank. In the upstream channel, a recirculating eddy develops, imposing rather significant changes on the incoming velocity. By changing the size of the stagnation zone, this also influences the mixing layer. In the downstream channel, the impinging flow causes stronger helicoidal cells, upwelling near the right bank and associated higher levels of TKE.

2. Confirmation of some of the known trends in confluence literature, such as increased three-dimensionality of the flow, and increasing dimensions of the separation zone. However, the new flow features can be regarded as deviations from the known trends.

3. Changes in the mixing layer, as the upstream channel inflow is influenced by the recirculating eddy. In addition, a new shear interface develops between the upwelling flow caused by impinging, and the non-impinging part of the tributary inflow follows a curved path to the downstream channel. The impinging flow enforces stronger helicoidal cells, though in the end, these do not result in faster flow recovery.

4. Upwelling events of much stronger upwelling flow, having an intermittent character. They seem to be linked to the height at which the tributary impinges on the opposing wall, thus they are associated with the small discharge ratios.

Further research is needed to fully understand the origin of the intermittent upwelling phenomenon. Moreover, an open question remains whether the observed changes occur gradually between $q = 0.25$ and $q = 0.05$, or if there is a distinct discharge ratio at which the flow features alter. For $q \gtrsim 0.10$, no changes in flow patterns have been reported, while Shumate [8] identified reflection at $q = 0.083$. In the present research, it was found that the recirculating eddy in the upstream channel is weak at $q = 0.05$, but is fully developed at $q = 0$. Therefore, it can be speculated that a gradual transition exists between confluences with important tributary discharges and those with dominant tributary discharges.

The current research was conducted for relatively low Froude numbers, representative of lowland rivers. Validating the conclusions for higher Froude numbers remains to be performed. Similarly, it should be emphasized that the present research only considered one geometrical confluence angle ($\theta = 90°$) and one value of the channel width to water depth ratio (W/h). Additionally, the confluent channels had a chamfered rectangular section, which might aid fluid motion in the vertical direction.

For movable bed situations, the changes in flow patterns might alter the bed morphology. Although the new shear layer and higher levels of TKE associated with the impinging flow are mainly located near the free surface, they might form a new effective way of sediment transport. Finally, referring to the finding of Rhoads *et al.* [15], *i.e.*, that the smaller discharge ratios occur on the rising limb of the inflow hydrographs (of a particular confluence), the results of the present research may aid understanding of the associated flow patterns, which are difficult to measure in field conditions because of the transient character of the flood.

ACKNOWLEDGMENTS

The first author is a PhD fellow of the research foundation Flanders (FWO), and the second author is a PhD fellow of the Special Research Fund (BOF) of Ghent University. The funding from these institutions is gratefully acknowledged. The computational resources (Stevin Supercomputer Infrastructure) and services used in this work were provided by the VSC (Flemish Supercomputer Center), funded by Ghent University, the Hercules Foundation and the Flemish Government, department EWI. The feedback of two anonymous reviewers is gratefully acknowledged.

REFERENCES

1. Best, J.L. Flow Dynamics and Sediment Transport at River Channel Confluences. Ph.D. Thesis, University of London, London, UK, 1985.
2. Taylor, E.H. Flow characteristics at rectangular open-channel junctions. *Trans. ASCE* **1944**, *109*, 893–902.
3. Webber, N.B.; Greated, C.A. An investigation of flow behaviour at the junction of rectangular channels. *ICE Proc.* **1966**,*34*, 321–334.
4. Hager, W.H. Transitional flow in channel junctions. *J. Hydraul. Eng.* **1989**, *115*, 243–259.
5. Ramamurthy, A.S.; Carballada, L.B.; Tran, D.M. Combining open channel flow at right angled junctions. *J. Hydraul. Eng.* **1989**, *114*, 1449–1460.
6. Gurram, S.K.; Karki, K.S.; Hager, W.H. Subcritical junction flow. *J. Hydraul. Eng.* **1997**, *123*, 447–455.
7. Hsu, C.; Wu, F.; Lee, W. Flow at 90° equal-width open-channel junction. *J. Hydraul. Eng.* **1998**, *124*, 186–191.
8. Shumate, E.D. Experimental Description of Flow at an Open-Channel Junction. Master's Thesis, University of Iowa, Iowa City, IA, USA, 1998.
9. Shakibainia, A.; Tabatabai, M.R.M.; Zarrati, A.R. Three-dimensional numerical study of flow structure in channel confluences. *Can. J. Civil Eng.* **2010**, *37*, 772–781.
10. Mignot, E.; Bonakdari, H.; Knothe, P.; Liperne Kouyi,

G.; Bessette, A.; Riviere, N.; Bertrand-Krajewski, J.L. Experiments and 3D simulations of flow structures in junctions and their influence on location of flowmeters. *Water Sci. Technol.* **2012**, *66*, 1325–1332.

11. Mignot, E.; Vinkovic, I.; Doppler, D.; Riviere, N. Mixing layer in open-channel junction flows. *Environ. Fluid Mech.***2014**, *14*, 1027–1041.
12. Schindfessel, L.; Creëlle, S.; Boelens, T.; De Mulder, T. Flow Patterns in An Open Channel Confluence with A Small Ratio of Main Channel to Tributary Discharge. In *River Flow 2014*; Schleiss, A.J., de Cesare, G., Franca, M.J., Pfister, M., Eds.; Taylor & Francis Group: London, UK, 2014; pp. 989–996.
13. Yoshimura, H.; Moriguchi, R.; Fujita, I. Numerical and Experimental Study of Three Dimensional Flow Structures and Water Surface Fluctuations at a Right-Angled Confluence. In Proceedings of the 36th IAHR World Congress, Hague, the Netherlands, 28 June–3 July 2015.
14. Leclair, S.F.; Roy, A.G. Variability of bed morphology and sedimentary structures at a discordant river confluence during low flows. *Géogr. Phys. Quat.* **1997**, *51*, 125–139. (In French).
15. Rhoads, B.L.; Riley, J.D.; Mayer, D.R. Response of bed morphology and bed material texture to hydrological conditions at an asymmetrical stream confluence. *Geomorphology* **2009**, *109*, 161–173.
16. Constantinescu, G.; Miyawaki, S.; Rhoads, B.; Sukhodolov, A.; Kirkil, G. Structure of turbulent flow at a river confluence with momentum and velocity ratio close to 1: Insight provided by an eddy-resolving numerical simulation. *Water Resour. Res.* **2011**, *47*.
17. Constantinescu, G.; Miyawaki, S.; Rhoads, B.; Sukhodolov, A. Numerical analysis of the effect of momentum ratio on the dynamics and sediment-entrainment capacity of coherent flow structures at a stream confluence. *J. Geophys. Res.***2012**, *117*, 1–21.
18. Leite Ribeiro, M.; Blanckaert, K.; Roy, A.G.; Schleiss, A.J. Flow and sediment dynamics in channel confluences. *J. Geophys. Res.* **2012**, *117*.
19. Khublaryan, M.G. Surface Waters: Rivers, streams, lakes and wetlands. In *Types and Properties of Water*, 1st ed.; EOLSS Publishers: Oxford, UK, 2009; pp. 1–16.
20. Goring, D.G.; Nikora, V.I. Despiking acoustic Doppler Velocimetry Data. *J. Hydraul. Eng.* **2002**, *128*, 117–126.
21. Wahl, T.L. Discussion of "Despiking acoustic Doppler Velocimetry Data". *J. Hydraul. Eng.* **2003**, *129*, 484–487.
22. Sagaut, P. *Large Eddy Simulation for Incompressible Flows*, 3rd ed.; Springer: Dordrecht, The Netherlands, 2006; pp. 1–556.
23. Keylock, C.J.; Constantinescu, G.; Hardy, R.J. The application of computational fluid dynamics to natural river channels: Eddy resolving *versus* mean flow approaches. *Geomorphology* **2012**, *179*, 1–20.
24. Rodi, W.; Constantinescu, G.; Stoesser, T. *Large-Eddy Simulation in Hydraulics*, 1st ed.; CRC Press/Balkema: London, UK, 2013; pp. 1–252.
25. Bradbrook, K.F.; Lane, S.N.; Richards, K.S.; Biron, P.M.; Roy, A.G. Large eddy simulation of periodic characteristics at river channel confluences. *J. Hydraul. Eng.* **2000**, *38*, 207–215.
26. Van Balen, W.; Blanckaert, K.; Uijttewaal, W.S.J. Analysis of the role of turbulence in curved open-channel flow at different water depths by means of experiments, LES and RANS. *J. Turbul.* **2010**, *11*, 1–34.
27. Stoesser, T. Large-eddy simulation in hydraulics: Quo vadis? *J. Hydraul. Eng.* **2014**, *52*, 441–452.
28. Fluent. In *Fluent 6.1 User's Guide*; Fluent Inc: Endicott, NY, USA, 2003.
29. Van Balen, W.; Uijttewaal, W.S.J.; Blackaert, K. Large-eddy simulations of a curved open-channel flow over topography. *Phys. Fluids* **2010**, *22*.
30. Creëlle, S.; De Mulder, T; Schindfessel, L; van Oyen, T. Influence of Hydraulic Resistance on Flow Features in An Open Channel Confluence. In Proceedings of the 3rd IAHR Europe Congress, Porto, Portugal, 14–16 April 2014.
31. Huang, J. Development and Validation of a Three-Dimensional Numerical Model for Application to River Flow. Ph.D. Thesis, University of Iowa, Iowa City, IA, USA, 30 August 2000.
32. Pope, S.B. Ten questions concerning the large-eddy simulation of turbulent flow. *New J. Phys.* **2004**, *6*.
33. Coussement, A.; Gicquel, O.; Degrez, G. Large eddy simulation of a pulsed jet in cross-flow. *J. Fluid Mech.* **2012**, *695*, 1–34.
34. Schindfessel, L.; Creëlle, S.; De Mulder, T. Influence of cross-sectional shape on flow patterns in an open-channel confluence. In Proceedings of the 36th IAHR World Congress, Hague, The Netherlands, 28 June–3 July 2015.
35. Best, J.L.; Reid, I. Separation zone at open-channel junctions. *J. Hydraul. Eng.* **1984**, *110*, 1588–1594.
36. Plesniak, M.W.; Metha, R.D.; Johnston, J.P. Curved two-stream turbulent mixing layers revisited. *Exp. Therm. Fluid Sci.***1996**, *13*, 190–205.
37. Lesieur, M. *Turbulence in Fluids*, 4th ed.; Springer: Dordrecht, The Netherlands, 2008; pp. 170–171.
38. Greated, C.A. *A Theoretical Study of Flow Behaviour at Open-Channel Junctions*; Report No. 6; University of Southampton: Southamption, UK, 1964.
39. Bergeron, N.; Roy, A.G. The effect of an ice jam on the morphology of a river channel confluence. *Géogr. Phys. Quat.***1988**, *2*, 191–196. (In French).
40. Roy, A.G.; Bergeron, N. Hydraulic geometry and changes in flow velocity at a river confluence with coarse bed material. *Earth Surf. Process. Landf.* **1988**, *13*, 583–598.
41. Roy, A.G.; Bergeron, N. Flow and particle paths at a natural river confluence with coarse bed material. *Geomorphology***1990**, *3*, 99–112.
42. Ashmore, P.E.; Ferguson, R.I.; Prestegaard, K.L.; Ashworth, P.J.; Paola, C. Secondary flow in anabranch confluences of a braided, gravel-bed stream. *Earth Surf. Process. Landf.* **1992**, *17*, 299–311.

CHAPTER 4

A NOVEL APPROACH FOR ESTIMATING THE RECURRENCE INTERVALS OF CHANNEL-FORMING DISCHARGES

Andy Ward , and Molly Moran

Keywords: Channel-forming discharges, Log-Pearson Type 3, annual series of peaks, probability density function, annual occurrences, river restoration

INTRODUCTION

Based on an analysis of the 37,099 projects in the National River Restoration Science Synthesis (NRRSS) database, Bernhardt et al. [1] estimated that annual expenditures on stream restoration in the United States exceeded one billion dollars and was increasing exponentially. They identified the most common goals of river restoration as enhancing water quality; managing riparian zones; and improving in-stream habitat and improving fish passage and stabilizing banks. A concern they identified a decade ago was that most small- to modest-sized projects had no post-construction monitoring or assessment of performance, therefore limiting our ability to know if our efforts approximated a river's remaining natural potential and to learn from successes and failures. Since the mid-1990s, the controversial Natural Channel Design approach that was developed by Dave Rosgen has seen widespread application in the United States [2,3,4]. The backbone of the approach is a classification system that depended on knowledge of bankfull discharges. Simon et al. [5] argued that a fundamental problem with the approach was that it was based on channel form and did not adequately consider channel processes. We will not jump into this debate, but we concur with Simon et al. that identifying and measuring the bankfull geometry in order to ascertain bankfull discharges is often difficult in an unstable stream system and that it is desirable to collect geotechnical data to help evaluate bank stability. Channel processes include the knowledge of channel-forming or dominant discharges and are often the foundation of river restoration projects [6]. In this manuscript, we use the term channel-forming discharge to designate either the bankfull (Q_b) or effective discharge (Q_e).

Wolman and Miller [7] described the bankfull discharge as the stream-flow that filled the main channel and that began to spill onto the active floodplain; while the effective discharge was the discharge that transported the most sediment over time. Goodwin [8] provided a useful study on what he called the *dominant discharge* that determined the cross-sectional and planform characteristics of a channel, and he noted that the dominant discharge could be based on measurements, the effective discharge, or the recurrence interval (RI) of the dominant discharge. Doyle et al. [6] stated that the bankfull discharge, or a discharge for a user-specified recurrence interval, was used widely in design procedures without consideration of the effective discharge. Although the bankfull and effective discharge were each considered to be channel-forming discharges, they are not identical. For example, Emmitt and Wolman [9] reported that the ratio of the effective discharge to bankfull discharge, Q_e/Q_b, ranged from 0.98 to 1.31 for five gravel-bed streams in the northern Rocky Mountains, USA. Torizzo and Pitlick [10], also in the Rocky Mountains, found that Q_e/Q_b ranged from 0.8 to 1.5 for 12 gravel-bed streams. Powell et al. [11] found that the Q_e/Q_b ratio ranged from 0.4 to 2.5 for 10 large rivers in Ohio. Channel-forming discharges have often been associated with a fixed recurrence interval such as 1.5 or 1.58 years [12], which were partially derived from the classic work of Leopold et al. [13]; however, studies also warn against using a fixed recurrence interval [14].

One method for determining the effective discharge is the Wolman–Miller Model [15], which Powell et al. [11] illustrated for the Maumee River in Ohio at a United States Geological Survey (USGS) gauge having a catchment area of 16,395 km^2(Figure 1). In this study, 43 years of discharge and suspended-sediment data were divided into 100-m^3/s discharge ranges (bins) starting with 0 to 100 m^3/s, 100+ to 200 m^3/s,

Citation: Andy Ward and Molly Moran, A Novel Approach for Estimating the Recurrence Intervals of Channel-Forming Discharges, Water 2016, 8(7), 269; doi:10.3390/w8070269.

200+ to 300 m^3/s, and so forth until a final bin was made that contained all data for discharges larger than 2700 m^3/s. The total sediment load per bin (solid line) was the product of the two dashed lines (days in a particular bin multiplied by the average daily load for discharges in the bin). Low discharges were ineffective in transporting sediment, and extreme events had very high sediment transport rates but occurred infrequently so the total sediment load they carried over a period of many decades was not the bin with the largest average discharge.

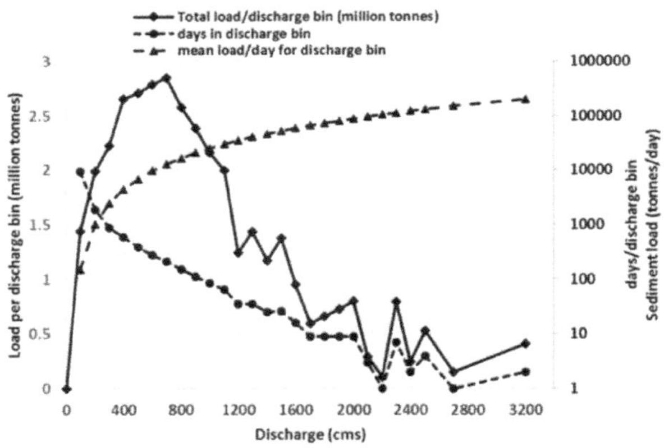

Figure 1: Illustration for the Maumee River, Ohio, that the effective discharge, 700 m^3/s, carries the largest total sediment load based on 14,415 days of discharge and suspended load data from 1950 to 1993 (cms = m^3/s). Powell et al. [11].

For the result shown in Figure 1, the effective discharge was about 700 m^3/s; however, there were two potential issues with this approach. First, at a particular gauge, there might be considerable variability in the calculated effective discharge, as it will vary with the flow range of the bin. Powell et al. [11] illustrated this concept for the Maumee River by reporting larger estimates of the effective discharge, in part, because wash loads were excluded and perhaps because the bins were sized based on stage increments rather than discharge increments. Simon et al. [16] stated that mean daily values of both flow and sediment loads, which were readily available from the USGS, tended to be biased towards lower flows, particularly in flashy basins. They noted that using data at shorter time increments was best but that type of measured data was seldom available. Second, in addition to the lack of a standard approach to determining bin sizes for an effective discharge analysis, the approach sometimes gave more than one peak (effective) discharge. This might be an anomaly of the analysis but also could be the geomorphic reality [17].

Useful guidelines on determining the effective discharge have been published by Biedenharn et al. [18]. Other approaches to determining the effective discharge included using a sediment transport threshold, such as the discharge that transported 50% of the sediment, to quantify the channel-forming discharge [19,20]. For large rivers in Ohio, Powell et al. [11] found the ratio of the threshold discharge, associated with transport of 50% of the sediment, to the bankfull discharge was 0.5 to 1.8. Powell et al. [11] reported that the effective discharge and bankfull discharge often had RIs of less than one year; however, they also found that even in rivers with RIs of at least a year the channel-forming discharges, or large discharges, occurred 1 to 24 days a year. Their results were not unique and were in contradiction not only with statistical expectations but also numerous studies in the literature [5,6,21]. Nolan et al. [22] found that discharges, which transported 50% to 90% of the sediment load, had recurrence intervals (RIs) of 0.27 to 16.1 years. Andrew and Nankervis [15] and Sichingabula [23] reported discharges that transported 50% to 80% of the sediment load to have flow durations of 11 to 80 days annually.

For headwater rivers in Idaho, Whiting et al. [24] reported RIs of 1.0 to 2.8 years for the effective discharge. In a study in the northern Rocky Mountains, Emmett and Wolman [9] reported RIs ranging from 1.5 to 1.7 years, while MacRae [25] reported RIs of 1.6 to 10 years for a study in British Columbia. Andrews and Nankervis [15] described the effective discharge as that which transports the largest portion of the mean annual bed-material load over time, and in the gravel bed rivers located in mountainous parts of Colorado, they found that 80% of the mean annual load was transported by flows between 0.8 to 1.6 times the bankfull discharge, which on average occurred as 15.6 events per year.

Gomez et al. [26], in a study of a gravel-bed river in New Zealand, estimated that the average RI of the effective discharge was 4 ± 2 years but it was equaled or exceeded three times a year. In a study of 12 urban river channels in southern Ontario, Canada, Annable et al. [27] reported that bankfull discharge occurred more than once a year in all of the rivers, often averaged from four to eight bankfull discharge or larger events per year, and in one case 18 events occurred in a year. Other examples of channel-forming discharges occurring many times annually include: Andrews [28], who reported effective discharge flow duration values of 1.5 to 11 days annually, with corresponding recurrence intervals of 3.26 and 1.18 years, respectively; and Wolman and Miller [7], who reported channel-forming discharge or larger of 6 to 11 days per year.

Numerous studies and methods have been used to develop relationships between discharge and RI, but much focus has been on applying these methods to extreme floods. Common approaches that are used to analyze river discharge data are based on: (1) an analysis of an annual series of peak values; (2) a partial duration or peaks over threshold approach; (3) an analysis of a probability density function; and (4) an annual frequency of exceedance approach. In the United States, the most commonly used method by state and federal agencies is the Log-Pearson Type 3 statistical method and procedures [28,29,30,31]. As the main purpose of the Log-Pearson Type 3 statistical method was to identify probable maximum discharges and flood elevations, it is usually used with an annual series of "instantaneous" peak values—a series of the highest discharge in each year that was recorded at an interval of 5 to 60 min depending on the river. The main disadvantage of using an annual series is that it excludes large floods when several occur in a single year [32].

Using a partial duration series, also called peaks over threshold approach, to determine the RI has also seen much application [33,34,35,36,37,38,39]. For frequent floods, it is the recommended method as it provides better estimates than the annual series of peaks approach [38]. The main problem with the peaks over threshold approach is determining the threshold to be used.

Logic and methods based on sediment transport, such as the effective discharge or half-load discharge [40], suggest that using a continuous series of discharges should be used to determine channel-forming discharges. If a continuous series of data is not available, such as discharge values every 15 min, an approach commonly used is to develop histograms and probability density functions (pdf) based on mean daily discharge data. There is much variability in how different scientists have developed strategies on how to perform this type of analysis. Doyle et al. [41] divided a histogram into 25 bins to represent the pdf of the daily discharge. Crowder and Knapp [42] used an iterative approach to reducing the number of bins from 25 until each bin had at least one discharge value. In addition to a lack of a standard approach to determining the bin size, the main drawback of a pdf approach is that the magnitude of the largest calculated RI is the number of years of available data. Determining the annual frequency of exceedance, for example, of specified mean daily discharge can be obtained from a histogram based on an annual series of mean daily discharges—or values associated with other time increments such as 15 min. In order to maximize the stability of channel reaches, channel-forming discharge is usually adopted as the design discharge in channel design and restoration [6].

We have cited just a few of the many studies that consider the annual frequency of occurrence of channel-forming discharges. The problems with the approach and its usefulness in river restoration are in knowing that a channel-forming discharge or larger occurs a certain number of times annually. Quader and Guo [43] warned that an inaccurate estimation of the design discharge in river restoration could adversely affect project performance. Amongst others, Doyle et al. [6] also noted that one should not just rely on return intervals or bankfull discharges in river restoration designs.

The goal of our study was to address statistical inconsistencies in the range of channel-forming discharges using 150 case studies across six states in the North Central Region of the United States, and to propose a novel, alternative approach to determining discharge versus RIs based on daily values rather than peak values, which we call the Full Daily Distribution (FDD). Specific objectives of our study were to: (1) determine the annual frequency of occurrence of daily discharges ranging from 2 to 100 years with particular focus on discharges with a RI of 10 years or less by comparing a simplified Log-Pearson Type 3 method to the more sophisticated method commonly used for USGS gauged sites; (2) determine if our method of calculating annual frequency of occurrences using mean daily values was consistent with statistical probability expectations; and (3) evaluate the appropriateness of the use of the new FDD approach to determine discharge versus RI relationships.

MATERIALS AND METHODS

Study Site Selection

The study was conducted by using USGS surface water gauge data for six states located in the North Central Region of the United States that are within the Mississippi River Basin and the Great Lakes Region (see Figure 2).

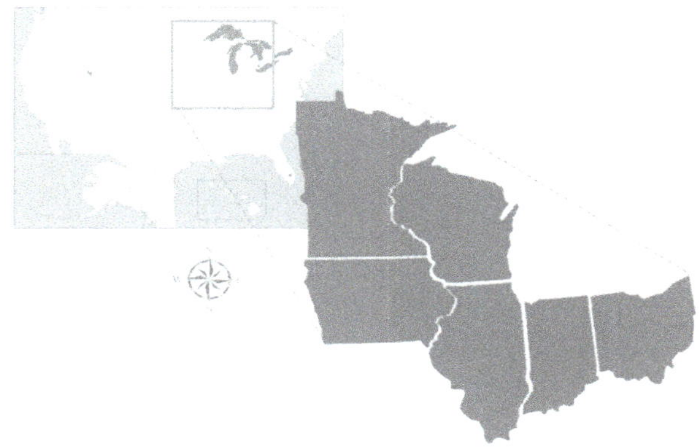

Figure 2: The six states from the North Central Region of the United States that were used in this study.

The data used in the study was obtained from the USGS-NWIS website (http://waterdata.usgs.gov/nwis). For each state, we searched gauges for mean daily and instantaneous peak surface water data that had a minimum of 14,610 days (or 40 years) in the period of record. We excluded the catchments where there was evidence on maps on the NWIS website of large lakes and dams. We then looked for the most recent published surface water discharge values for unregulated streams in each state (Table 1). In order to have a common comparison between sites, we included only rivers that had published estimates using the Log-Pearson Type 3 method in the data set. Published unregulated river discharge values are reported as peak discharges and were calculated using a Log-Pearson Type 3 method that considers the skew [33]. We compared unregulated sites listed in the publications with sites we selected visually. If we had a site in our data set that was not listed in the publication it was removed. The most recent publication for Indiana did not report the 2-year and 5-year recurrence interval discharge values so those values were not included in our study.

Table 1: Publications with discharge versus recurrence interval data for the 150 rivers evaluated.

State	Authors	Scientific Report
Illinois	Soong et al. [44]	USGS 2004-5103
Indiana	Knipe and Rao [45]	FHWA/IN/JTRP-2005/1 **
Iowa	Eash et al. [46]	USGS 2013-5086
Minnesota	Lorenz et al. [47]	USGS 2009-5250
Ohio	Koltun [48]	USGS 03-4164
Wisconsin	Walker and Krug [49]	USGS 03-4250

Note: ** Joint Transportation Research Program (Federal, Indiana and U.S. Highway Administration).

We used a second approach to determine if a gauge site was regulated or unregulated to help further reduce the data set. In that approach, we determined the number of times the daily discharge event occurred or was exceeded for a given recurrence interval for the period of record and normalized the occurrence of daily discharge events so that every site provided calculated results for 100-year time periods. For example, if a recurrence interval event was met or exceeded 32 times for a gauge with 40 years of data the result was normalized to 80 times in 100 years (32 times 100 divided by 40). If 100-year events at a site occurred five times more often than the expected statistical probability, that was an indicator that it might have been a regulated river and it was removed from the data set.

Data Analysis

Comparison of Discharge versus Recurrence Interval Relationships

Using Microsoft Excel spreadsheet software and analysis tools we developed, we performed Objectives 1 and 2 using the following 6-part method for each catchment. First, for each catchment, we downloaded annual series of "instantaneous" peak discharges and daily mean discharge data for the period of record. We then used a pivot table to determine the peak daily mean discharge in each year of the period of record. Third, we developed discharge versus recurrence interval relationships for the period of record using HydroToolBox, a Microsoft Excel add-on that provides specialized functions for commonly used hydrology analyses including determining recurrence intervals from an annual series of discharge data using a simplified Log-Pearson Type 3 method. We used the HydroToolBox to develop annual "instantaneous" and daily discharge versus recurrence interval relationships for the 2-year, 5-year, 10-year, 25-year, 50-year, and 100-year recurrence intervals at each site.

Fourth, we determined the number of times discharges associated with a particular recurrence interval were exceeded. Fifth, because periods of record varied between sites we chose to normalize the data to provide an estimate of the number of days in a 100-year time period a particular discharge might be exceeded. To do this, the calculated number of exceedances were multiplied by 100 divided by the period of record expressed in years.

Sixth, a second Excel analysis tool was used to calculate the number of times during the period of record that 75% and 125% of the annual peak daily mean discharge values, associated with a particular recurrence interval were exceeded. To do this, we took the annual peak daily mean values of each site from Step 1 and multiplied them by 75% and 125%, respectively. We then repeated Steps 2 and 3 to obtain our final calculated values. The purpose of Step 6 was to account for estimation errors associated with the period of record and because our simplified Log-Pearson Type 3 model did not vary the skew on the data. Based on the work of McCuen and Galloway [33], the plus or minus 25% range should account for the expected uncertainty in most estimates of the smaller 2-year, 5-year and 10-year recurrence interval discharges.

Last, we checked for errors in our simplified Log-Pearson Type 3 discharge estimates (herein referred to as "calculated") by comparing the results to results published in the documents listed earlier in Table 1.

Determining Full Daily Distribution (FDD)

We developed an Excel analysis tool to determine a Full Daily Distribution (FDD) for each gauge. The premise of the FDD is that it uses the whole period of record for a given site, but adjusts the bin size as a function of both the period of record and the maximum discharge of the data series. The FDD approach we developed provided a method for determining a bin size whose magnitude was related to the period of record. The FDD tool required importing mean daily discharge data from the USGS-NWIS website for each study site, ranking the discharge values from highest to lowest, and then dividing the data into bin sizes determined with the following equation:

$$Q_{bin} = Q_{max}/KY \tag{1}$$

where Q_{bin} is the bin discharge range, Q_{max} is the highest daily discharge in the series, Y is the number of years in the period of record, and K is a user-defined number. We developed this approach so the bin size increment for each gauge would vary as a function of the maximum daily discharge rather than using a fixed discharge such as 1 m³/s. The number of bins, n, for a catchment was Q_{max}/Q_{bin} or KY. Non-numeric discharge entries, such as an "ice" notation and missing data in the downloaded daily mean discharges were assigned a discharge value of 0 m³/s. Typically, there were 0 to 150 days where we made this correction and most catchments did not have any non-numeric entries. In this study, a value of 20 was used for K.

The Excel tool used the following procedure: (1) the downloaded data were ranked from highest to lowest; (2) Equation (1) was used to determine the bin size increment and then n bins were created with sizes incrementally increasing from Q_{bin} to nQ_{bin}; (3) the histogram function in Excel was used to populate the bins and to determine the number of discharge occurrences in each bin; (4) the cumulative number of occurrences was determined starting with the bin with the largest discharges and ending with the bin with the smallest discharges; and (4) the occurrence interval of each bin size was calculated as the period of record Y divided by the cumulative number of occurrences. The statistical number of occurrences, Os, was determined as Y/RI and calculated results were then determined for recurrence intervals that were closest to 0.8-years, 1-year, 1.4-years, 2-years, 5-years, 10-years, 25-years, 50-years, and 100-years.

Analysis of Results

The focus for Objectives 1 and 2 was evaluating whether the Log-Pearson Type 3 approach provided reasonable estimates of discharges associated with recurrence intervals up to 100 years. Of particular interest were discharges for recurrence intervals less than 10 years. There is no scientific method to measure these relationships so the analysis was based on calculated versus expected number of occurrence of different magnitude, or larger, discharges in 100 years. For the 2-year, 5-year, 10-year, 25-year, 50-year, and 100-year RIs the number of occurrences in 100 years would be 50, 10, 4, 2, and 1, respectively. For example, a 10-year or larger discharge is expected to occur on average 10 times. The expected value of 10 would consist of all discharges equal or larger than the 10-year discharge. This analysis was done with the results from the Log-Pearson Type 3 approach with the annual series of peak daily discharges. A meaningful analysis could not be done with the annual series of "instantaneous" peak data because only one value per year is available in the USGS-NWIS database.

The calculated number of occurrences were divided by the expected number of occurrences for each catchment and the recurrence intervals previously listed. Mean and standard deviations of these ratios were then determined for all catchments in the dataset and for the catchments located in each state. Step 5 in Section 2.2.1 was used in a similar way to determine means and standard deviations for ratios based on 75% and 125% of the discharges obtained with the Log-Pearson Type 3 approach. Regression relationships were determined between the occurrence ratios and recurrence intervals for the mean of the catchments for each state and the mean value for all catchments in the dataset.

Last, we compared our calculated Log-Pearson Type 3 discharge versus recurrence interval results, based on the annual "instantaneous" peak discharge series, to results published in the documents listed in Table 1. In this case, "instantaneous" peak data were used because comprehensive results based on an annual series of daily peak data were not found in the literature.

For Objective 3, ratios of the discharge versus recurrence interval values, determined with the FDD method, were compared to discharges calculated using the Log-Pearson Type 3 method with both the annual series of "instantaneous" peak discharges and the annual series of daily peak discharges. To calculate these ratios it was first necessary to obtain logarithmic regression relationship between discharge and recurrence intervals for values obtained with the Log-Pearson Type 3 results. This was done because, with the FDD approach, it was not possible to calculate results for the specific recurrence intervals of 2-year, 5-year, 10-year, 25-year, 50-year and 100-year recurrence intervals used for the Objective 1 and 2 analysis. Also, with the FDD approach, there was variability in the calculated recurrence intervals for each catchment. The form of these relationships was:

$$Q = a \ln RI + b \tag{2}$$

where Q is the discharge (m^3/s), RI is the recurrence interval (years) and a and b are regression coefficients.

The use of logarithmic regression relationships with the Log-Pearson Type 3 data facilitated obtaining calculated results for recurrence intervals less than 1 year.

RESULTS

Site Selection

The average discharge record used in the analysis ranged from 68 years in Indiana to 79 years in Illinois (Table 2). There was one gauge in Minnesota with a recurrence interval less than 40 years. This was due to difficulty in finding rivers with long discharge records that did not contain numerous lakes and dams. The final dataset consisted of 150 catchments with 25 catchments located in each state. Catchment areas ranged from 29 km^2 to 6475 km^2. Of the 150 gauges used in the analysis, 12 had a catchment area smaller than 200 km^2, while 76 and 15 had catchment areas larger than 1000 km^2 and 3000 km^2, respectively.

Table 2: Average Annual Precipitation, Discharge and Catchment Area Data for 25 rivers in each of the six states.

State	Annual Rainfall (mm)	Average Discharge Record (Years)	Minimum Discharge Record (Years)	Maximum Discharge Record (Years)	Average Catchment Area (km^2)	Minimum Catchment Area (km^2)	Maximum Catchment Area (km^2)
Illinois	996	80	46	101	1520	29	5416
Indiana	1060	68	52	86	1104	222	2222
Iowa	864	69	43	102	1772	66	6475
Minnesota	693	69	39	93	1657	363	4869
Ohio	993	74	51	100	896	158	3240
Wisconsin	829	77	40	101	982	47	2678

Objective 1: Annual Frequency of Occurrences of Daily Discharges

A summary of the calculated number of daily occurrences in a 100-year time period for different recurrence interval discharge is shown in Table 3. The large standard deviations for all states and all recurrence intervals were an indicator of the large variability between catchments. For example, the range in the daily occurrences in 100 years, for the two-year recurrence interval discharge, were 77–382, 62–748, 110–834, 83–254, 102–612 and 97–1854 times for Illinois, Indiana, Iowa, Ohio, Wisconsin and Minnesota, respectively.

Table 3: Calculated number of daily occurrences in a 100-year time period for different recurrence interval discharge events for each state. Standard deviation reported in parentheses. The expected number of occurrences and the number associated with discharges that are 75% and 125% of the calculated Log-Pearson Type 3 values are also reported.

State	Number of Daily Occurrences					
	2-Year	5-Year	10-Year	25-Year	50-Year	100-Year
Illinois	174 (93)	40 (18)	15 (6.3)	5.1 (3.1)	2.1 (1.9)	1.0 (1.5)
Indiana	209 (175)	58 (49)	24 (22)	8.2 (6.1)	3.8 (2.8)	1.4 (1.6)
Iowa	245 (190)	48 (35)	18 (11)	5.2 (2.5)	2.8 (2.2)	0.9 (1.1)
Ohio	133 (53)	34 (12)	14 (5.4)	6.2 (3.0)	2.9 (1.9)	1.4 (1.1)
Wisconsin	210 (128)	47 (22)	17 (9.1)	6.3 (4.3)	2.6 (1.9)	1.6 (1.7)
Minnesota	494 (351)	94 (64)	35 (22)	9.4 (5.1)	4.1 (2.5)	2.0 (1.9)
Mean	244 (165)	54 (33)	21 (13)	6.7 (4.0)	3.1 (2.2)	1.4 (1.5)
Expected	50	20	10	4.0	2.0	1.0
75% Log P Mean	510 (193)	144 (45)	70 (24)	31 (15)	18 (11)	11 (7.3)
125% Log P Mean	133 (83)	22 (12)	7.1 (2.8)	1.8 (0.7)	0.7 (0.3)	0.4 (0.2)

Objective 2: Analysis of Calculated versus Expected Occurrences

As the recurrence interval became larger the difference between the expected number of times and the calculated number of times becomes less. This is shown in Table 4, where the mean ratios of the calculated to expected daily discharges for each state vary from 2.6 to 10.5 for the two-year event compared to a ratio of 0.9 to 1.7 for the 100-year event.

Table 4: The ratio of the calculated to expected number of daily occurrences in a 100-year time period for different recurrence interval discharges.

Attribute	Ratio of Calculated to Expected Number of Daily Occurrences					
	2-Year	5-Year	10-Year	25-Year	50-Year	100-Year
Illinois	3.5 (1.8)	2.0 (0.9)	1.5 (0.6)	1.3 (0.8)	1.0 (1.0)	1.0 (1.5)
Indiana	4.2 (3.5)	2.9 (2.5)	2.4 (2.2)	2.1 (1.5)	1.9 (1.4)	1.4 (1.6)
Iowa	4.8 (3.8)	2.4 (1.8)	1.8 (1.1)	1.3 (0.6)	1.5 (1.1)	0.9 (1.1)
Ohio	2.6 (1.1)	1.7 (0.6)	1.4 (0.5)	1.5 (0.7)	1.4 (1.0)	1.4 (1.1)
Wisconsin	4.2 (2.6)	2.3 (1.1)	1.7 (0.9)	1.6 (1.1)	1.3 (0.9)	1.6 (1.7)
Minnesota	10.5 (7.0)	4.9 (3.2)	3.6 (2.2)	2.3 (1.3)	1.8 (1.2)	1.7 (1.9)
Mean	4.9 (4.5)	2.7 (2.2)	2.1 (1.6)	1.7 (1.1)	1.5 (1.1)	1.3 (1.5)
75% Log P	10.2 (9.1)	7.2 (6.6)	7.0 (7.5)	7.8 (11.1)	9.2 (16.9)	11.3 (25.9)
125% Log P	2.7 (2.9)	1.1 (1.1)	0.7 (0.6)	0.4 (0.4)	0.3 (0.5)	0.3 (0.8)

The analysis of reducing the annual peak daily mean discharges by 25% or increasing them by 25% showed that if the calculated discharges were under-estimated they deviated even further from the expected discharge and gave unlikely mean ratios of 7 to 11.3 across the range of recurrence interval evaluated (Table 4). In contrast, if the daily discharge were 25% larger than calculated the ratios become unrealistically low for the more extreme recurrence interval of 10-year to 100-year. However, the two-year and five-year occurrence intervals still had mean ratios larger than one.

Objective 3: Evaluation of Statistical Relationships between the Expected and Calculated Frequency of Occurrences

Regression relationships for each state that are based on the calculated mean ratios in Table 4 are reported in Figure 3. The expected ratio of 1.0 and mean results when the discharge was 25% larger than the Log-Pearson Type 3 estimates are also shown. Power relationships had R-squared values that ranged from 0.64 to 0.97. The high R-squared values are misleading as they are based on mean values. This is illustrated in Figure 4 for an analysis with all of the six states and an analysis with the 25 catchments in the state of Indiana. Figure 4A has an R-squared value of 0.9 for the relationship based on the mean of the mean value for all states combined. However, in Figure 4B where the relationship was based on the mean values for each state, the R-squared was reduced to 0.6. There was an even more marked reduction in the R-squared value when a relationship was developed with data for each catchment evaluated in a state. Figure 4C shows the relationship with the mean data for Indiana (R-squared = 0.97) while Figure 4D shows the relationship based on the 25 individual catchments in Indiana (R-squared = 0.05).

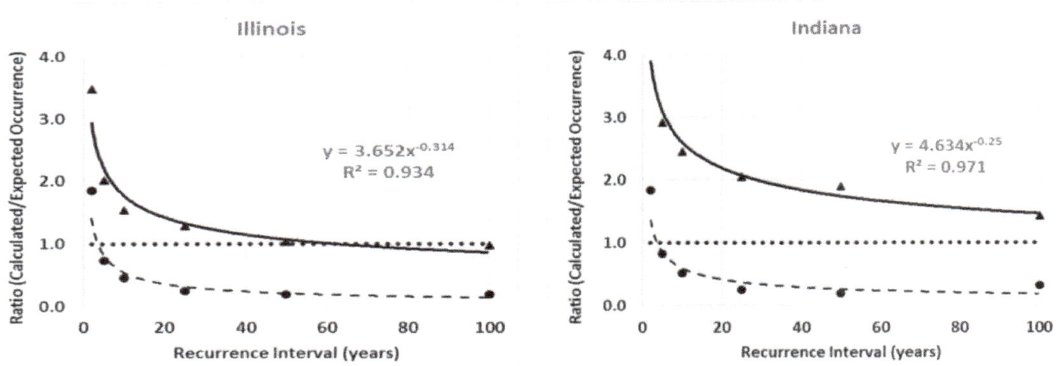

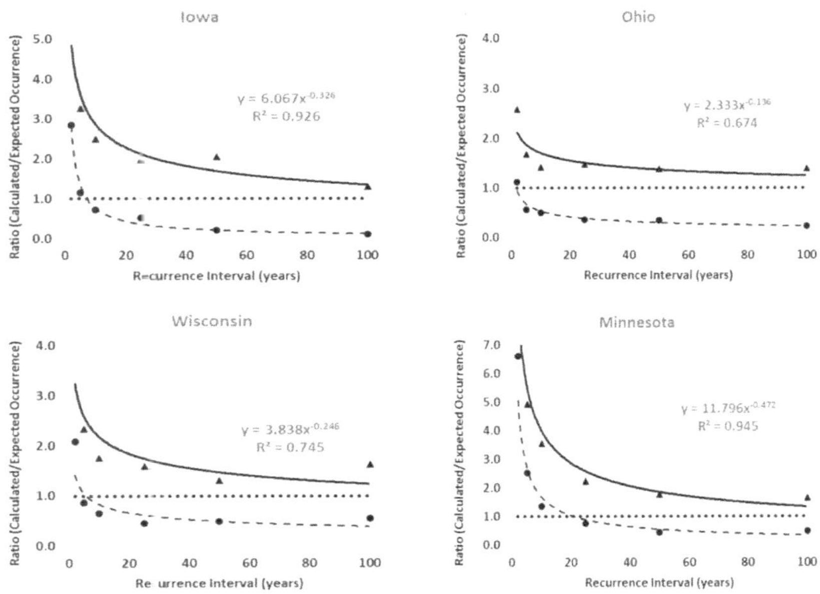

Figure 3: Plots of the ratios of the calculated versus expected occurrences based on the mean discharge of all 25 catchments evaluated for each state (solid triangles). A regression relationship through this data is shown as a solid line together with the relationship and R-squared value. Solid circles and dashed lines are associated with occurrence ratios based on 125% of the discharges calculated with the Log-Pearson Type 3 analysis.

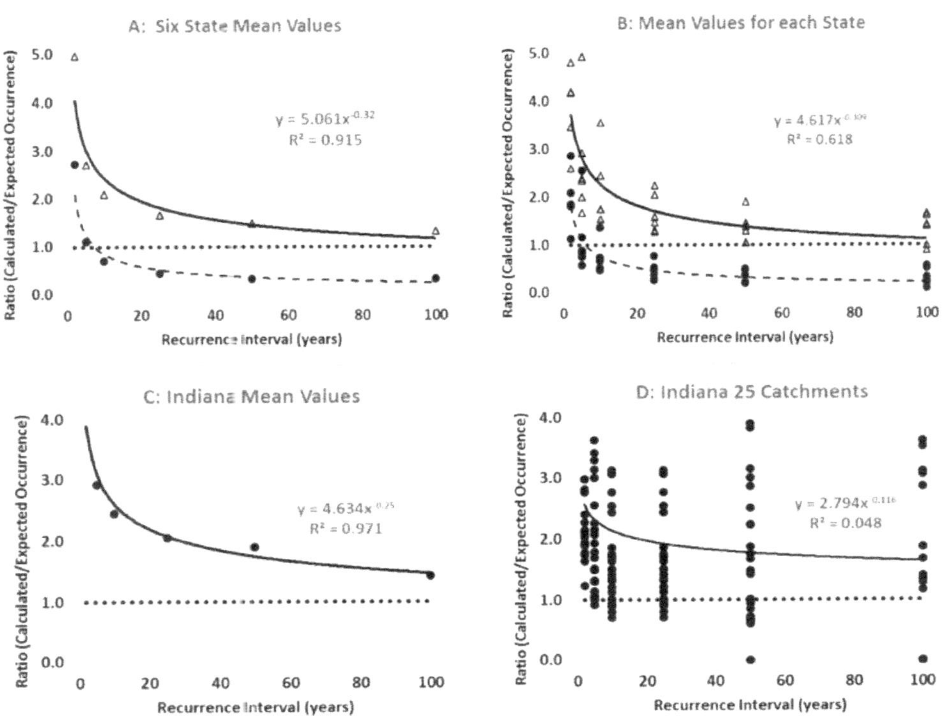

Figure 4: For the Log-Pearson Type 3 approach, the ratio of the calculated versus expected occurrences based on the mean discharge of all 150 catchments evaluated (**A**); and the mean discharges for the 25 catchments in each state (**B**). Solid circles and dashed lines are associated with occurrence ratios based on 125% of the discharges calculated with the Log-Pearson Type 3 analysis. Also shown is the ratio of the calculated versus expected occurrences based on the mean discharge for the 25 catchments evaluated in Indiana (**C**) and the ratios determined with the discharge for each of the 25 catchments evaluated in Indiana (**D**).

When the calculated discharge was decreased by 25% the mean ratio of calculated to expected occurrences for the 150 catchments and across all recurrence intervals was 10. There was only one catchment with a ratio less than 1 for the 50-year recurrence interval and four catchments with a ratio less than 1 for the 100-year recurrence interval. In contrast, when the calculated discharges were increased by 25% the mean ratio of calculated to expected occurrences, for the 150 catchments, and across all recurrence intervals, was 0.9. However, the expected occurrence of the two-year discharge was exceeded by 87% of the catchment.

The results of the evaluation of the accuracy of the calculated discharges compared to published values are presented in Table 5. A minor difference would be expected, as there were small differences in the published periods of record and those used in our study. The overall mean absolute difference between the 850 (six states times 25 catchment times six recurrence intervals less 50 missing values for Indiana) calculated and published discharges was 6.2%. If the overall mean difference was not based on absolute values then the mean difference between the calculated and published discharges was 0.2%. The mean absolute difference for each recurrence interval increased from 5.2% for the 2-year recurrence interval to 8.5% for the 100-year recurrence interval. These results suggest that there is little difference in the simplified Log-Pearson Type 3 approach used in this study and the more rigorous Log-Pearson Type 3 approaches often used in published studies conducted by federal agencies.

Table 5: Absolute values of the mean percent difference between the calculated and published discharges for 2-year to 100-year recurrence interval for each state. For Indiana, values for the two-year and five-year values are not reported [45].

	State Mean [1]	Mean Absolute Difference [2]					
		2-Year	5-Year	10-Year	25-Year	50-Year	100-Year
Illinois	6.2	3.4	4.5	5.2	6.5	8.2	9.2
Indiana	4.5	-	-	3.8	3.8	4.5	5.4
Iowa	7.5	6.0	6.7	6.2	7.1	8.6	10.3
Minnesota	7.0	5.5	5.5	5.3	6.8	8.7	10.4
Ohio	6.7	6.0	5.6	5.9	6.6	6.9	9.5
Wisconsin	5.3	5.3	4.5	5.9	5.2	4.7	6.3
Mean	6.2	5.2	5.4	5.4	6.0	6.9	8.5

Notes: [1] Reference information for the published discharges is reported in Table 1. Reported in this column is the mean absolute difference, in each state, between the calculated and published discharges for all six recurrence intervals and the 25 catchments in a state. The absolute mean values for all 125 catchments and all recurrence intervals are also reported; [2] Mean absolute difference for each recurrence interval.

Objective 4: Evaluation of the Full Daily Distribution (FDD) Approach

The FDD approach results, based on the mean of the mean results for the six states, are presented in Table 6. The first column in Table 6 reports the number of catchments used in the analysis. Although the R-squared values for the logarithmic regression relationships, fitted to the calculated Log-Pearson Type 3 discharges, almost always exceeded 0.96, some of the calculated discharges for the frequent recurrence intervals were negative and not included in the analysis. Ten catchments had periods of record that were too short to provide results for the largest recurrence interval.

Table 6: Recurrence Interval versus discharge ratios of the full daily distribution (FDD) discharges to those based on an annual series of instantaneous peaks (Peak) and an annual series of peak mean daily discharges (Daily)—based on an analysis of mean results for the six states. Also reported are ratios of the mean daily peaks to the instantaneous peaks. Standard deviations are reported in parentheses.

Number of Catchments	Recurrence Interval Years (std. dev.)	FDD/Peak Ratio (std. dev.)	FDD/Daily Ratio (std. dev.)	Daily/Peak Ratio (std. dev.)
106	0.79 (<0.01)	1.73 (0.29)	1.98 (0.45)	0.89 (0.16)
124	0.99 (<0.01)	1.72 (0.35)	1.96 (0.53)	0.90 (0.15)
143	1.38 (<0.01)	1.54 (0.41)	1.78 (0.48)	0.88 (0.11)
150	1.92 (0.09)	1.31 (0.32)	1.61 (0.32)	0.81 (0.06)
150	4.7 (0.22)	1.00 (0.18)	1.23 (0.14)	0.81 (0.06)
150	9.02 (0.48)	0.90 (0.14)	1.13 (0.09)	0.79 (0.06)
150	20.81 (1.07)	0.84 (0.12)	1.09 (0.07)	0.77 (0.07)
150	36.58 (2.67)	0.82 (0.08)	1.09 (0.04)	0.76 (0.07)
140	74.36 (4.57)	0.82 (0.05)	1.10 (0.07)	0.75 (0.09)

The results in Table 6 and Figure 5 show that for recurrence intervals larger than about 10 years the discharges calculated by the FDD method and those based on an annual series of peak daily means were about 10% larger. In contrast, for recurrence intervals larger than about 10 years the discharges calculated by the FDD method and those based on an annual series of "instantaneous" peak means were 16% to 18% smaller. However, in the recurrence interval range commonly associated with channel-forming discharges (0.8 to 2 years), the FDD discharges were 30% to 98% larger than discharges obtained by using the Log-Pearson Type 3 method. The results reported in Figure 6 show that the state-to-state variability (Figure 6B) and the catchment-to-catchment variability (Figure 6D) were too large for the regression equations to have any general application.

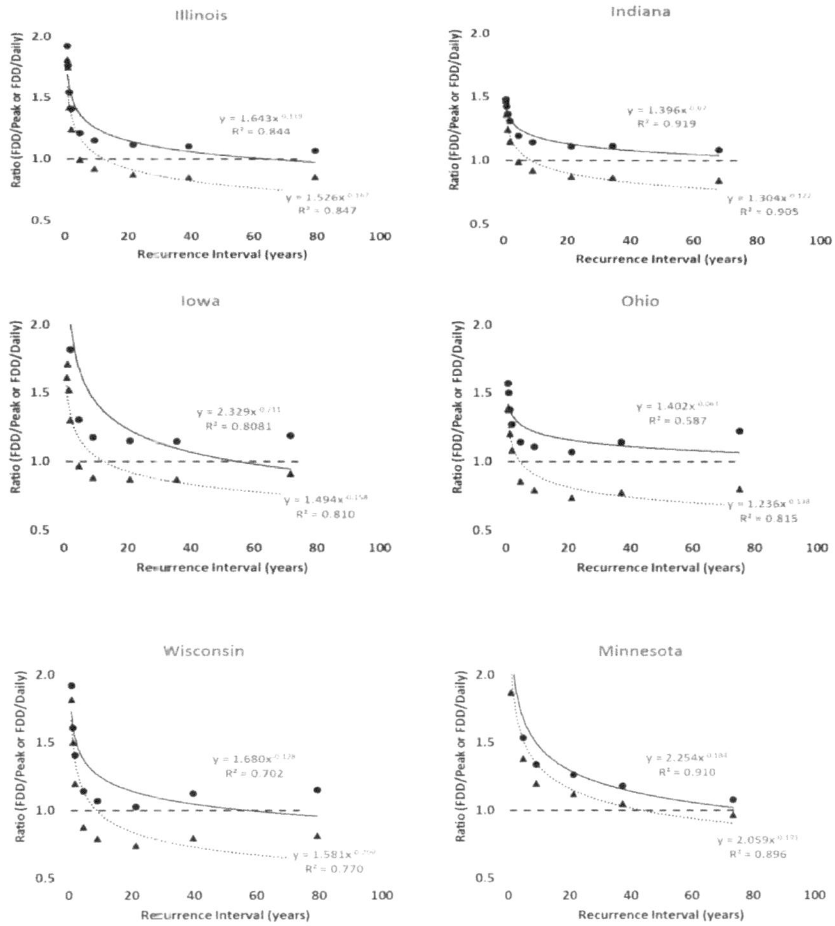

Figure 5: For different recurrence intervals, the ratio of calculated full daily distribution (FDD) discharges to the Log-Pearson Type 3 results for an annual series of instantaneous peaks (solid circles) and peak daily means (solid triangles). The values are based on data from Table 6. The solid and dotted lines are regressions lines associated with the solid circles and solid triangles, respectively.

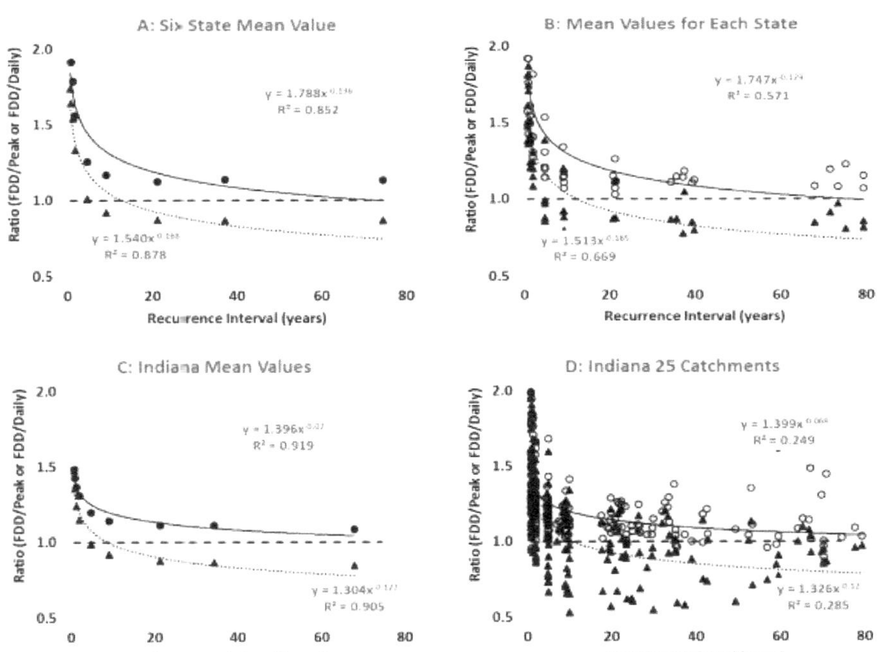

Figure 6; For the FDD approach, the ratio of the calculated versus expected occurrences based on the mean discharge of all 150 catchments evaluated (**A**); and the mean discharges for the 25 catchments in each state (**B**). In (**A**,**B**), the circles and solid lines are associated with the

FDD/Peak ratios and the solid triangles and dashed lines are associated with the FDD/Daily ratios. Also shown are the ratio of the calculated versus expected occurrences based on the mean discharge for the 25 catchments evaluated in Indiana (**C**); and the ratios determined with the discharge for each of the 25 catchments evaluated in Indiana (**D**). Data symbols and lines are the same as in (**A**,**B**).

DISCUSSION

The evidence from this study suggests that the frequency of channel-forming discharges that have a recurrence interval less than five years are likely to occur much more frequently than statistical expectations. The calculated smallest number of occurrences in 100 years for the two-year recurrence interval daily discharge was 62 times for one of the catchments, while the highest number of occurrences for the two-year event was 1854 days in 100 years. On average, a two-year or larger discharge was calculated to occur about 2.5 days annually.

Based on an annual series of peak daily means, the ratio of the calculated to the expected mean number of occurrences were highly correlated to the recurrence interval with the exception of the results for Ohio (R-squared = 0.64, see Figure 3). However, when the state-to-state or catchment-to-catchment variability was considered (Figure 3) the correlation was too low for scientific purposes. The analysis based on discharges 25% larger or smaller than those obtained from the Log-Pearson Type 3 analysis indicated that the discharges were generally unlikely to be much smaller or larger than the calculated values.

The results obtained with the full daily distribution (FDD) approach were in general agreement with a study conducted by Endreny [32] with data for 46 bankfull surveyed gauges in Maryland, North Carolina, and New York. Using a flow duration approach with a daily peak series he found that on average the bankfull discharge occurred 2.1 days annually and a maximum mean number of times of 13.2 days annually. He determined that by using either an annual instantaneous peak series or an annual daily peak series the recurrence interval of the bankfull discharge was about 1.4 years and concluded that use of a flow duration analysis helped to describe that bankfull discharges occur more than one time or one day annually. Navratil et al. [50], in a study of 16 gravel-bed rivers in France, evaluated five different methods to identify bankfull and determined that the mean annual duration of occurrence varied from 4.5 days to 11.5 days depending on the method. One of their conclusions was that the annual maximum flood (AMF) analysis should be avoided.

Using the FDD method for each catchment the maximum recurrence interval was limited by the number of years in the period of record. For example, for the results reported in Table 7, the maximum recurrence interval was 89 years.

Table 7: Comparison of FDD method calculated exceedance and expected occurrences (reported as days) for different recurrence intervals. The difference between the calculated exceedances and expected occurrences were the number of discharge values in the bin with the maximum discharge shown. Reported results are for the Sandusky River, near Fremont Ohio, by using K values of 0.5 and 20 in Equation (1).

K = 20 and Bin Size of 0.53 m^3/s				K = 0.5 and Bin Size of 21.25 m^3/s			
RI (years)	Discharge (m^3/s)	Calculated (days)	Expected (days)	RI (years)	Discharge (m^3/s)	Calculated (days)	Expected (days)
0.59	380	146	152	0.56	435	133	160
0.79	413	108	112	0.67	458	94	133
0.98	439	86	91	0.95	481	77	94
1.46	462	59	61	1.48	527	48	60
1.98	487	44	45	1.85	550	40	48
4.94	598	17	18	4.94	641	14	18
9.89	652	8	9	9.89	710	7	9
22.25	765	3	4	22.25	779	2	4
44.50	946	1	2	44.50	962	1	2
89.00	1020	0	1	89.00	1031	0	1

The FDD results were a function of the K value of 20 used in the study. In Table 7, calculated values are the number of exceedances and expected values are the number of occurrences (days) for a RI. The difference between the calculated exceedances and expected occurrences were the number of discharge values in the bin with the maximum discharge shown. For example, with K equal to 20, there were 146 discharge values larger than 380 m^3/s and six discharge values (152–146) in the bin containing discharges in the range 379.47 to 380 m^3/s. The expected value of 152 is 89 years/0.586 years (0.586 is round to 0.59 in Table 7). Using a K value of 20, as shown in Figure 7B, resulted in smaller bin discharge ranges than for a K value of 0.5 (Figure 7A) and a better representation of the frequency distribution of different magnitude discharges. For example with K values of 20 and 0.5 the 4.94-year recurrence interval discharges are 598 m^3/s and 641 m^3/s, respectively. For the Sandusky River catchment, with a period of record of 89 years, a K value of 20 gave 1780 bins (20 times 89) while a K value of 0.5 only gave 45 bins. For a K value of 20, the average number of values in each bin was 18.3 for all the catchments while a K value of 0.5 had an average of 722 values.

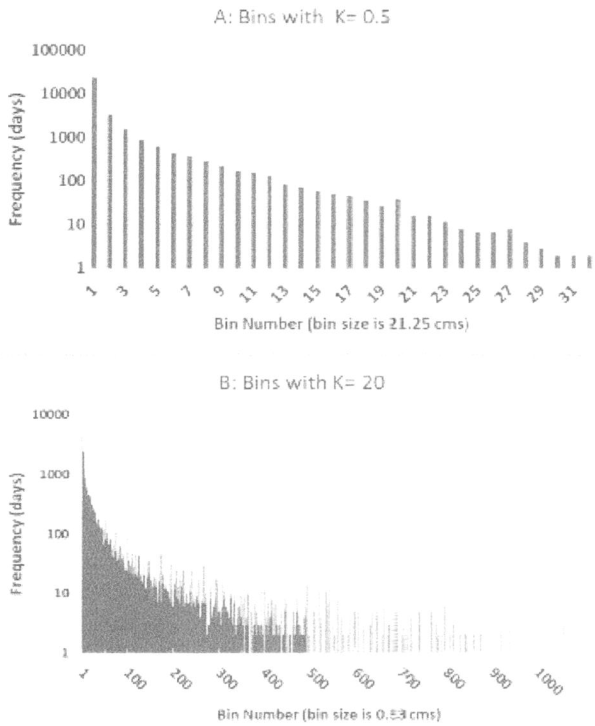

Figure 7: Histograms with K = 0.5 (**A**) and K = 20 (**B**) for the Sandusky River, near Fremont Ohio. A bin with a bin number of 1 has all discharges less than 21.25 m^3/s (K = 0.5) and 0.53 m^3/s (K = 20). A bin with a bin number of 10 (K = 0.5)) contains discharges between approximately 229.0 and 229.07 m^3/s. A bin with a bin number of 400 (K = 20) contains discharges between approximately 229.07 and 229.64 m^3/s.

The strength of the FDD approach was that the calculated number of occurrence were always in close agreement with the expected number of occurrences. The FDD approach does, therefore, provide a useful approach for discharge versus recurrence interval relationships in both the range of channel-forming discharges and also for extreme floods. However, it is less useful for extreme floods as the maximum recurrence interval that can be calculated by the method is the number of years in the period of record.

The number of occurrences of a discharge associated with a recurrence interval raises the question as to what is the meaning of the term channel-forming discharge. It might be more appropriately called a channel-forming threshold discharge. Discharges below the threshold are sufficient to transport sediment and to shape the channel geometry—geomorphic work associated with scour and deposition. The function of discharges above the threshold will depend on floodplain connectivity and the size of the floodplain. What is important to note is that, statistically, all discharges higher than the threshold are included in the statistical definition. Therefore, in mountain streams where the banks will generally not be flat and in locations where the floodplain is small, the functional usefulness of basing a restoration design just on a channel-forming discharge must be questionable. A number of approaches based on sediment transport, hydraulics, and dynamics of the system have more merit [1,51] than using an uncertain estimate of the channel-forming discharge.

Other implications of the statistical meaning and statistical uncertainty of channel-forming discharges for river restoration are unclear. For example, He and Wilkerson [52] determined that in some cases regional bankfull geometry relationship can be better estimated as a function of the 2-year recurrence interval discharge or this discharge and the drainage area. If that approach is used it would be desirable to have a standard method for determining the 2-year recurrence interval discharge.

In the United States, a commonly used Internet tool for ungauged sites is the USGS StreamStats Method (http://water.usgs.gov/osw/streamstats/) that, based on regional regression equations, can provide discharge versus recurrence interval estimates for any location in the more than 30 states where StreamStats is operational. A problem with the approach is that reported errors in a discharge estimate usually range from 30% to 50%. Like other prediction methods, it is necessary to calibrate the method with data from the nearest suitable river gauge [33,53]. Typically, this is done by developing calibration variables based on first using discharge data at the river gauge and then relating the results to a Log-Pearson Type 3 analysis of an annual series of maximum instantaneous peaks as that data are readily available from the USGS-NWIS state websites. The results of this study indicated that calibration with an FDD analysis of mean daily discharges might be more appropriate. However, this creates an inconsistency in methodology as hydrologic prediction methods usually estimate peak discharges not mean daily discharges.

CONCLUSIONS

Using a Log-Pearson Type 3 statistical method with an annual series of instantaneous peaks or peak daily mean values provided discharge versus recurrence interval results that were consistent with flood events with recurrence intervals larger than 25 years. However, for more frequent recurrence intervals in the range of channel-forming discharges (less than five years) calculated discharges occured several times more frequently than statistical expectations.

With the FDD approach, the calculated number of occurrences associated with a recurrence interval were similar to the expected number of occurrences. This was not the case with the Log-Pearson Type 3 method. The FDD approach can easily be used where long periods of record of daily discharge data are available as it uses analysis functions commonly available in spreadsheets and statistical software. A strength of the method is that it adjusts the number of bins and discharge range of the bins as a function of the period of record and the magnitude of the measured discharges.

Further work is needed to evaluate how skew values in the Log-Pearson Type 3 method could be modified to better account for statistical inconsistencies in the range of channel-forming discharges. In addition, where discharge data are not available, work is needed to determine how peak flow prediction methods might be modified. The importance of the current inconsistencies in determining recurrence intervals or discharges associated with channel-forming discharges is not clear and requires further work. However, we speculate, based on the literature that suggests a high frequency of river restorations do not perform as expected, that incorrect estimates of the channel-forming discharge could be a contributing factor.

ACKNOWLEDGMENTS

The work presented in this manuscript was funded in part by the Ohio Water Development Authority (OWDA), the Ohio Agricultural Research and Development Center (OARDC), the College of Food Agricultural and Environmental Sciences and the Department of Food, Agricultural and Biological Engineering at The Ohio State University. Jessica D'Ambrosio assisted with the editing of the manuscript.

REFERENCES

1. Bernhardt, E.S.; Palmer, M.A.; Allan, J.D.; Alexander, G.; Barnas, K.; Brooks, S.; Carr, J.; Clayton, S.; Dahm, C.; Follstad-Shah, J.; et al. Restoration of U.S. Rivers: A national Synthesis. *Science* **2005**, *308*, 636–637.
2. Hey, R. Fluvial Geomorphological Methodology for Natural Stable Channel Design. *J. Am. Water Resour.* **2006**, *42*, 357–372.
3. Rosgen, D.L. The Rosgen Geomorphic Approach for Natural Channel Design. In *Stream Restoration Design. National Engineering Handbook Part 654*; U.S. Department of Agriculture, National Resources Conservation Service: Washington, DC, USA, 2007.
4. Lave, R. The controversy over natural channel design: Substantive explanations and potential avenues for resolution.*J. Am. Water Resour.* **2009**, *45*, 1519–1532.
5. Simon, A.; Doyle, M.; Kondolf, M.; Shields, F.D., Jr.; Rhoads, B.; McPhillips, M. Critical evaluation of how the Rosgen classification and associated "natural channel design" methods fail to integrate and quantify fluvial processes and channel response. *J. Am. Water Resour.* **2007**, *43*, 1117–1131.
6. Doyle, M.W.; Shields, F.D.; Boyd, K.F.; Skidmore, P.B.; Dominick, D. Channel-forming discharge selection in river restoration design. *J. Hydraul. Eng.* **2007**, *133*, 1–7.
7. Wolman, M.G.; Miller, J.P. Magnitude and frequency of forces in geomorphic processes. *J. Geol.* **1960**, *68*, 54–74.
8. Goodwin, P. Analytical solutions for estimating effective discharge. *J. Hydraul. Eng.* **2004**, *130*, 729–738.
9. Emmett, W.W.; Wolman, M.G. Effective discharge and gravel-bed rivers. *Earth Surf. Proc. Landf.* **2001**, *26*, 1369–1380.
10. Torizzo, M.; Pitlick, J. Magnitude-frequency of bed load transport in mountain streams in Colorado. *J. Hydrol.* **2004**,*290*, 137–151.
11. Powell, G.E.; Mecklenburg, D.; Ward, A. Evaluating channel-forming discharges: A study of large rivers in Ohio.*Trans. ASABE* **2006**, *49*, 35–46.
12. Dury, G.H. Discharge prediction, present and former, from channel dimensions. *J. Hydrol.* **1976**, *30*, 219–245.
13. Leopold, L.B.; Wolman, M.G.; Miller, J.P. *Fluvial Processes in Geomorphology*; W.H. Freeman: San Francisco, CA, USA, 1964; p. 522.
14. Williams, G.; Arnett, P. Bank-full discharge of rivers. *Water Resourc. Res.* **1978**, *14*, 1141–1154.
15. Andrews, E.D.; Nankervis, J.M. Effective discharge and the design of channel maintenance flows for gravel-bed rivers. In *Natural and Anthropogenic Influences in Fluvial Geomorphology*; Costa, J.E., Miller, A.J., Potter, K.W., Wilcock, P.R., Eds.; Geophysical Monograph Series; American Geophysical Union: Washington, DC, USA, 1995; Volume 89, pp. 151–164.
16. Simon, A.; Dickerson, W.; Heins, A. Suspended-sediment transport rates at the 1.5-year recurrence interval for ecoregions of the United States: Transport conditions at the bankfull and effective discharge? *Geomorphology* **2004**, *58*, 243–246.
17. Lenzi, M.A.; Mao, L.; Comiti, F. Effective discharge for sediment transport in a mountain river: Computational approaches and geomorphic effectiveness. *J. Hydrol.* **2006**, *326*, 257–276.
18. Biedenharn, D.S.; Thorne, C.R.; Soar, P.J.; Hey, R.D.; Watson, C.C. Effective discharge calculation guide. *Int. J. Sediment Res.* **2001**, *16*, 445–459.
19. Vogel, R.M.; Stedinger, J.R.; Hooper, R.P. Discharge indices for water quality loads. *Water Resour. Res.* **2003**, *39*.
20. Ferro, V.; Paolo, P. Identifying a dominant discharge for natural rivers in southern Italy. *Geomorphology* **2012**, *139–140*, 313–321.
21. USGS. *The "100-Year Flood"*; Fact Sheet 229-96; USGS Washington: Tacoma, WA, USA, 1996; p. 2.
22. Nolan, K.M.; Lisle, T.E.; Kelsey, H.M. Bankfull discharge and sediment transport in northwestern California. In*Erosion and Sedimentation in the Pacific Rim*; IAHS Publication 165; International Association of Hydrological Sciences: Wallingford, UK, 1987; pp. 439–450.
23. Sichingabula, H.M. Magnitude-frequency characteristics of effective discharge for suspended sediment transport, Fraser River, British Columbia, Canada. *Hydrol. Process.* **1999**, *13*, 1361–1380.
24. Whiting, P.J.; Stamm, J.F.; Moog, D.B.; Orndorff, R.L. Sediment transporting flows of headwater streams. *GSA Bull.***1999**, *111*, 450–466.
25. MacRae, C.R. Experience from morphological research on Canadian streams: Is control of the 2-year frequency runoff event the best basis for stream channel protection? In *Effects of Catchment Development and Management on Aquatic Systems*; Roesner, L., Ed.; American Society of Civil Engineers: Reston, VA, USA, 1996; pp. 144–160.
26. Gomez, B.; Coleman, S.E.; Sy, V.W.K.; Peacock, D.H.; Kent, M. Channel change, bankfull and effective discharges on a vertically accreting, meandering, gravel-bed river. *Earth Surf. Proc. Landf.* **2007**, *32*, 770–785.
27. Annable, W.K.; Lounder, V.G.; Watson, C.C. Estimating channel-forming discharge in urban watercourses. *River Res. Appl.* **2011**, *27*, 738–753.
28. Andrews, E.D. Effective and bankfull discharges of streams in the Yampa River basin, Colorado and Wyoming. *J. Hydrol.* **1980**, *46*, 311–330.
29. Interagency Committee on Water Data (IACWD). *Guidelines for Determining Flood Flow Frequency*; Bulletin 17B of the Hydrology Subcommittee;

U.S. Department of the Interior Geological Survey, Office of Water Data Coordination: Reston, VA, USA, 1982.

30. Water Resources Council Hydrology Committee. *A Uniform Techniques for Determining Flood Flow Frequencies*; Bulletin No. 15; Water Resources Council: Washington, DC, USA, 1967.
31. Stedinger, J.R.; Griffis, V.W. Flood Frequency Analysis in the United States: Time to Update. *J. Hydrol. Eng.* **2008**, *13*, 199–204.
32. Endreny, T.A. Estimation of channel bankfull occurrence from instantaneous discharge data. *J. Hydrol. Eng.* **2007**, *12*, 524–531.
33. McCuen, R.H.; Galloway, K.E. Record length requirements for annual maximum flood series. *J. Hydrol. Eng.* **2010**, *15*, 704–707.
34. Ward, A.D.; Trimble, S.W.; Burckhard, S.R.; Lyon, J.G. *Environmental Hydrology*, 3rd ed.; CRC Press, Inc.: Boca Raton, FL, USA, 2015; p. 663.
35. Petit, F.; Pauquet, A. Bankfull discharge recurrence interval in gravel-bed rivers. *Earth Surf. Proc. Landf.* **1997**, *22*, 685–693.
36. Keast, D.; Ellison, J. Magnitude frequency analysis of small floods using the annual and partial series. *Water* **2013**, *5*, 1816–1829.
37. Madsen, H.; Rasmussen, P.F.; Rosbjerg, D. Comparison of annual maximum series and partial duration series methods for modeling extreme hydrologic events: 1. At-site modeling. *Water Resour. Res.* **1997**, *33*, 747–757.
38. The Institution of Engineers Australia. *Australian Rainfall and Runoff: A Guide to Flood Estimation*; Institution of Engineers Australia: Canberra, Australia, 1987.
39. Langbein, W.B. Annual floods and the partial-duration flood series. *Trans. Am. Geophys. Union* **1949**, *30*, 879–881.
40. Claps, P.; Laio, F. Can continuous streamflow data support flood frequency analysis? An alternative to the partial duration series approach. *Water Resour. Res.* **2003**, *39*.
41. Doyle, M.W.; Stanley, E.H.; Strayer, D.L.; Jacobson, R.B.; Schmidt, J.C. Effective discharge analysis of ecological processes in streams. *Water Resour. Res.* **2005**, *41*, 1–16.
42. Crowder, D.W.; Knapp, H.V. Effective discharge recurrence intervals in Illinois streams. *Geomorphology* **2005**, *64*, 167–184.
43. Quader, A.; Guo, Y. The relative importance of hydrological and sediment-transport characteristics affecting the effective discharge of small urban streams in southern Ontario. *J. Hydrol. Eng.* **2009**, *14*, 698–710.
44. Soong, D.T.; Ishii, A.L.; Sharpe, J.B.; Avery, C.F. *Estimating Flood-Peak Discharge Magnitudes and Frequencies for Rural Streams in Illinois*; Scientific Investigations Report 2004-5103; U.S. Geological Survey: Reston, VA, USA, 2004; p. 147.
45. Knipe, D.; Rao, A.R. *Estimation of Peak Discharges of Indiana Streams by Using Log-Pearson (III) Distribution*; FHWA/IN/JTRP-2005/1; School of Civil Engineering, Purdue University: West Lafayette, IN, USA, 2005; p. 194.
46. Eash, D.A.; Barnes, K.K.; Veilleux, A.G. *Methods for Estimating Annual Exceedance-Probability Discharges for Streams in Iowa, Based on Data through Water Year*; Scientific Investigations Report 2013-5086; U.S. Geological Survey: Reston, VA, USA, 2010; p. 63.
47. Lorenz, D.L.; Sanocki, C.A.; Kocian, M.J. *Techniques for Estimating the Magnitude and Frequency of Peak Flows on Small Streams in Minnesota Based on Data through Water Year 2005*; U.S. Geological Survey Scientific Investigations Report 2009-5250; U.S. Geological Survey: Reston, VA, USA, 2010; p. 54.
48. Koltun, G.F. *Techniques for Estimating Flood-Peak Discharges of Rural, Unregulated Streams in Ohio*, 2nd ed.; Water-Resources Investigations Report 2003-4164; U.S. Geological Survey: Denver, CO, USA, 2003; p. 75.
49. Walker, J.F.; Krug, W.R. *Flood-Frequency Characteristics of Wisconsin Streams*; USGS Water-Resources Investigations Report 03-4250; U.S. Geological Survey: Denver, CO, USA, 2003; p. 37.
50. Navratil, O.; Albert, M.-B.; Hérouin, E.; Gresillon, J.-M. Determination of bankfull discharge magnitude and frequency: comparison of methods on 16 gravel-bed river reaches. *Earth Surf. Process. Landf.* **2006**, *31*, 1345–1363.
51. Wohl, E.; Angermeier, P.L.; Bledsoe, B.; Kondolf, G.M.; MacDonnell, L.; Merritt, D.M.; Palmer, M.A.; Poff, N.L.; Tarboton, D. River restoration. *Water Resour. Res.* **2005**, *41*, W10301.
52. He, L.; Wilkerson, G.V. Improved bankfull channel geometry prediction using two-year return-period discharge. *J. Am. Water Resour.* **2011**, *47*, 1298–1316.
53. Moore, R.J.; Bell, V.A.; Cole, S.J.; Jones, D.A. *Rainfall-Runoff and Other Modelling for Ungauged/Low-Benefit Locations*; R&D Technical Report SC030227/SR1; Environment Agency: Bristol, UK, 2007; p. 247.

CHAPTER 5

THE EFFECT OF THE HYDRAULIC RETENTION TIME ON THE PERFORMANCE OF AN ECOLOGICAL WASTEWATER TREATMENT SYSTEM: AN ANAEROBIC FILTER WITH A CONSTRUCTED WETLAND

María L. Merino-Solís, Edgardo Villegas, José de Anda and Alberto López-López

Keywords: Constructed wetland, up-flow anaerobic filter, organic matter removal, nutrient removal, hydraulic retention time (HRT)

INTRODUCTION

Untreated wastewater discharged by cities and small rural communities contains nutrients that alter lakes and streams by accelerating the eutrophication process and thus change the quality of freshwater bodies [1,2,3]. A convenient ecological alternative that works well for small rural communities is to treat the municipal wastewater through constructed wetlands [4,5].

An artificial or constructed wetland is a natural treatment system designed and constructed to imitate hydrodynamic mechanisms as well as biogeochemical pollutant degradation processes that occur in natural wetlands, but with a greater degree of control over the hydraulic regime of the system [6]. The goal is to improve the quality of an effluent by optimizing the treatment systems [7]. A horizontal subsurface flow constructed wetland (HSSCW) includes low depth excavations or channels, emergent vegetation planted on a layer of soil of a porous or granular media that allows for the horizontal subsurface flow of the effluent. The permeable bed (formed by the granular media, the rhizomes and plant roots) is elevated and the water level is below the surface of the bed. Construction includes the placement of a plastic membrane for groundwater and soil protection and adequate inlet and outlet hydraulic structures to control water level [8,9,10,11,12]. Low energy requirements, easy operation and maintenance, modest installation and maintenance costs, minor sludge production, and the creation of a visually pleasing landscape are some advantages of this technology [5,13,14]. The clogging of granular media is one of the main operation problems in subsurface flow constructed wetlands [15]. It is linked to media type and it, in turn, is linked to hydraulic retention time since the hydraulic conductivity of porous media is very sensitive to media size and is influenced by particle size distribution and particle shape [16]. An adequate selection of granular media could decrease the estimated wetland area and improve the removal efficiencies [4,17].

Constructed wetlands are commonly used for secondary treatment in small communities [8]. However, they need a large land area [18] and therefore, it is sometimes necessary to add a previous treatment stage to the natural treatment system [18,19]. Anaerobic processes are a treatment technology that can be effectively integrated with constructed wetlands because they require a very small input of energy, minimum investment and operational costs, generate less biosolids when compared to conventional aerobic systems, and they can prevent clogging [15,18,20,21]. The up-flow anaerobic filter (UAF), as part of this technology, is able to treat wastewater with different pollutant loadings and it is successfully used, specifically for small communities in warm climates, to achieve the desired

Citation: María L. Merino-Solís, Edgardo Villegas, José de Anda, and Alberto López-López, The Effect of the Hydraulic Retention Time on the Performance of an Ecological Wastewater Treatment System: An Anaerobic Filter with a Constructed Wetland, Water 2015, 7, 1149-1163; doi:10.3390/w7031149.

water quality. In general, the UAF removes dissolved and suspended solids through close contact with anaerobic bacteria attached to the filter media [22,23].

Both the wetlands and UAF wastewater treatment processes have been individually studied with good results. The UAF removes mainly organic matter in wastewater with low and high organic loads, while the HSSCW is capable of removing organic matter and nutrients and is commonly used to obtain secondary or tertiary effluent concentration levels [8,23,24,25]. In other words, combining an anaerobic reactor with a constructed wetland brings important benefits to the constructed wetlands system, such as the reducing of the required planted area, reducing of the retention time, and increasing of their life cycle [18,20]. Few studies have been conducted on this coupled treatment process.

The aim of this study was to determine the effect of the hydraulic retention time on the performance of the coupled treatment system, integrated with an up-flow anaerobic filter and a horizontal subsurface constructed wetland, operated under subtropical climate conditions. Moreover, it was to evaluate the constructed wetland efficiency considering its area reduction. Provisions were included that allowed the control of the hydraulic retention time (HRT) to assess the impact on the reduction of organic matter and nutrients.

MATERIALS AND METHODS

Treatment System Description

The pilot-scale constructed wetland was built on the site of the municipal wastewater treatment plant for Chapala, Jalisco, Mexico. This town is located on Lake Chapala (20°07'–20°21' N and 102°40'45"–103°25'30" W) at an altitude of 1524.6 m above sea level. This area has a semi-warm sub-humid climate with summer rain, dry and semi-warm winters and springs. According to Köppen–Geiger climate classification, it is defined as a subtropical humid climate (Cwa) making this proposal possible in this type of climate around the world [26,27].

Wastewater was diverted from the grit removal chamber at the wastewater treatment plant. Raw wastewater was pretreated with a UAF with a filling height of 90% and with the dimensions of 1.75 m deep, 4 m long and 2.3 m wide. UAF effluent flowed to the HSSCW with the dimensions of 0.6 m deep, 9.2 m long and 7.6 m wide. The system was designed to operate with a water flow rate of 11 $m^3 \cdot day^{-1}$, in order to provide for 110 person equivalents (PE). Both, the UAF and the HSSCW use tezontle as biofilm support. Tezontle is the Indian name of a volcanic red-orange extrusive rock classified as part of the feldspar family. This material has a high porosity (up to 65% pore volume for grain size of 20–25 mm), and this characteristic gives it a high surface area to volume ratio (390 $m^2 \cdot m^{-3}$) compared to synthetic media (164 $m^2 \cdot m^{-3}$). It has been tested and used in a number of environmental applications [28], and it is inexpensive, since it is abundant in Mexico [29]. Furthermore, the material is considered inert, with pH values close to neutral, low cation exchange capacity, good ventilation and capacity in humidity retention depending on particle diameter [30]. While the UAF contains tezontle of 1 inch diameter, the HSSCW was backfilled with two layers: The first with 50 cm of tezontle of 1 inch diameter and the second with 10 cm of tezontle of 0.25 inch diameter. The porosity of clean tezontle, tested prior to the experiments was determined to be 58% (1 inch) and 60% (0.25 inch), respectively. As part of the construction, an impermeable plastic liner of 1 mm thick HDPP (High Density Polypropylene) was installed at the bottom of the treatment unit to prevent groundwater contamination. In addition, an adjustable outlet was installed at the end of the bed, which permitted better water level control. The HSSCW was planted with two plant species, the first half with *Canna hybrids* and the second half with *Strelitzia reginae.* Around one hundred plants from each species were planted in a geometric pattern of quincunx with a distance of 0.5 m between each plant of *Canna hybrids* and 0.55 m between each *Strelitzia reginae*, which provided an initial plant density of three plants per m^2. The warm climate enables the use of non-conventional species; Zurita *et al.*, (2009) suggest that it is possible to produce commercial flowers in constructed wetlands without reducing the efficiency of the treatment system. Therefore, these plants can acclimate well and grow, and also give a pleasing appearance [29,31].

Treatment System Monitoring

Once construction was completed, a start-up period of six months was used before beginning the monitoring stage in order to ensure greater plant coverage, which suggests a stable performance for pollutant removal [32], since it provides microorganisms the opportunity to grow and to adhere to the surface roots of the planted species and to the solid-porous substrate. Figure 1 shows the cover plant in the constructed wetland.

Figure 1: Wastewater Treatment System (field view).

Monitoring was conducted for 66 weeks, divided into three periods, to test the following hydraulic retention times: HRT1 corresponds to 18 h in the UAF and two days in the HSSCW, HRT2 was for 28 h in the UAF and three days in the HSSCW, and HRT3 was for 48 h in the UAF and four days in the HSSCW. In HSSFCW, the hydraulic retention time can vary between three and fifteen days to achieve an adequate treatment depending on pollutants to be removed [33]. Shorter retention times were chosen in order to test the response of the constructed wetland performance considering the previous addition of an anaerobic filter. The samples were collected weekly at four monitoring points along the treatment system (Figure 1): Influent (M1), outlet UAF/inlet HSSCW (M2), middle part of HSSCW (M3) and effluent (M4). For sampling points M2 and M4 (constructed wetland stage) it is important to point out that the first part of HSSCW was planted with different ornamental plants than the second stage. The sampling point M3 marked the separation between the two ornamental plants used.*Strelitzia reginae* was placed in the second part of the HSSCW because it has been proven that *Canna hybrids* adapt better to these conditions, showing high growth rates [31,34].

Water Quality Analysis

Water samples were taken to measure five-day biochemical oxygen demand (BOD), chemical oxygen demand (COD), total nitrogen (N_{tot}), ammonia nitrogen (NH_4-N), nitrite (NO_2), nitrate (NO_3) and total phosphorous (P_{tot}). The analytical methods used for the water quality analyses were in accordance with "Standard Methods for the Examination of Water and Wastewater" [35]. Total inorganic nitrogen was estimated as the sum of NH_4-N, NO_2 and NO_3. Some parameters were measured in the field, including Electrical conductivity (EC), pH (waterproof PC 300 hand-held pH/conductivity/TDS/ Temperature meter OAKTON®,Vernon Hills, IL, USA), temperature (Temp.), relative humidity (RH) (4184 Traceable®, Jumbo Humidity/Temperature meter Friendswood, TX, USA) and dissolved oxygen (DO) (YSI 58 DO meter, Yellow Springs, OH, USA) in the constructed wetland with at a depth of 20 cm below the water table.

Statistical Analyses

A statistical model was utilized to determinate how pollutant concentration varied between sampling points and for different HRT. Data was assessed with a Nested Design with a crossover factor using Statgraphics Centurion XVI®(Warrenton, VA, USA) with a confidence level of 95% [36]. The experiment was designed to compare the response (pollutant concentration) relative to categorical factors, including sampling points, HRT and time.

RESULTS AND DISCUSSION

In-Situ Field Conditions

The major parameters that influence the removal mechanisms for nitrogen and organic matter in constructed wetlands are pH, dissolved oxygen (DO) and temperature [17]. This is because organisms present in biological wastewater treatments are sensitive to these parameters. Table 1 shows the mean values and standard deviations for temperature, RH and pH. The UAF stage of the process is represented by sampling points M1 (influent) to M2 (effluent), while the HSSCW stage is between sampling points M2 (influent) and M4 (effluent). A significant difference ($p \leq 0.05$) between treatment stages was observed for pH, but the DO did not show a significant difference between the sampling points M3 with a mean value of 0.71 mg·L^{-1} (± 0.2) and M4 with a mean value of 0.8 mg·L^{-1} (± 0.2) in the three tested HRTs.

Table 1: Mean values (± SD [a] n = 40) of temperature (Temp.) and relative humidity (RH) for environmental conditions and pH in every sampling point.

Parameter		HRT1	HRT2	HRT3
Temp. (°C)		23.2 ± 1.9	21.7 ± 1.8	24.6 ± 1.4
RH (%)		49.8 ±10.5	50.3 ± 9.8	42.9 ± 9.4
pH	M1	7.4 ± 0.2	7.1 ± 0.3	7.2 ± 0.2
	M2	6.9 ± 0.2	7.1 ± 0.3	6.9 ± 0.2
	M3	7.2 ± 0.2	7.4 ± 0.2	7.2 ± 0.1
	M4	7.2 ± 0.3	7.4 ± 0.3	7.1 ± 0.1

Note: [a] SD: Standard deviation.

The pH reduction at sampling point M2 is due to anaerobic conditions that are predominate in the UAF ($p \leq 0.05$). After M2, in almost all cases, a slight increase in pH was observed at the next sampling points (M3 and M4) because conditions in HSSCW are generally considered facultative. A straight-line correlation ($p \leq 0.05$) was established between inlet and outlet pH, for every stage that indicates a dependence on this parameter in the influent. In addition, the BOD removal efficiency and N_{tot} removal efficiency of UAF showed dependence on the inlet pH ($p \leq 0.05$) with 40% and 35%, respectively, demonstrating that the efficiency of an anaerobic process depends mainly on pH [24].

The microorganisms responsible for nitrogen removal depend on temperature. The range of temperature favorable for nitrification is between 16.5 and 32 °C and for denitrification, the favorable temperature range is between 20 and 25 °C [17]. During the experiment, the recorded average temperature was 23 °C (± 2.1) and the relative humidity was 48% (± 10%). The warm environmental conditions and the wastewater characteristics boosted the degradation of organic matter and nutrient removal.

Performance and Removal Efficiencies

Initially, organic matter decomposition is determined by the balance between organic matter load and oxygen transfer rate, which defines the predominant conditions of anaerobic and aerobic systems [9]. The organic matter, expressed as BOD and COD, is decomposed by aerobic and anaerobic microbial processes and also by physical processes, such as sedimentation and filtration. The Figure 2 shows the mean values for COD and BOD while the Figure 3a,b show all the COD and BOD data during the monitoring period. At each sampling point, the results showed a reduction across the treatment system, the BOD/COD global ratio registered was 0.52. It is possible that the UAF is responsible for the removal of most part of the organic matter (from 50% to 80%). At sampling point M2, the lowest BOD/COD ratio (0.498) was found, indicating the presence of more oxidizable species that are biodegradables. In the beginning, the monitoring plan was scheduled to assess from three to five days, but once the treatment system behavior was observed and evaluated, the retention time of five days was changed to two days as it is shown in the Figure 3.

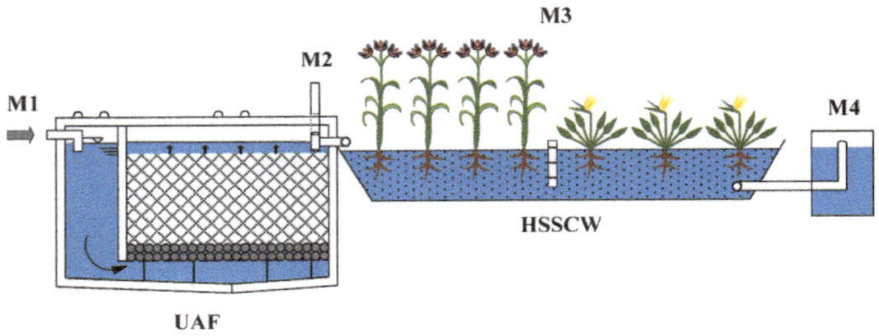

Figure 2: Sampling points in the wastewater treatment system. M1-influent; M2-outlet UAF/inlet HSSCW; M3-middle part of HSSCW; and M4-effluent.

It was observed that the greatest portion of the BOD was removed in the first half, from 42% to 53%, while the second half of the HSSCW was between 8% and 19%. This could be because the majority of the easily biodegradable BOD is removed in the first stage. The complete treatment system reached a BOD removal of 89.5%, 79.3% and 86.1% for the HRT1, HRT2 and HRT3, respectively. The COD removal values obtained were 84.2% for HRT1, 85.6% for HRT2 and 81.7% for HRT3. In theTable 2 is possible to observe the mean values for COD and BOD obtained in every sampling point and HRT. Significant differences ($p \leq 0.05$) were found between the treatment stages (evaluated in the sampling points M1, M2 and M4), and between the HRTs as followed: HRT1 and HRT2, and HRT1 and HRT3.

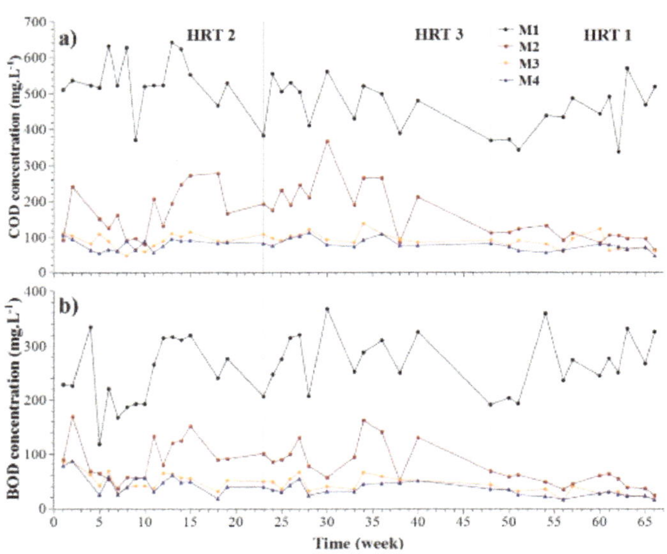

Figure 3: COD (**a**) and BOD (**b**) concentration for every sampling.

Table 2: Mean values (±SD, $n = 40$) of COD and BOD ($mg \cdot L^{-1}$) in every sampling point and HRT tested.

Sampling Point	HRT1		HRT2		HRT3	
	COD	BOD	COD	BOD	COD	BOD
M1	436.54 (72.42)	261.50 (54.70)	537.77 (68.81)	243.84 (63.90)	478.90 (62.68)	279.58 (49.58)
M2	98.76 (19.88)	48.73 (14.10)	167.80 (68.16)	90.35 (40.31)	217.58 (66.97)	100.25 (33.92)
M3	75.04 (21.65)	28.25 (9.06)	84.82 (21.21)	52.03 (18.01)	100.72 (16.35)	47.25 (11.45)
M4	66.86 (12.92)	25.83 (7.47)	76.88 (16.49)	47.63 (18.85)	86.25 (14.45)	38.33 (9.20)

Most of the nitrogen forms present in municipal wastewater are organic nitrogen at about 60% and ammonium at about 40% [37]. In this study, the average N_{Org} and NH_4-N concentrations in the affluent (M1) of 14 ± 7 and 26 ± 6.4 $mg \cdot L^{-1}$, respectively, were found, which corresponds to 65% and 35.5% of the total nitrogen, respectively. Nitrogen has a complex biogeochemical cycle, despite the multiple mechanisms involved, nitrogen removal is only achieved by means of denitrification, ammonia volatilization, nitrogen assimilation in vegetal tissue, sedimentation of dead plant material and ammonia adsorption to the bed [6,38]. Figure 4 shows the change of nitrogen concentration along the treatment system. The N_{Org} is converted into NH_4-N by ammonification (mineralization). During this process, the organic matter is biologically degraded in aerobic and anaerobic conditions with an optimum pH range between 6.5 and 8.5 [6,38,39]. Therefore, the N_{Org} decrease along the entire treatment system with a global efficiency of 75.7% in HRT1, 60.5% in HRT2 and 80.8% in HRT3. In the second part of HSSFCW (from M3 to M4), a small increment was found that attributed to the accumulation of detritus [6,11,39]. The next step is the nitrification process; whereby the compounds with reduced nitrogen are sequentially oxidized to nitrate with the formation of nitrite as an intermediate compound, which means that the oxygen is the terminal electron acceptor [6,39,40]. The anaerobic conditions of UAF (M2) promoted the conversion of organic nitrogen to ammonia with a result of 20.7% and 79.1% final concentrations, respectively, with a significant difference ($p \leq 0.05$), thus reducing the nitrification and the formation of nitrates. Removal of total nitrogen in UAF occurs mostly due to the reduction of nitrate under the anaerobic conditions. Besides, municipal wastewater is characterized by low concentrations of nitrates and nitrites [39]. The average nitrate concentration obtained at sampling point M1, 0.086 ± 0.042 $mg \cdot L^{-1}$, and the average nitrite concentration, 0.022 ± 0.007 $mg \cdot L^{-1}$, confirmed this. During the next sampling points, the nitrate continued to decrease; the mean value found at M2 was 0.037 ± 0.024 $mg \cdot L^{-1}$, M3 0.017 ± 0.11 $mg \cdot L^{-1}$ and M4 0.007 ± 0.007 $mg \cdot L^{-1}$. Finally, the nitrate could be reduced in anoxic conditions to gaseous nitrogen by means of denitrification, which is considered the main mechanism for nitrogen removal [6,41]. In general there are significant differences between the treatment stages for N_{tot}, N_{Org} and NH_4-N; regarding the retention time, the N_{tot} and N_{Org} showed significant differences between HRT1 and HRT3 as well as between HRT2 and HRT3.

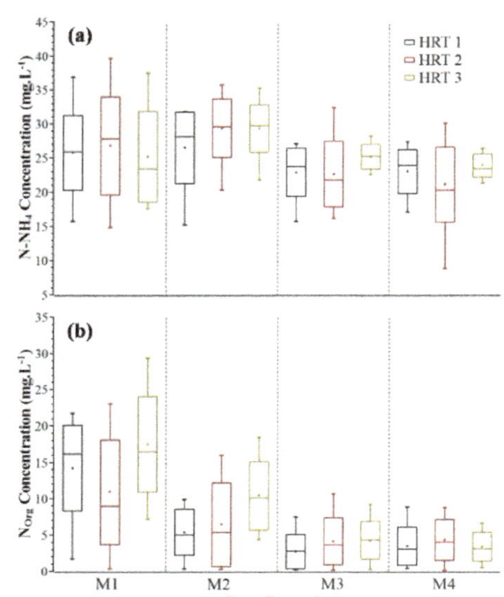

Figure 4: Plot of N-NH_4 ($mg \cdot L^{-1}$) (**a**) and N_{Org} ($mg \cdot L^{-1}$) (**b**), with

maximum and minimum values and standard deviation.

When the objective is to remove nutrients from municipal wastewater, several authors report that 10 days of retention time is required [25,42] to remove any constituent. The hydraulic retention time recommended as a prescriptive scaling factor to wetland design is between 10 and 13 days [8]. In this study, the treatment system removal of nitrogen and phosphorus showed a removal efficiency of 32.5% and 43.8%, respectively, in the HRT1; 30% and 24.2% in the HRT2; and, finally, 32.9% and 30.2% in the HRT3.

Phosphorous removal is governed mainly by sorption processes on the media that is used, and phosphorus can be stored in the accumulated sediments [43]. The largest decrease of P_{tot} was observed in the UAF at sample points M1 and M2 ($p \leq 0.05$), as is shown in the Figure 5; since the retention of phosphorus in previous units is mostly due to sedimentation of particulate fraction [39]. Removal of up to 37% was measured and is attributed, primarily, to adsorption on the biofilm formed on tezontle. The biofilm begins to form by a natural process on the tezontle surface. Then, the oldest layers are detached while a newer layer is formed, and finally the biofilm settles on the bottom of the UAF. This process is repeated constantly. When the sediment reaches a determined volume, it is removed with the sludge purge where the phosphorus contained in the biofilm is removed [24].

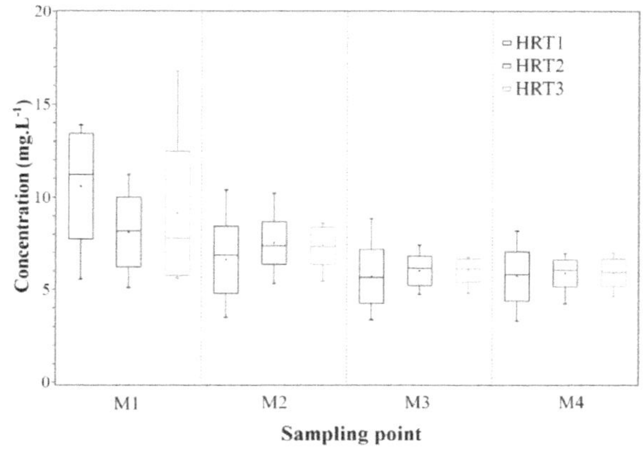

Figure 5: Plot of P_{tot} ($mg{\cdot}L^{-1}$), with maximum and minimum values and S.D.

The classic approach for treating wastewater from small communities has generally been with septic tanks [44]. Vymazal and Kröpfelová [39] report an average efficiency for previous treatment stages, such as septic and Imhoff tanks of 40.1% in BOD_5, 37% in COD, 25.8% in N_{Org}, −1.8% in NH_4-N, 8.8% in N_{tot} and 10.5% in P_{tot}. Population growth and environmental conditions influence the factors used for the selection of biological wastewater treatment. The efficiency of treatment required as well as the size and cost of systems have led to implementation of new anaerobic technologies as shown in Table 3. This Table reflects the best removal efficiencies achieved by this study and compares the results with other studies where anaerobic technology treatment was used previous to a horizontal subsurface constructed wetland in order to treat municipal wastewater.

Table 3: Pollutant removal efficiency comparison (%).

Case Study	Anaerobic Treatment						Constructed Wetland (CW)						Global Efficiency				
	1	2	3	4	6	7	1	2	3	5	6	7	1	2	3	6	7
Wastewater Treatment	UAF[a]	UAF	UASB[b]	TSUAR[c]	ABR[d]	UASB	HSSCW[e]	HSSCW	HSSCW	HSSCW	SFCW[f]	SFCW HSSCW	UAF HSSCW	ST[g] UAF HSSCW	UASB HSSCW	ABR SFCW HSSCW	UASB SFCW HSSCW
BOD	79.8 ± 9.0	56.6 ± 2.1	63.2	70	–	44.8	38.4 ± 35.8	50.6 ± 21.7	78	79–82	–	76.6	89.5 ± 4.4	78.1	92	–	87.1
COD	76.1 ± 7.5	49.6 ± 2.6	≈61	80	30.3 ± 1.4	49.4	44.7 ± 29.5	–	78	78–82	78.1 ± 2.3	70.9	84.3 ± 4.0	81.1	91	81.2 ± 2.3	85.3
N_{tot}	23.2 ± 10.7	–	–	–	31.3 ± 1.3	–	30.0 ± 8.0	–	≈24	–	83.4 ± 2.5	–	33.0 ± 17.7	–	≈27	82.3 ± 1.9	–
P_{tot}	35.4 ± 15.5	–	35	–	34.4 ± 2.2	–	21 ± 9.9	–	38	15	61.2 ± 2.1	–	43.8 ± 12.3	–	60	67.2 ± 2.3	–
HRT (h)	18	<8	8	48	110	5.5–10	72	<24	120	8.9–9.0	7	–	–	–	–	–	–

Notes: 1 This study, 2 [44], 3 [46], 4 [49], 5 [47], 6 [45]. 7 [21]; [a] UAF: Up-flow anaerobic filter, [b] UASB: Up-flow anaerobic sludge blanket; [c] TSUAR: Two-step upflow anaerobic reactor; [d] ABR: Anaerobic baffled reactor; [e] HSSCW: Horizontal subsurface flow constructed wetland, [f] SFCW: Surface flow constructed wetland, [g] ST: Septic tank.

With other anaerobic technologies, the UAF process has demonstrated a good performance for removal of organic matter and phosphorus. Nitrogen removal was slightly lower compared to the result of Ye *et al* [45], where it used a full-scale anaerobic baffled reactor (ABR) followed by a constructed wetland with a total retention time of 124 h, with a temperature between 20 and 25 °C, an HSSFCW area of 297 m^2, under a hydraulic loading of 50 $m^3{\cdot}(m^2{\cdot}day)^{-1}$, and with the start-up phase of nine months; however, the HRT tested in the HSSFCW was nine times shorter. Constructed wetlands coupled with anaerobic treatments in these studies, had some differences with regard to the wetland type, the type of plants used, granular media, climate and the retention time or the system configuration. The BOD removal efficiency found in this study was similar to the efficiency reported by Villegas *et al.* [44], who also worked in tropical conditions, with a UAF with an HRT of 8 h while the HSSCW used an HRT shorter than one day, an area of approximately of 107 m^2 and a mean water flow of 1.85 ± 0.4 $L{\cdot}s^{-1}$. While the N_{tot} removal efficiency was slightly higher in the study of El-Khateeb and El-Gohary [46], who assessed the capability of a system consisting of an up-flow anaerobic sludge blanket (UASB) reactor followed by HSSCW. This system was implemented in an arid and semi-arid zone, the HSSCW was planted with a density of three rhizomes per m^2, in an area of 2 m^2, and operated an HRT of 5 days while the UASB retention time was of 8 h, meaning a longer HRT compared to that used in the present study. The P_{tot} removal efficiency was higher than that found by El Hamuri *et al.* [47]

where limestone aggregates were used as filling material. In this study the behavior of a HSSCW used as post-treatment behind a two-step upflow anaerobic reactor (TSUAR) in Mediterranean conditions is evaluated. The HSSCW had an area of 28 m^2, a water flow of 9.5 $m^3 \cdot day^{-1}$ and HRT of 9 h. The global efficiencies reached for organic matter removal (up to 80%) were as expected, and they were comparable with to treatment systems with combined high rate anaerobic reactors and constructed wetlands as surface flow constructed wetlands (SFCW) with HSSCW. Better nitrogen removal has been found when constructed wetlands are combined with anaerobic treatment than for HSSCW units without a treatment stage previous. In addition, the area decrease is noticeable when the P-k-C* kinetic removal model [8] is applied to the design data, considering an influent BOD concentration (M1) of 259 ± 58 $mg \cdot L^{-1}$ and a target concentration of 60 $mg \cdot L^{-1}$ according to Mexican regulations [48], making it possible to calculate a wetland area decrease of 27%.

CONCLUSIONS

Mean global efficiencies found for the complete system for the lower HRT (two days) were: 89.5% ± 4.4% for BOD, 84.2% ± 4.0% for COD, 32.5% ± 13.2% for total nitrogen (N_{tot}) and 43.8% ± 12.3% for total phosphorous (P_{tot}). Mean removal efficiencies for the higher HRT (four days), were 86.1% ± 3.1% for BOD, 81.7% ± 3.9% for COD, 32.9% ± 17.7% for N_{tot} and 30.2% ± 16.0% for P_{tot}. Finally, for the intermediary HRT (three days), the results were: 79.3% ± 8.7% for BOD, 85.6% ± 3.1% for COD, 30.0% ± 16.8% for N_{tot} and 24.2% ± 17.7% for P_{tot}. The best mean efficiencies reached in the UAF stage for COD, BOD, N_{tot} and P_{tot} were 79.8%, 76.1%, 23.2% and 35.4%, respectively, at the HRT1. The best removal for nitrogen in the HSSCW was reached with a detention time of three and four days, with the longer time allowing further progress in the nitrogen cycle. The HRT recommended for phosphorus removal for UAF is two days, while for HSSCW an HRT of at least three days is recommended. In other words, two days is adequate to remove organic matter, but when the objective is to remove organic matter and nutrients a three-day HRT is recommended. Moreover, it would be interesting to test HRTs longer than four days for HSSCW and to conduct a more comprehensive investigation of the interactions on the support media. The UAF was responsible for the organic matter reduction while nitrogen removal was performed by the HSSCW. The goal of reducing the land surface required by the wetland was achieved, since the organic matter removal by the UAF decreased the land requirements moderately, and could lessen early clogging in the constructed wetland.

ACKNOWLEDGMENTS

The authors would like to thank to State Water Commission (Comisión Estatal del Agua, CEA-Jalisco) for its support during the experimental stage and to National Council of Science and Technology (Consejo Nacional de Ciencia y Tecnología, CONACYT) for the scholarship granted to student María de la Luz Merino Solís and the support of the Project 128209 (FOMIX CONACY-Veracruz). Also, a special thanks to MSc Leonel Hernández Mena for his constructive comments and to Dana Erickson and Walter Meyer for their linguistic contribution.

REFERENCES

1. Cui, L.; Zhu, X.; Ma, M.; Ouyang, Y.; Dong, M.; Zhu, W.; Luo, S. Phosphorus sorption capacities and physicochemical properties of nine substrate materials for constructed wetland. *Arch. Environ. Contam. Toxicol.* 2008, *55*, 210–217.
2. Smith, V.H. Eutrophication. In *Encyclopedia of Inland Waters*; Likens, G.E., Ed.; Elsevier: Oxford, UK, 2009; Volume 3, pp. 61–73.
3. Yates, C.R.; Prasher, S.O. Phosphorus reduction from agricultural runoff in a pilot-scale surface-flow constructed wetland. *Ecol. Eng.* 2009, *35*, 1693–1701.
4. Akratos, C.S.; Tsihrintzis, V.A. Effect of temperature, HRT, vegetation and porous media on removal efficiency of pilot-scale horizontal subsurface flow constructed wetland. *Ecol. Eng.* 2007, *29*, 173–191.
5. Puigagut, J.; Villaseñor, J.; Salas, J.J.; Bécares, E.; García, J. Subsurface-flow constructed wetlands in Spain for the sanitation of small communities: A comparative study. *Ecol. Eng.* 2007, *30*, 312–319.
6. Vymazal, J. Review: Removal of nutrients in various types of constructed wetlands. *Sci. Total Environ.* 2007, *380*, 48–65.
7. Langergraber, G. Modeling of processes in subsurface flow constructed wetlands: A review. *Vadose Zone J.* 2008, *7*, 830–842.
8. Kadlec, R.; Wallace, S. *Treatment Wetlands*, 2nd ed.; CRC Press, Taylor & Francis Group: Boca Raton, FL, USA, 2009; pp. 3–10, 715–734.
9. Matamoros, V.; Bayona, J.M. Behavior of emerging pollutants in constructed wetlands. *Handb. Environ. Chem.* 2008, *5*, 199–217.
10. Reed, C.; Crites, R.W.; Middlebrooks, E.J. *Natural Systems for Waste Management and Treatment*; Mc Graw Hill Co.: New York, NY, USA, 1995; pp. 173–178.
11. U.S. EPA. *Manual Constructed Wetlands Treatment of Municipal Wastewater*; Office of Research and Development: Cincinnati, OH, USA, 2000.
12. Wang, X.; Bai, X.; Qiu, J.; Wang, B. Municipal wastewater treatment with pond-CW system: A case study. *Water Sci. Technol.* 2005, *51*, 325–329.
13. Sim, C.; Yusoff, M.; Shutes, B.; Ho, S.; Mansor, M. Nutrient removal in a pilot and full scale constructed wetland, Putrajaya city, Malaysia. *J. Environ. Manag.* 2008, *88*, 307–317.
14. Ye, F.; Li, Y. Enhancement of nitrogen removal in towery hybrid constructed wetland to treat domestic wastewater for small rural communities. *Ecol. Eng.* 2009, *35*, 1043–1050.
15. De la Varga, D.; Díaz, M.A.; Ruiz, I.; Soto, M. Avoiding clogging in constructed wetlands by using anaerobic digesters as pre-treatment. *Ecol. Eng.* 2013, *52*, 262–269.
16. Knowles, P.R.; Griffin, P.; Davies, P.A. Complementary methods to investigate the development of clogging within a horizontal sub-surface flow tertiary treatment wetland. *Water Res.* 2010, *44*, 320–330.
17. Saeed, T.; Sun, G. A review on nitrogen and organics

removal mechanisms in subsurface flow constructed wetlands: Dependency on environmental parameters, operating conditions and supporting media. *J. Environ. Manag.* 2012, *112*, 429–448.

18. Barros, P.; Ruiz, I.; Soto, M. Performance of an anaerobic digester-constructed wetland system for a small community.*Ecol. Eng.* 2008, *33*, 142–149.
19. Ansola, G.; González, J.M.; Cortijo, R.; de Luis, E. Experimental and full-scale pilot plant constructed wetlands for municipal wastewater treatment. *Ecol. Eng.* 2003, *21*, 43–52.
20. Álvarez, J.A.; Ruíz, I.; Soto, M. Anaerobic digesters as a pretreatment for constructed wetlands. *Ecol. Eng.* 2008, *33*, 54–67.
21. Ruiz, I.; Díaz, M.A.; Crujeiras, B.; García, J.; Soto, M. Solids hydrolysis and accumulation in a hybrid anaerobic digester-constructed wetlands system. *Ecol. Eng.* 2010, *36*, 1007–1016.
22. Morel, A.; Diener, S. *Greywater Management in Low and Middle-Income Countries, Review of Different Treatment Systems for Households or Neighbourhoods*; Swiss Federal Institute of Aquatic Science and Technology (Eawag): Dubendorf, Switzerland, 2006; pp. 27–30.
23. Manariotis, I.; Grigoropoulos, S. Restart of anaerobic filters treating low-strength wastewater. *Bioresour. Technol.* 2008,*99*, 3579–3589.
24. López-López, A.; Albarrán-Rivas, M.G.; Hernández-Mena, L.; León-Becerril, E. An assessment of an anaerobic filter packed with a low-cost material for treating domestic wastewater. *Environ. Technol.* 2013, *34*, 1151–1159.
25. Trang, N.; Konnerup, D.; Schierup, H.; Chiem, N.; Tuan, L.; Brix, H. Kinetics of pollutant removal from domestic wastewater in a tropical horizontal subsurface flow constructed wetland system: Effects of hydraulic loading rate.*Ecol. Eng.* 2010, *36*, 527–535.
26. Ramsar Information Sheet, Version 2006–2008. Available online: http://ramsar.conanp.gob.mx/docs/sitios/FIR_RAMSAR/Jalisco/Lago%20de%20Chapala/LAGO_DE_CHAPALA.pdf (accessed on 13 March 2015).
27. Rubel, F.; Kottek, M. Observed and projected climate shifts 1901–2100 depicted by world maps of the Köppen-Geiger climate classification. *Meteorol. Z.* 2010, *19*, 135–141.
28. López-López, A.; Vallejo-Ramírez, R.; Méndez-Romero, D.C. Evaluation of a combined anaerobic and aerobic system for the treatment of slaughterhouse wastewater. *Environ. Technol.* 2010, *31*, 319–326.
29. Zurita, F.; de Anda, J.; Belmont, M.A. Treatment of domestic wastewater and production of commercial flowers in vertical and horizontal subsurface-flow constructed wetland. *Ecol. Eng.* 2009, *35*, 861–869.
30. Trejo-Téllez, L.I.; Ramírez-Martínez, M.; Gómez-Merino, F.C.; García-Albarado, J.C. Physical and chemical evaluation of volcanic rocks and its use for tulip production. *Rev. Mex. Cienc. Agríc.* 2013, *5*, 863–876.
31. Konnerup, D.; Koottatep, T.; Brix, H. Treatment of domestic wastewater in tropical, subsurface flow constructed wetlands planted with Canna and Heliconia. *Ecol. Eng.* 2009, *35*, 248–257.
32. Lin, Y.F.; Jing, S.R.; Lee, D.Y.; Wang, T.W. Nutrient removal from aquaculture wastewater using a constructed wetlands system. *Aquaculture* 2002, *209*, 169–184.
33. Kincanon, R.; McAnally, S. Enhancing commonly used model predictions for constructed wetland performance: As-built design considerations. *Ecol. Model.* 2004, *174*, 309–322.
34. Zurita, F.; de Anda, J.; Belmont, M.A. Performance of laboratory-scale wetlands planted with tropical ornamental plants to treat domestic wastewater. *Water Qual. Res. J. Can.* 2006, *41*, 410–417.
35. American Public Health Association (APHA). *Standard Methods for the Examination of Water and Wastewater*, 20th ed.; American Public Health Association, American Water Works Association, Water Environment Federation: Washington, DC, USA, 1999.
36. *Statgraphics Centurion, version XVI*; StatPoint Tecnologies, Inc.: Warrenton, VA, USA, 2011.
37. Sedlak, R.I. *Phosphorus and Nitrogen Removal from Municipal Wastewater: Principles and Practice*, 2nd ed.; CRC Press, Lewis publishers: Boca Raton, FL, USA, 1991; p. 3.
38. Wallace, S.D.; Knight, R.L. *Small-Scale Constructed Wetland Treatment Systems: Feasibility, Design Criteria and O&M Requirements*; Water Environment Research Foundation (WERF): Alexandria, VA, USA, 2006; pp. 4–21.
39. Vymazal, J.; Kröpfelova, L. *Wastewater Treatment in Constructed Wetlands with Horizontal Sub-Surface Flow*; Alloway, B.J., Trevors, J.T., Eds.; Springer: Berlin, Germany, 2008; Volume 14, pp. 40, 277–278.
40. Vymazal, J. Review: Horizontal sub-surface flow and hybrid constructed wetlands systems for wastewater treatment.*Ecol. Eng.* 2005, *25*, 478–490.
41. Maltais, G.; Maranger, R.; Brisson, J.; Chazarenc, F. Nitrogen transformations and retention in planted and artificially aerated constructed wetlands. *Water Res.* 2009, *43*, 535–545.
42. Yousefi, Z.; Mohseni-Bandpei, A. Nitrogen and phosphorus removal from wastewater by subsurface wetlands planted with Iris pseudacorus. *Ecol. Eng.* 2010, *36*, 777–782.
43. Green, M.; Safray, I.; Agami, M. Constructed wetlands for river reclamation: experimental design, start-up and preliminary results. *Bioresour. Technol.* 1995, *55*, 157–162.
44. Villegas, J.D.; Guerrero, J.; Castaño, J.M.; Paredes, D. Septic tank (ST)-Up flow anaerobic filter (UFAF)-subsurface flow constructed wetland (SSF-CW) systems aimed at wastewater treatment in small localities in Colombia. *Rev. Tec. Fac. Ing. Univ. Zulia* 2006, *29*, 269–281.
45. Ye, C.; Li, L.; Zhang, J.; Yang, Y. Study on ABR stage-constructed wetland integrated system in treatment of

rural sewage. In Proceedings of the 2011 International Conference on Environmental Science and Engineering, Bali Island, Indonesia, 1–3 April 2011.

46. El-Khateeb, M.A.; El-Gohary, F.A. Combining UASB technology and constructed wetland for domestic wastewater reclamation and reuse. *Water Sci. Technol. Water Supply* 2003, *3*, 201–208.

47. El Hamouri, B.; Nazih, J.; Lahjouj, J. Subsurface-horizontal flow constructed wetland for sewage treatment under Moroccan climate conditions. *Desalination* 2007, *215*, 153–158.

48. Secretaría de economía, Norma Oficial Mexicana NOM-001-ECOL-1996. Available online: http://www.economia-noms.gob.mx/normas/noms/1997/001-ecol.pdf (accessed on 12 December 2014).

49. El Hafiane, F.; El Hamouri, B. Anaerobic reactor/high rate pond combined technology for sewage treatment in the Mediterranean area. *Water Sci. Technol.* 2005, *52*, 125–132.

CHAPTER 6

REHABILITATION PRIORITY DETERMINATION OF WATER PIPES BASED ON HYDRAULIC IMPORTANCE

Do Guen Yoo, Doosun Kang , Hwandon Jun and Joong Hoon Kim

Keywords: Rehabilitation priority, water pipes, hydraulic importance, decision making

INTRODUCTION

A water distribution system is an important part of the social infrastructure, facilitating water transport, distribution and supply. Such a system is a highly complicated network that combines pipelines, pumps and valves. Hence, the facilities in any such system should be continuously improved and updated based on specific plans to maintain the stability and safety of the water supply. As the importance of maintaining and managing this water distribution system has increased, projects for repairing and replacing deteriorating water pipes are currently being undertaken throughout Korea. However, the current methods used to deal with deteriorating pipelines involve an evaluation of the degree of deterioration based on empirical means, as well as reactive rehabilitation projects undertaken after accidents, leading to economic losses and failure to improve system functions. The current approach to determining the rehabilitation priority order for pipelines is based only on the year of installation of the pipes, with no clear criteria for evaluating the degree of deterioration. To address these problems, a new approach to determine the rehabilitation priority order for a water distribution system should be developed that overcomes the drawbacks of the existing methods. In addition, the rehabilitation priority order should be determined according to not only the physical deterioration rate of individual pipes, but also the relative importance of those pipes, to increase the overall stability and safety of the system.

Early studies of the methods to determine the rehabilitation priority order for water distribution systems were conducted using rehabilitation models based on empirical determination that use only general guidelines. Subsequent studies on the determination of the rehabilitation priority order can be broadly classified into the categories of analysis of the deterioration rate and failure probability, regression analysis coupled with failure probability analysis and priority order estimation based on economic feasibility analysis. The Guidance Manual-Water Mains Evaluation for Rehabilitation/Replacement published by the American Water Works Association Research Foundation [1] proposes physical strength, water quality in the pipes, hydraulic conditions and water leakage as criteria for evaluating pipeline functionality. Accordingly, the manual notes that comprehensive management that considers these four criteria is required to effectively maintain and manage water distribution pipelines. The Water Research Centre [2] of the United Kingdom states that it is reasonable to prioritize the repair and rehabilitation of water distribution pipes that are frequently associated with accidents. K-water [3] determined pipeline deterioration rates and developed a weighting system for pipeline repair and rehabilitation by collecting and

Citation: Do Guen Yoo, Doosun Kang, Hwandon Jun and Joong Hoon Kim, Rehabilitation Priority Determination of Water Pipes Based on Hydraulic Importance, Water 2014, 6, 3864-3887; doi:10.3390/w6123864.

summarizing domestic and international literature, as well as data on large-diameter water distribution pipes in metropolitan water distribution systems. Subsequently, K-water developed a model to estimate the deterioration rate of water pipes using a point-based evaluation method that considers the estimated pipe condition and weight, as well as a model for prioritizing the replacement and rehabilitation of water pipes. However, this model is limited in that the point-based evaluation method does not reflect the hydraulic characteristics of water distribution systems. Kim *et al.* [4] re-calculated the cost function by modifying and complementing the failure rate function proposed by Shamir and Howard [5]. In addition, they estimated the deterioration rate using a probabilistic neural network (PNN) and proposed a rehabilitation model that prioritizes the rehabilitation and replacement of pipes using a shortest-path model. Studies in which regression analysis and failure probability analysis have been applied to estimating the rate of deterioration of water pipes have been conducted by Marks and Clark [6,7], Agbenowosi [8] and Park and Loganathan [9,10]. Marks and Clark [6,7] proposed a method for failure modeling based on the deterioration rate of the water distribution pipes. After classifying the deterioration status of water distribution pipes into "early stage", with a small number of failures, or "later stage", with a large number of failures, their method applies these stages to each case. Based on their results, they proposed that the deterioration status of a water distribution pipe could be represented by a proportional hazards model in the early stage and a Poisson-type model in the later stage. Deb *et al.* [11] classified water distribution pipes according to their installation year, pipe material, diameter and backfilled soil type. They also developed a probabilistic model called "KANEW" based on this classification to estimate the number of pipes in a water distribution system that should be replaced annually. Agbenowosi [8] represented the factors related to water distribution pipes in a pipe load model and a pipe break model, so that the factors could be analyzed mathematically and the most economical time at which to replace a water distribution pipe could be determined. Park and Loganathan [9,10] proposed a failure estimation model for water distribution pipes using a threshold break rate. Using this model, the economically optimal replacement time is determined as a function of the costs of pipe replacement and rehabilitation, a discount rate and the length and diameter of the pipes.

In general, these models, which determine the rehabilitation priority order of pipes in water distribution systems using regression and failure probability analyses, predict pipe failure through complex formulas and determine the most economical replacement and rehabilitation times through economic feasibility analysis. While these models can predict pipe failure through failure probability analysis, they require a large amount of basic data and complex formulas to produce results. Models for determining the rehabilitation priority order of pipes in water distribution systems on the basis of economic feasibility, which have been examined by Shamir and Howard [5], Walski [12] and Luong and Fujiwara [13], achieve maximum efficiency at minimum cost. Alvisi and Franchini [14] proposed a near-optimal rehabilitation scheduling method based on a multi-objective genetic algorithm. With reference to a fixed time horizon, the goal is to minimize the overall costs of repairing and/or replacing pipes and to maximize the hydraulic performances of the water network. However, given that these models also require a large amount of basic data and that water distribution pipes are part of social infrastructure networks that directly affect public welfare, cost optimization cannot be an essential condition.

In recent years, computer-aided models and decision support tools, such as Care-W [15,16,17,18,19,20,21], CASSES [22] and AWARE-P [23], have been developed. The CARE-W project aimed to develop methods and software that would enable engineers of the water undertaking to define and implement an effective management of their water supply networks, rehabilitating the right pipelines at the right time. This project was organized in eight work packages (WP), which were the construction of a control panel of performance indicators (WP1), description and validation of technical tools (WP2), elaboration of a decision support system for annual rehabilitation programs (WP3), elaboration of long-term strategic planning and investment (WP4), elaboration of the CARE-W prototype (WP5), testing and validation of the CARE-W prototype (WP6), dissemination (WP7) and project management (WP8). CARE-W used four probabilistic forecast failure models (proportional hazard model, Markov model, Poisson analysis, and non-homogeneous Poisson process) and three mathematical hydraulic reliability assessment models (Aquarel, Failnet-Reliab, Relnet). Among these WPs, WP3 and WP4 focused on the determination of priority and scheduling for rehabilitation according to the target time period (one year and long-term plan). They used multi-criteria techniques, such as scoring and ELETRE TRI methods, to determine relevant procedures during annual rehabilitation programs. In case of long-term rehabilitation strategies, extended KANEW is used. In the continuation of CARE-W, a new break prediction model, called linear extension of the Yule process (LEYP), was developed. It involved a statistical model based on a counting process and relied not only on the pipe's characteristics and environment, but also its age and previous breaks. This new model was chosen to be used in the development of the break prediction software, CASSES. The main result from this software was the number of breaks for each pipe for a period in the future. The AWARE-P project has been performed for providing water and wastewater utilities with the know-how and the tools needed for efficient decision-making in infrastructural asset management (IAM) of urban water services. The key point of this project is IAM as a management process, based on plan-do-check-act (PDCA) principles and requiring alignments between the organization's strategic objectives and targets and the actual priorities and actions implemented.

Water distribution pipes can be divided into hydraulically significant or insignificant pipes depending on the installation location, valve location, base demand for water at nodes, variation in demand, pipe flow rates and the population served. Accordingly, the importance of individual water distribution pipes can be termed "hydraulic importance." This study was conducted to develop a method to determine the rehabilitation priority order of pipes in water distribution systems on the basis of both the deterioration rate of water pipes and their hydraulic importance. Using the method proposed in this paper, pipes are first classified according to their deterioration rate and hydraulic importance according to two different attributes. A multi-criteria decision-making method called a weighted utopian approach, which combines a weighting method and a distance measurement method, is then used to prioritize the pipes' rehabilitation needs.

METHODOLOGIES

In this study, the deterioration rate under a normal situation and hydraulic importance under abnormal conditions of water distribution pipes were used to determine the rehabilitation priority order of pipes in a water distribution system, as shown in Figure 1. The deterioration rate was divided into two components: the deterioration rate due to internal factors and the deterioration rate due to installation environment factors (external factors). The hydraulic importance was also divided into two components: the importance related to single pipe failure (which accounts for most pipe accidents) and the importance related to multiple pipe failures due to disasters, such as earthquakes.

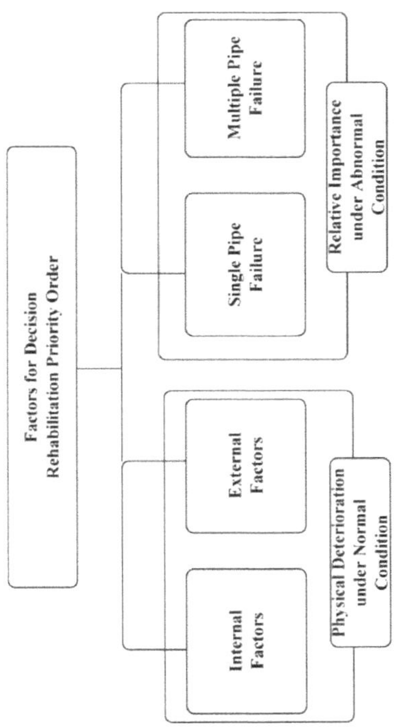

Figure 1: Attributes for rehabilitation priority order.

Pipe Deterioration

In this study, the deterioration rate calculation method and factors proposed by Kim *et al.* [4] were considered, and a total of 12 factors were classified as either internal or external factors. Eight internal factors were identified: pipe material, pipe diameter, interior corrosion rate, elapsed time since installation, type of joint, record of leakage and breakage, record of civil appeals regarding water quality and pressure and maximum pipe pressure. Four external factors were identified: exterior corrosion rate, backfilled soil type, road width and installation district. To quantify the deterioration rate, weights for the factors should be taken into consideration. From among the methods available for determining such weights, an eigenvector method employed for weight calculation in an analytic hierarchy process (AHP) [24], a decision-making method, was selected for use in this study.

A PNN model proposed by Specht [25] was employed to take the weight of each factor into consideration in calculating the degree of deterioration. To apply this PNN model to the prioritization method developed in this study, the conditions of the factors should be classified according to conditional values. For example, the factors that are most heavily influenced by the deterioration rate are assigned a value of 1.0, and those that are influenced the least are assigned a value of 0.0. The conditions that have the greatest influence on a factor are assigned a value of 1.0, and those with the least influence are assigned a factor of 0.0. The conditions of factors, such as the exterior corrosion rate, backfilled soil type, road width and installation district, which are used to calculate the possibility of water distribution pipes failing due to installation environment/external factors, were classified according to a total of five conditional values. The relative weight of each factor is then calculated using the eigenvector method. The PNN model applied in this study consists of input, pattern, summation and output layers, as illustrated in Figure 2. In the input layer, a conditional value defined according to conditions for standard pipes is assigned to each neuron, while five large neurons are placed in the pattern layer, with one large neuron being responsible for a certain boundary condition value. A large neuron is one of the main components when we make the structure of a PNN. In this study, a large neuron works as a dividing criterion, which is divided into five deterioration groups among the pipes. That is, the first large neuron includes neurons that have a conditional value of 1, while the second, third, fourth, and fifth large neurons consist of small learning neurons with conditional values of 0.75, 0.5, 0.25 and 0.0, respectively. The four small learning neurons correspond to the exterior corrosion rate, backfilled soil type, road width and installation district, respectively. The pattern layer plays a critical role in identifying input data and performing classification, by measuring the distance between the input data and learning neurons, and includes the calculated distance in the activation function. In the model used in this study, a Gaussian function was used to determine the width and area of the data in the activation function. The summation layer has a single summation neuron that simply adds data learned from the pattern layer. The output layer has five neurons that have a simple classification function. When classification is performed in the output layer, classification is not done based on a percentage value, but by a simple summation of all values obtained in the summation layer, once trained, after the input data are entered. When output values are defined as P1, P2, P3, P4 and P5; P1 means a probability value that is to be included in the neuron responsible for the conditional value of 1. From among P1, P2, P3, P4 and P5, obtained through the process above, the largest value is selected; if P1 is the largest value, the deterioration rate of the corresponding pipe is classified into Group 1. For example, if P1, P2, P3, P4 and P5 have values of 0.45, 0.15, 0.2, 0.15 and 0.05, respectively, as a result of calculating the deterioration rate, P1, which has the largest probability, is selected, and the deterioration rate of the corresponding pipe becomes that corresponding to Group 1. The groups and probability values of the pipes are obtained by using the PNN algorithm to assess the deterioration rate of the pipes based on internal factors and installation environment/external factors. In addition, values are distributed between 0 and 1 using an interpolation method to compare the values for each pipe.

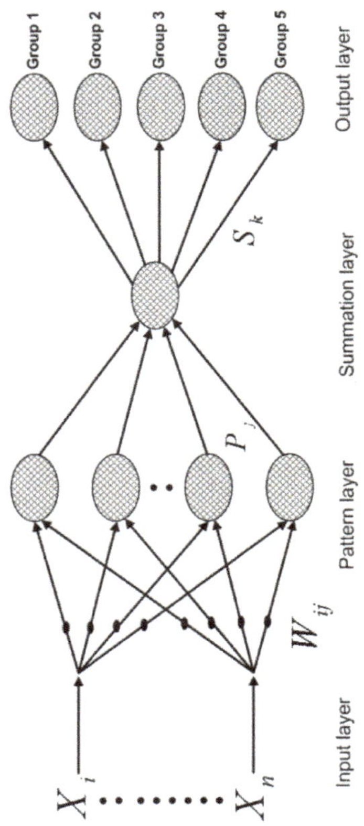

Figure 2: Probabilistic neural networks for assessing pipe deterioration.

Hydraulic Importance: Single Pipe Failure

Whereas most existing methods restrict the range of damage due to the failure of a water pipe to the corresponding water pipes, Jun *et al.* [26] introduced the concept of "unintended isolation," in addition to the concept of partial segments in water distribution networks, as proposed by Walski [27], which are used to estimate the damage area due to the failure of water pipes. According to Walski [27], a segment can be defined as a set of surrounding pipes that are closed by water control valves, along with a failed pipe, in the event of a failure. That is, when a failed pipe and the surrounding pipes should be isolated, a segment can be defined as the number of separated water pipes and nodes. When the area of damage is estimated using only failed water pipes, it is limited to those cases in which only the failed water pipes are closed. However, considering the practicalities, adjacent pipes may also have to be closed, depending on the number and locations of the water control valves. For example, Figure 3 shows that when P2 fails, such that the water control valves have to be closed to enable repairs, P4 is also closed. Consequently, the water supply to node N1 within the closed segment is also stopped. This damage area estimation method can produce more realistic estimates of the areas of damage, because it considers the number and locations of the water control valves. When a segment that includes a failed water pipe is closed to enable repair, the water supply to other sub-pipes connected to the failed pipes is also shut off if the segment constitutes the only path from the water source. That is, some pipes can experience disconnection of the water supply due to the segment closure. Jun *et al.* [26] defined this event as unintended isolation and developed an updated breadth-first search algorithm based on a node-arc matrix, which is more advanced than a conventional breadth-first search algorithm, to identify the unintentionally isolated pipes. In the event of unintended isolation, consumers of water in the area are deprived of their supply until the failed water pipe is fixed. That is, even if the supply in the area is not shut down by the water control valves, consumers in the area are isolated from the water supply being provided to those living in the segment. In Figure 3, a segment consisting of P6 and N6 generates unintended isolation.

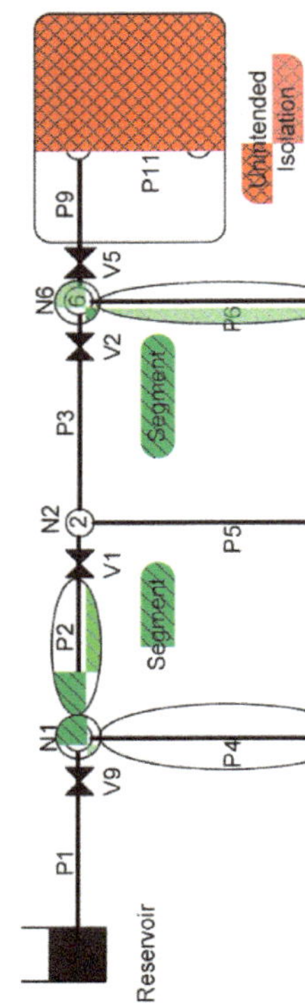

Figure 3: Segment and unintended isolation [26]. P, probability value; N, node.

To determine the importance of individual water pipes in the event of a single pipe failure, it is necessary to assume the following: the failure probability of each pipe is the same, and no more than two pipes can fail at any one time. The probability of pipe failure can differ among pipes according to the deterioration rate of the pipe, size, material, *etc.* In the case of hydraulic importance calculations of single pipe failure simulation, components of failures include not only unintentional failure (e.g., breakage and leakage), but also intentional failure (e.g., cut off the water). In this case, we cannot exactly predict intentional failure. Because of that, we assumed that the failure probability of each pipe is the same. It is also assumed that all of the valves are functioning properly, so valve malfunctions do not have to be considered. Based on these assumptions, the segments and instances of unintended isolation that are generated as a result of a single pipe failure in a water distribution system are calculated, and this calculation process is executed for all pipes. Finally, the importance of an individual water pipe in the event of a single pipe failure is calculated for each of the water pipes, based on Equation (1). The greater the $HISPF_i$ (hydraulic importance by single pipe failure when pipe i fails) value of an individual pipe, the higher the relative importance of the pipe in the event of a failure. Conversely, the smaller the $HISPF_i$ value, the lower the relative importance of the pipe in the event of a failure.

$$HISPF_i = \frac{Q_{i,S} + Q_{i,UI}}{Q} \quad (1)$$

where $HISPF_i$ = hydraulic importance by single pipe failure when pipe i fails; Q = total pipe flow under normal conditions; $Q_{i,s}$ = segment pipe flow when pipe i has failed; $Q_{i,UI}$ = unintended isolation pipe flow when pipe i has failed.

Hydraulic Importance: Multiple Pipe Failures

Because earthquakes typically lead to major damage and multiple pipe failures, the importance of water pipes as evaluated in the event of an earthquake was considered in this study. Yoo *et al.* [28,29] developed the reliability evaluation model of seismic hazards for water supply networks (REVAS.NET) for use in evaluating seismic risks to water distribution systems. This model is intended to prevent damage from earthquakes by calculating the reliability of a water distribution system in the event of an earthquake. This model was employed in this study. The process by which REVAS.NET was employed and the reliability factors used in this study are as follows:

- Step (1) Establishment of basic topological structure of water supply network in the area of application;
- Step (2) Construction of fragility curves for each component of a water distribution system;
- Step (3) Simulated earthquake generation (seismic location and magnitude);
- Step (4) Seismic wave attenuation;
- Step (5) Determination of component conditions;
- Step (6) Execution of EPANET hydraulic analysis;
- Step (7) Negative pressure processing;
- Step (8) Calculation of hydraulic reliability of systems/components.

Once the states of the water pipelines, tanks and pump facilities after the simulated earthquake have been determined, these states can be simulated via hydraulic analysis, and the hydraulic analysis results for the water distribution system can be derived. To quantify the results, a reliability factor is used. In general, system reliability, which is a widely-used concept in many fields, refers to the probability of a service running continuously within a system. In REVAS.NET, nodal serviceability (N_S) and system serviceability (S_S) [30,31,32], which are known to be relevant to a water distribution system, are used as reliability factors.

In this study, the hydraulic importance of water pipelines was calculated using the nodal serviceability (N_S), expressed by Equation (2). In the case of N_S, the hydraulic importance of a node for which nodal demand is required is calculated from the ratio of the available nodal demand to the required nodal demand. The usability of a node at which the required nodal demand is not present is assessed by evaluating the ratio of the pressure after the earthquake to the allowable minimum nodal pressure. This reliability index, which represents the serviceability that each node can provide, is calculated from the required nodal demand and minimum pressure, so it can be used to assess the relative supply ability per node.

$$\text{Nodal Serviceability } (N_{S,i}) = \begin{cases} \dfrac{Q_{avl,i}}{Q_{in\,i,i}} & \text{when } Q_{in\,i,i} \neq 0 \\ \sqrt{\dfrac{Min(P_i, P_{min})}{P_{min}}} & \text{when } Q_{in\,i,i} = 0 \end{cases} \quad (2)$$

where $Q_{avl,i}$ = available nodal demand at node i; $Q_{ini,i}$ = required nodal demand at node i; P_i = nodal pressure at node i; P_{min}= allowable minimum nodal pressure.

Finally, the importance of a water pipeline in the event of multiple pipe failures is calculated using Equation (3). The importance can be calculated by subtracting the average reliability value of each node connected to each water pipe from value 1. The higher the resulting value is, the lower the serviceability of the water pipelines related to the corresponding water pipeline as a result of a corresponding pipeline breakage followed by multiple pipe failures.

$$HIMPF_i = 1 - \frac{UN_{S,i} + DN_{S,i}}{2} \quad (3)$$

where $HISPF_i$ = hydraulic importance of pipe i by multi-pipe failure; $UN_{S,i}$ = upstream nodal serviceability of pipe i; $DN_{S,i}$ = downstream nodal serviceability of pipe i.

Weights of Attributes

Many studies have been conducted to examine methods for determining the weighting factors of attributes used in multi-criteria decision-making methods. These studies have shown that common usage standards do not exist among the conventional methods and that no one method stands out as better than the others. Although some studies have proposed weighting factor methods based on the type of decision-making domain, the theoretical validity of these methods has not been confirmed. Thus, it is crucial to determine appropriate weighting methods on the basis of the types and attributes of the decision-making domain and to make a reasonable decision accordingly. In this study, the five weighting methods listed in Table 1 were employed.

Table 1: Weighting methods.

No.	References	Method
1	-	Same weighting
2	[24]	Eigenvector
3	[33]	Churchman–Ackoff
4	[34]	Rating
5	[35,36]	Entropic

Eigenvector

The eigenvector method proposed by Saaty [24] has been used as a weighting calculation method for AHP, which is one of the main multi-criteria decision methods. Weighting factor values can be calculated using Equation (4) below. This is a widely used method, because it is easily applied and suitable for problem analysis.

$$A\varpi = \lambda_{max}\varpi \tag{4}$$

where A = positive pairwise comparison reciprocal matrix; λ_{max} = maximum eigenvalue of A; ϖ = weighting factor of the attribute.

Churchman–Ackoff Method

The Churchman–Ackoff method [33] is based on rankings, whereby weightings are calculated via Equation (5). For example, when a problem consists of five attributes, n is 5 and $\sum_{j=1}^{5} j$ is 15. When Attributes 1 to 5 are ranked from 1st to 5th, respectively, the weights of the attributes will be 0.33, 0.27, 0.20, 0.13 and 0.07, in sequence.

$$\omega_i = \frac{J}{\sum_{j=1}^{n} j} = \frac{2J}{n(n+1)} \tag{5}$$

where j = attribute (j=1,2,…,n); n = number of attributes; J = rank of the importance of the j-th attribute.

Rating

The rating method [34] calculates weights based on ratings received from a respondent, and calculation is performed as indicated in Equation (6). The scale of the rating can be arbitrarily established from 0 to 10.

$$\omega_i = \frac{\sum_{k=1}^{l} \omega_{jk}}{\sum_{j=1}^{n}\sum_{k=1}^{l} \omega_{jk}} \tag{6}$$

where j = attribute (j=1, 2,…, n); n = number of attributes; k = respondent (k=1, 2,…, l); l = number of respondents; $\omega_{jk} = \frac{r_{jk}}{\sum_{j=1}^{n} r_{jk}}$; r_{jk} = rating of respondent k.

Entropic Method

The entropic method was proposed by Hwang and Yoon [35] and Soofi [36]. It is an objective method that is used to calculate weights using only data in the decision-making domain and not using the subjective judgments of decision makers. The procedure for the entropic method starts by estimating the weight vectors from an alternative-attribute matrix. From the viewpoint of entropy, an alternative-attribute matrix contains information that can be used to estimate weights for criteria. That is, a criterion that has a large difference between alternatives is an important criterion, whereas a criterion that has a small difference between alternatives is a less important criterion. It is assumed that a decision-making domain can be represented by matrix D, shown in Equation (7) below.

$$D = \begin{bmatrix} x_{11} \cdots x_{1j} \cdots x_{1n} \\ \vdots \cdots \vdots \cdots \vdots \\ x_{i1} \cdots x_{ij} \cdots x_{in} \\ \vdots \cdots \vdots \cdots \vdots \\ x_{m1} \cdots x_{mj} \cdots x_{mn} \end{bmatrix} \tag{7}$$

If p_{ij} is a normalized result with respect to all attributes, p_{ij} can be represented by Equation (8).

$$p_{ij} = \frac{x_{ij}}{\sum_{i=1}^{m} x_{ij}} \ (i=1,2,\ldots,m; j = 1,2,\ldots,n) \tag{8}$$

Assuming that the entropy of each attribute is E_j, the entropy can be calculated using Equation (9).

$$E_j = -k\sum_{i=1}^{m} p_{ij} \cdot \log p_{ij} \ (k = \frac{1}{\log m}) \tag{9}$$

To calculate the weight of an attribute, a level of diversity, d_j, is used that can be calculated using Equation (10). When these values are normalized with respect to every attribute, this represents the weight of the corresponding attribute (Equation (11)).

$$d_j = 1 - E_j \tag{10}$$

$$\omega_i = \frac{d_j}{\sum_{j=1}^{n} d_j} \tag{11}$$

Weighted Utopian Approach

The utopian approach proposed by Xanthopulos *et al.* [37] can be applied whenever there are multiple attributes and offers the advantage of convenience in application. However, it suffers the drawback that calculations are performed assuming the same weights for each attribute. In general, the weights of the attributes cannot be the same during decision making. If the same weights were to be used, it would produce results and judgments that differed from those of the decision makers.

Yoo *et al.* [38] proposed a weighted utopian approach (WUA) that combines a weighting method for multi-criteria decision making. A schematic view of the WUA is shown in Figure 4. Assuming that there are two attributes that enter into making a decision, namely A1 and A2, a coordinate plane with two axes, can be visualized. The normalization of the attribute values of every alternative between 0 and 1 over the coordinate plane can be seen in the graph on the left-hand side of Figure 4.

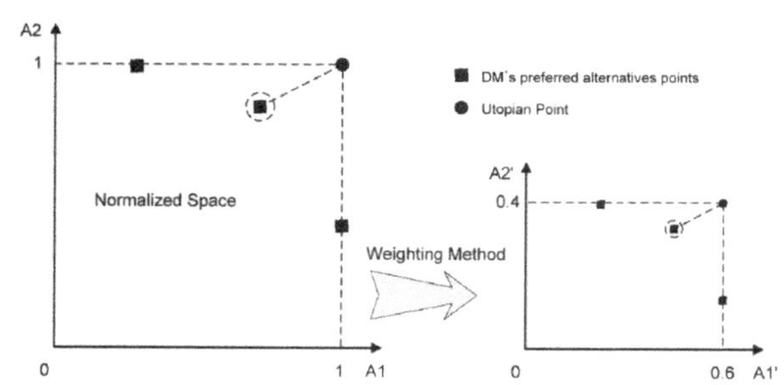

Figure 4: Weighted utopian approach.

Once the weights of the attributes have been calculated using the weighting method for multi-criteria decision making, they are multiplied for each axis, so that alternative attribute and utopian point values, distributed between 0 and 1, are redistributed on a coordinate plane. For example, if the weights of A1 and A2, which were calculated using the weighting method, are 0.6 and 0.4, respectively, the newly created utopian point on a graph changes from (1, 1) to (0.6, 0.4). Finally, using the re-distributed graph, the distance between the utopian point and the alternative is calculated. The distance measurement method used to calculate a distance employs the commonly used Euclidean distance, illustrated in Figure 5. Finally, the priority of the alternatives is determined from the values of the Euclidean distance, starting with the shortest distance.

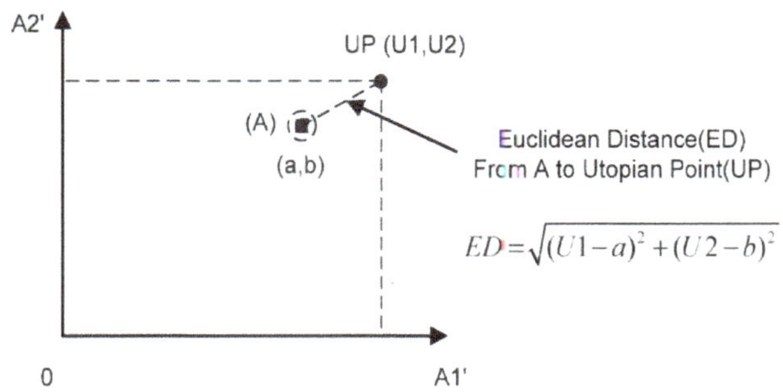

Figure 5: Euclidean distance.

That is, the priority of each alternative is determined using a weighting method, and the final priority ranking of the alternatives is determined by sorting the average ranking values produced using the weighting method in an ascending order.

Figure 6 shows a flow chart for the model proposed in this paper for determining the rehabilitation priority order for water pipelines.

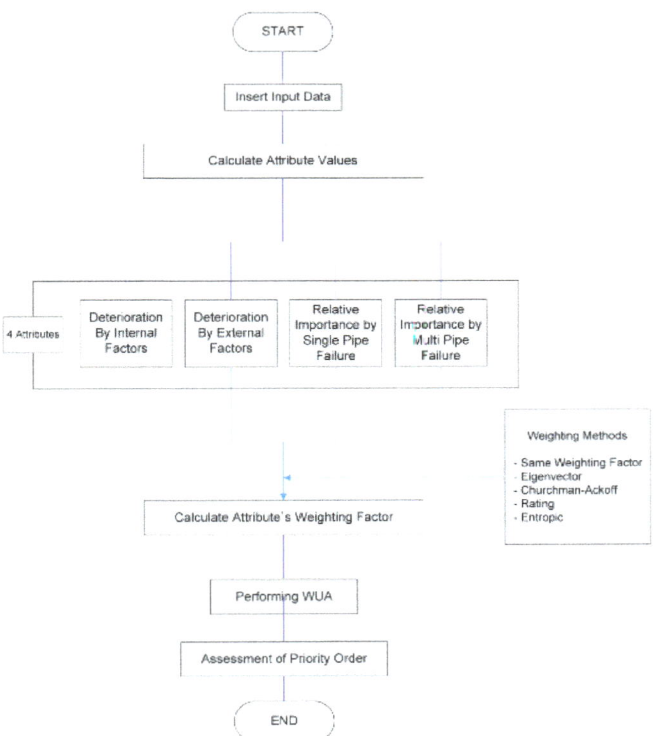

Figure 6: Flow chart for determination of rehabilitation order. WUA, weighted utopian approach.

APPLICATION RESULTS

The model proposed in this study was applied to the KA network in City J, which divides its overall water network to enable area isolation, with 10 large district meter areas and 130 small district meter areas. City J has established main project plans to replace and rehabilitate deteriorating water pipelines in 50 small district meter areas and to maintain the water distribution facilities. There are 350 pipelines in the KA network, which is one of the 10 large district meter areas, with a total pipeline length of 35,951 km. The average demand per day is 16,058.4 cubic meters per day (CMD). The number of nodes is 230, and the number of installed valves is 160. Figure 7 shows the locations of pipelines and nodes in the KA network. Within the network, 24 pipes have diameters greater than 300 mm, while most of the pipes have diameters between 80 and 300 mm.

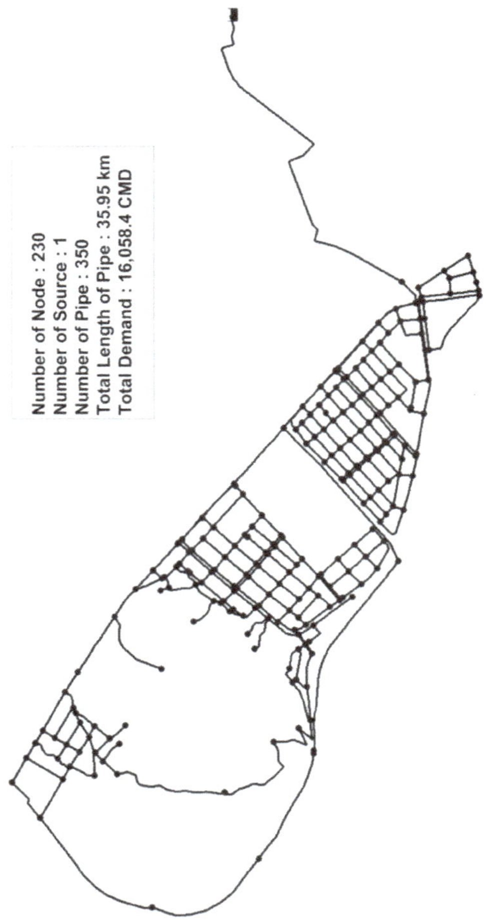

Figure 7: KA network in City J. CMD, cubic meters per day.

Results: Deterioration Rate

Input data consisting of internal factors and external factors for each pipe were acquired to calculate the deterioration rates of the water pipelines using the PNN algorithm. Scores for each factor, to be applied in the PNN algorithm, were determined according to Table 2 and Table 3. In addition, the weights for each factor were obtained by applying the eigenvector method to the results of surveys of 15 experts (the results are summarized in Table 4). Of the 15 experts, five were from academia, five were from corporations and five were from research centers and government agencies. The estimates of the deterioration rates with respect to the internal factors indicated that the number of Group 1 water pipes was 250 of the 350 pipes and that the number of Group 5 water pipes was 100. The reason for such a distinctive difference can be discerned from the years of installation of the water pipes. The average number of years that had elapsed from the installation date of the Group 1 water pipes was 18.5 years, which is considerably longer than that of the Group 5 water pipes, which was four years, many of the Group 5 water pipes having recently been replaced or rehabilitated. This means that those water pipes for which many years had elapsed since the installation date had a greater inside corrosion rate and more frequent records of civil appeals and repairs than more recently installed water pipes. The classification of deterioration rates by installation environment/external factors placed 10 pipes in Group 2, 81 pipes in Group 3 and 259 pipes in Group 5. The main reason that most pipes were in Group 5 was that no water pipes with a significantly large exterior corrosion rate (a factor to which a large weight applied) existed. Another reason was that most of the water pipes were installed in sandy soil, which tends to inhibit deterioration. As a result, the deterioration rate associated with the installation environment/external factors was not high.

Table 2: Scores resulting from internal factors.

Factor	Detailed Classification	Scaled Score	Factor	Detailed Classification	Scaled Score
Pipe material	CI *, CIP *	1.00	Pipe diameter (mm)	Below 80	1.00
	PE *, PVC *	0.75		80–100	0.75
	SP *	0.50		100–150	0.50
	HI-3P *	0.25		150–250	0.25
	DCIP *	0.00		Above 250	0.00
Inside corrosion rate (mm)	Below −2	1.00	Installation year (Installation duration, y)	Above 20	1.00
	−2–0.5	0.75		15–20	0.75
	−0.5–0.3	0.50		10–15	0.50
	−0.3–0	0.25		5–10	0.25
	Above 0	0.00		Below 5	0.00
Record of leakage or valve replacement (No. of cases/(y·m))	Above 10	1.00	Maximum pressure of pipes (kg/cm^2)	Above 5	1.00
	7–10	0.75		4–5	0.75
	3–7	0.50		3–4	0.50
	0–3	0.25		2–3	0.25
	0	0.00		Below 2	0.00
Joint type	Welding	1.00	Record of a civil appeal (water quality or pressure)	Yes	1.00
	Rubber ring	0.50			
	Mechanic	0.00		No	0.00

Notes: * CI/CIP: Cast Iron Pipe; PE: Poly-Ethylene Pipe; PVC: Poly-Vinyl Chloride Pipe; SP: Steel Pipe; HI-3P: High Impact (3-Layer) Pipe; DCIP: Ductile Cast Iron Pipe.

Table 3: Scores resulting from external factors.

Factor	Detailed Classification	Scaled Score	Factor	Detailed Classification	Scaled Score
Outside corrosion rate (mm)	More than 9	1.00	Road width	Industrial road	1.00
	6–9	0.75		4-lane road	0.75
	4–6	0.50		2-land road	0.50
	2–4	0.25		Street	0.25
	Less than 2	0.00		Sidewalk and bare ground	0.00
Installation district	Factory	1.00	Refilled soil type	Clay	1.00
	Roadside	0.75		Silt	0.50
	Commercial district	0.50			
	Apartment	0.25		Sand	0.00
	Residential area, farmland	0.00			

Table 4; Weighting values resulting from internal and external factors.

Factor	Weight
Pipe material	0.452
Pipe diameter	0.286
Inside corrosion rate	1.864
Installation year	1.252
Joint type	0.515
Record of leakage or valve replacement	1.848
Record of a civil appeal	1.140
Maximum pressure of pipes	0.643
Outside corrosion rate	2.015
Refilled soil type	0.680
Road width	0.610
Installation district	0.694

Results: Hydraulic Importance of Single Pipe Failure

The number of segments, as determined by the locations of valves, calculated to estimate the importance of each water pipe in the event of a single pipe failure was found to be 50. The average calculated importance of a water pipe when each segment failed was found to be 0.083. This result indicates that the KA network has a low importance value in the event of a single pipe failure. This low importance value is attributable to the proper placement of the valves. However, several major segments had considerably higher levels of importance. For example, if a segment consisting of water pipes 975, 1104 and 1105 were to fail, unintended isolation would occur for as many as 174 water pipes. As a result, the importance of the corresponding pipes was calculated to be 0.6340, which is extremely high. As such, a high relative importance, such as that calculated for the above example, was generated in a total of four cases, except for those water pipes for which the water is received directly from the water source. These four segments are described in Table 5 and Figure 8. Water pipes that belong to such segments can present a major risk when breakage or emergency situations occur. Accordingly, they must always be carefully maintained and monitored. However, the importance of the other segments, which had a small number of water pipes and which did not give rise to unintended isolation (UI), was low.

Table 5: Value of importance of segment in the event of single pipe failure. UI, unintended isolation.

Segment	Pipe ID	No. of Pipes by UI	Importance
1	975, 1104, 1105	174	0.6340
2	976	174	0.5868
3	1005,1118	98	0.4582
4	1059	184	0.4156

Figure 8; Importance of segments by single pipe failure.

Results: Hydraulic Importance of Multiple Pipe Failures

The REVAS.NET model was used in this study to calculate the importance of water pipes in the event of multiple pipe failures. As shown in Table 6, a previous earthquake of magnitude M7 that had occurred in Province J, which includes the KA network, was considered. The execution number for the Monte Carlo simulation was set to 10,000, while the minimum pressure in the KA network was considered to be 15 m (KWWA [39]).

Table 6: Reliability evaluation scenario and results for the KA network.

Seismic Hazard		Nodal Serviceability (Average Value) (N_S)
Historical Location Data (Number of Data Points)	Magnitude	
J Province (29)	Specific Magnitude (M = 7)	0.662

The results obtained with REVAS.NET indicate that the average system serviceability (S_S) and nodal serviceability (N_S) were 0.533 and 0.662, respectively. Figure 9 depicts the spatial distribution of the nodal serviceability and shows that the reliability was relatively low at the nodes of the terminal area, which are far from the water source and which are branched into a single path. One of the key features of this figure is that the reliability of the system appears to be almost as high as in the case of completely looped segments. This means that looped segments can secure various paths by which water is supplied, so that rapid service degradation, even in the event of the breakage of some pipes, can be prevented.

To confirm the importance of the water pipes more accurately, 20 major water pipelines were marked as shown in Figure 10. The figure shows that the serviceability of the water pipes in the terminal area, which ultimately branches into a single path, was very low. Therefore, these areas require rehabilitation work to improve the durability or diversification of the water supply paths.

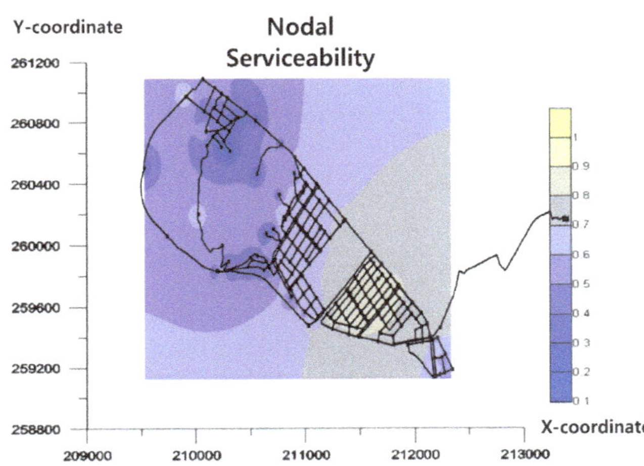

Figure 9: Nodal serviceability of the KA network.

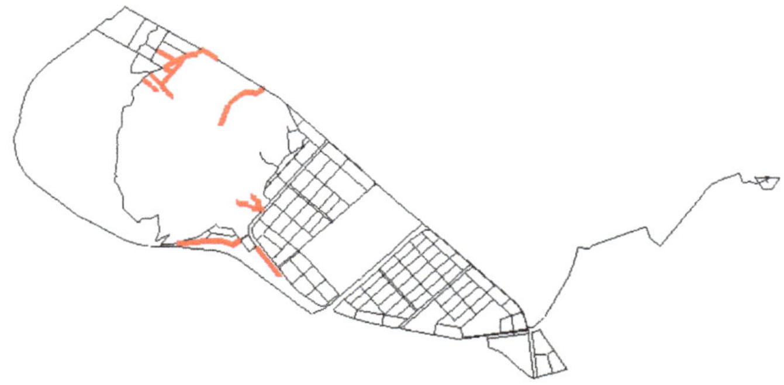

Figure 10: Importance of pipes by multi-pipe failure.

Results for Final Rehabilitation Priority Order

Table 7 summarizes the statistical values of four attributes. Based on the deterioration rate, the relative importance, the weighted utopian approach was used to calculate the rehabilitation priority order of individual water pipes.

Table 7: Statistical value of attributes of deterioration and importance.

Statistical Indicator	Deterioration Rate		Importance	
	Internal Factor	Installation Environments/External Factor	Single Pipe Failure	Multiple Pipe Failure
Average	0.7036	0.1470	0.0832	0.3179
Variance	0.1963	0.0556	0.0090	0.0232
Max value	1.0000	0.7111	0.6340	0.8126
Min value	0.0039	0.0039	0.0001	0.0744

Five weighting methods were used to apply the weighted utopian approach. The weighting factors of the attributes used in four of the methods (excluding the entropic method) were based on the results of the survey of 15 experts in the field of water distribution systems (the results are listed in Table 8). As with the survey concerning the weighting factors for the deterioration rate, the 15 experts consisted of five from academia, five from corporations and five from research centers and government agencies.

Table 8: Weighting factors of attributes depending on weighting methods.

Weighting Method	Weighting Factor			
	Deterioration by Internal Factors	Deterioration by External Factors	Relative Importance by Single Pipe Failure	Relative Importance by Multi Pipe Failure
Same Weighting	0.250	0.250	0.250	0.250
Eigenvector	0.344	0.097	0.349	0.210
Churchman-Ackoff	0.400	0.200	0.300	0.100
Rating	0.351	0.220	0.268	0.162
Entropic	0.155	0.549	0.246	0.050

Table 9 lists the weighting methods used and the resulting priority orders obtained for the rehabilitation of the top 20 water pipes, after applying the WUA. The results obtained using the five weighting methods indicate that the eigenvector, Churchman-Ackoff and rating methods produced similar results, whereas the same weighting and entropic methods produced results that were somewhat different from those obtained with the other three methods. Accordingly, the redundancy of the top 20 water pipes was investigated, and the results confirmed that a total of five water pipes were identified as priorities by all five methods and that a total of 14 water pipes overlapped in more than three of the methods. As described above, the rehabilitation priorities changed slightly as the weights were adjusted. These results indicate that although the superiority of any one of the weighting methods (whereby each of their axiomatic systems has been verified both logically and independently) cannot be determined, the sensitivity to any change in the weighting factors should be taken into account when a decision or rehabilitation prioritization is made. Hence, the proposed model offers not only the advantage of being representative, but also generality among multi-criteria decision-making methods, as it considers several methods of determining weighting factors with diverse characteristics. The proposed method can also provide final total results, as well as the results for each weighting method for consideration by final decision makers, thereby providing a wide range of alternatives depending on the context. That is, if there are attributes that must be considered depending on the decision-making environment and the characteristics of the issues, a specific weighting method can also be used to produce a priority ranking other than the priority ranking produced by the model developed in this study.

In this study, the final rehabilitation priority order was determined based on the average ranking for each water pipe, selected using the five weighting methods. The final results for the top 20 water pipes are shown in Figure 11 and Table 10. Although the results do not exhibit a clear trend, the rehabilitation priority order of the water pipes that are directly connected to the water source or that connect networks tended to receive high priority rankings. Water pipes other than those described above also generally ranked highly in terms of rehabilitation priority because of the effect of the deterioration rate determined from internal factors, for which the weighting values were set high as a result of the survey results. The deterioration rate values resulting from the internal and external factors showed that most of the values for the top 20 water pipes were high and that their variation was not considerable. However, the importance of single and multiple failures showed that the variation in the deterioration rates for the top 20 water pipes was relatively large. This result reflects the characteristics of the distribution of low factor values in most cases, except for some segments, in the case of a single failure. However, in the case of multiple failures, because the weighting values used to calculate the final rehabilitation priority order were not sufficiently large, the effect on the final rehabilitation priority order was minimal, even if the importance value resulting from multiple failures is large, which is why the variation for the top 20 values was large and unevenly distributed.

Table 9: Priority order for rehabilitation according to weighting methods.

Rehabilitation Priority Order	Same Weighting	Eigenvector	Churchman-Ackoff	Rating	Entropic
1	1105	1105	1105	1105	1088
2	1197	975	977	977	1073
3	1236	1104	1104	1197	1236
4	1220	976	975	1220	1105
5	1018	1118	976	1104	1040
6	1153	1005	1118	1018	977
7	909	977	1005	975	986
8	977	1197	1197	1236	1153
9	1121	1220	1220	976	909
10	1241	1009	1018	1088	1132
11	1213	1013	1088	1073	1128
12	1001	1015	1073	1153	1181
13	1132	1014	1206	1118	1197
14	999	1018	937	1206	1206
15	1000	1011	959	909	937
16	1002	1017	950	959	959
17	1226	1019	931	937	950
18	997	1236	930	1005	931
19	995	1206	1097	950	930
20	1199	959	912	930	1097

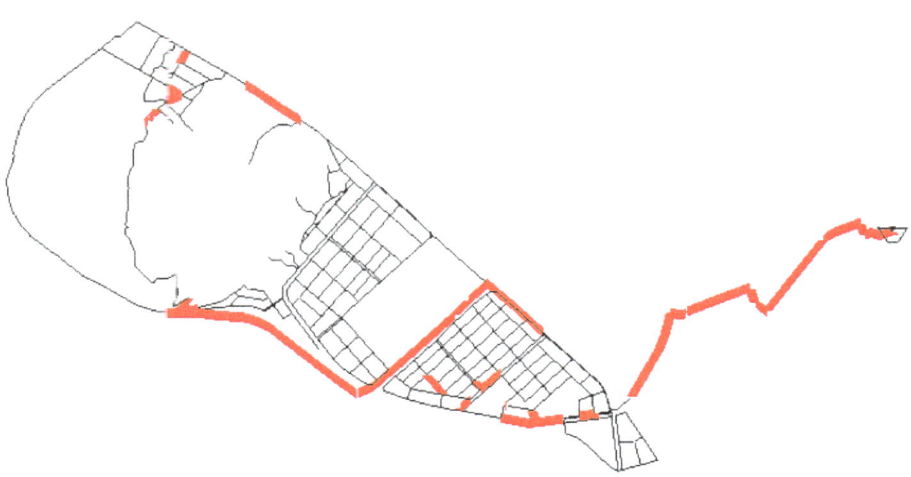

Figure 11: Top 20 pipes to be rehabilitated.

Table 10; Values of four attributes according to rehabilitation order.

Rehabilitation Priority Order	Deterioration by Internal Factors	Deterioration by External Factors	Relative Importance by Single Pipe Failure	Relative Importance by Multi Pipe Failure
1105	0.9997	0.5187	0.6340	0.2524
977	0.9905	0.5158	0.5538	0.0744
1197	0.9997	0.5332	0.1572	0.6167
1220	0.9975	0.5158	0.1572	0.5729
1236	0.9975	0.7111	0.0624	0.6172
1018	0.9995	0.5158	0.1572	0.4845
1088	0.9978	0.7111	0.1292	0.1686
1073	0.9978	0.7111	0.1292	0.1668
1206	0.9642	0.5332	0.1292	0.2620
959	0.9642	0.5332	0.1292	0.2476
937	0.9997	0.5332	0.1292	0.2426
950	0.9940	0.5332	0.1292	0.2122
930	0.9642	0.5332	0.1292	0.1995
931	0.9905	0.5332	0.1292	0.1960
1153	0.9996	0.7111	0.0039	0.6181
909	0.9998	0.7111	0.0030	0.5858
1104	0.9990	0.0569	0.6340	0.2232
975	0.9998	0.0069	0.6340	0.2908
1097	0.9940	0.5332	0.1292	0.1676
912	0.9940	0.5332	0.1292	0.1670

CONCLUSIONS

To determine the rehabilitation priority order for pipes in a water distribution system, we developed a new method to address the limitations of existing methods that use only the pipeline deterioration rate. We aimed to determine the priority of pipes for rehabilitation consider normal and abnormal conditions using multi-criteria decision making methods (using compromise of some weighting methods). In this study, we also added single and multiple pipe failures simulation in addition to the deterioration of pipes. These are key points of this manuscript. The proposed method considers the hydraulic importance in addition to the deterioration rate in determining the rehabilitation priority order with higher reliability. The results of this study are significant in that they show how a rehabilitation priority order model can be combined with the concept of the hydraulic importance of water pipes in applying existing deterioration rate calculation methods to determine the rehabilitation priority order. The proposed method was applied to the KA water distribution network of City J to determine the rehabilitation priority order for the pipes in the KA network. The results confirm that the proposed method provides a more realistic determination of the rehabilitation priority order that considers not only the deterioration rate, but also the relative hydraulic importance of each pipe. This model is able to determine the rehabilitation priority order for pipes in a water distribution network and can be applied more easily than existing rehabilitation priority order models that require large amounts of data and involve complex failure probabilities and mathematical models.

Because large-scale infrastructure networks, such as water distribution systems, require constant maintenance and rehabilitation, the design and reinforcement of water distribution networks to guard against multiple failures in the event of events, such as earthquakes, place huge financial and physical burdens on the water supplier. Korea has never experienced large-scale earthquakes, and most Koreans believe that their country has little to fear in this respect. However, although the probability of a disaster, such as an earthquake in Korea, is low, such an event would cause significant damage if it were to occur. Thus, Korea needs to be prepared for such disasters. Given these circumstances, a method that can determine the level of reliability of a water distribution system in the face of a disaster capable of causing immense damage is necessary within the current framework of maintenance and rehabilitation. The proposed model is also advantageous from this perspective.

The results of this study suggest that further research on this subject is warranted. The proposed method should be compared with other recent methodologies, such as CARE-W, CASSES and AWARE-P. Therefore, a detailed comparative study can be one of the further studies in the near future. The possible malfunction of valves was not considered in calculating the relative importance of structural failures. Because the area in which damage may occur varies depending on the ability to operate the relevant valve, the effect of valve malfunctioning should be studied. In this study, given that water distribution systems are essential social infrastructure networks that directly affect public welfare, an economic feasibility analysis for cost optimization purposes was not performed. Therefore, further studies on rehabilitation prioritization with consideration of economic constraints should be undertaken.

ACKNOWLEDGMENTS

This work was supported by a grant from the National Research Foundation (NRF) of Korea, funded by the Korean government (Ministry of Science, ICT and Future Planning, MSIP) (No. 2013R1A2A1A01013886).

REFERENCES

1. American Water Works Association Research Foundation (AWWARF). *Guidance Manual Water Mains Evaluation for Rehabilitation/Replacement*; AWWARF: Denver, CO, USA, 1986.
2. Water Research Centre (WRC). *Planning the Rehabilitation of Water Distribution Systems*; WRC: Marlow, UK, 1989.
3. K-Water. *Development of Decision-Making System for Pipe Rehabilitation*; K-Water: Daejeon, Korea, 1995.
4. Kim, E.S.; Baek, C.W.; Kim, J.H. Estimate of Pipe Deterioration and Optimal Scheduling of Rehabilitation. *Water Sci. Technol. Water Supply* **2005**, *5*, 39–46.
5. Shamir, U.; Howard, C.D. An Analytic Approach to Scheduling Pipe Replacement. *J. Am. Water Works Assoc.* **1979**, *71*, 248–258.
6. Marks, D.H.; Clark, M.R. A New Methodology for Modeling Break Failure Patterns in Deteriorating Water Distribution Systems: Theory. *Adv. Water*

Resour. **1987**, *10*, 2–10.

7. Marks, D.H.; Clark, M.R. A New Methodology for Modelling Break Failure Patterns in Deteriorating Water Distribution Systems: Applications. *Adv. Water Resour.* **1987**, *10*, 11–20.
8. Agbenowosi, N.K. A Mechanistic Analysis Based Decision Support System for Scheduling Optimal Pipeline Replacement. Ph.D. Thesis, Virginia Polytechnic Institute and State University, Blacksburg, VA, USA, September 2000.
9. Park, S.; Loganathan, G.V. Methodology for Economically Optimal Replacement of Pipes in Water Distribution Systems: 1. Theory. *KSCE J. Civil Eng.* **2002**, *6*, 539–543.
10. Park, S.; Loganathan, G.V. Methodology for Economically Optimal Replacement of Pipes in Water Distribution Systems: 2. Application. *KSCE J. Civil Eng.* **2002**, *6*, 545–550.
11. Deb, A.K.; Hasit, Y.J.; Grablutz, F.M.; Herz, R.K. *Quantifying Future Rehabilitation and Replacement Needs of Water Mains*; Report to the American Water Works Association Research Foundation; Roy F. Weston, Inc.: West Chester, PA, USA, 1997.
12. Walski, T.M. Economic Analysis of Rehabilitation of Water Mains. *J. Water Resour. Plan. Manag.* **1982**, *108*, 296–304.
13. Luong, H.T.; Fujiwara, O. Fund Allocation Model for Pipe Repair Maintenance in Water Distribution Networks. *Eur. J. Oper. Res.* **2002**, *136*, 403–421.
14. Alvisi, S.; Franchini, M. Near Optimal Rehabilitation Scheduling of Water Distribution Systems based on Multi-objective Genetic Algorithms. *Civil Eng. Environ. Syst.* **2006**, *23*, 143–160.
15. Saegrov, S. *Computer Aided Rehabilitation of Water Networks*; IWA Publishing: London, UK, 2005; ISBN 1843390914.
16. National Civil Engineering Laboratory. *CARE-W WP1 D1 Construction of Control Panel for Performance Indicators for Rehabilitation*; EVK1-CT-2000–00053, No. 1.1; Research and Technological Development Project of European Community, European Commission: Brussels, Belgium, 2001.
17. National Civil Engineering Laboratory. *CARE-W WP1 D2 Description of Technical Tools for Failure Forecasting and Network Reliability*; EVK1-CT-2000–00053, No. 1.2; Research and Technological Development Project of European Community, European Commission: Brussels, Belgium, 2002.
18. Cemagref Bordeaux. *CARE-W WP2 D3 Description and Validation of Technical Tools- Models Description*; EVK1-CT-2000–00053, No. 2.1; Research and Technological Development Project of European Community, European Commission: Brussels, Belgium, 2002.
19. INSA Lyon-URGC. *CARE-W WP3 D6 Decision Support for Annual Rehabilitation Programs: Criteria for the Prioritization of Rehabilitation Projects*; EVK1-CT-2000–00053, No. 3.1; Research and Technological Development Project of European Community, European Commission: Brussels, Belgium, 2002.
20. INSA Lyon-URGC. *CARE-W WP3 D6 Decision Support for Annual Rehabilitation Programs: Survey of Multi-Criteria Techniques and Selection of Relevant Procedures*; EVK1-CT-2000–00053, No. 3.2; Research and Technological Development Project of European Community, European Commission: Brussels, Belgium, 2002.
21. Technique University Dresden. *CARE-W WP4 D9 Development of the "Rehab Strategy Manager" Software*; EVK1-CT-2000–00053, No. 4.2; Research and Technological Development Project of European Community, European Commission: Brussels, Belgium, 2002.
22. Cemagref (2008) CASSES, User Manual. Available online: http://casses.irstea.fr/wp-content/uploads/2013/02/CassesManual_2.0.0_b.pdf (accessed on 1 November 2014).
23. AWARE-P (2012). The AWARE-P project—Infrastructure asset management of urban water services. Impact report for application to the 2012 Muelheim Water Award. Available online: http://www.aware-p.org/np4/?newsId=13&fileName=MWA_AWARE_P_Application_compressed.pdf (accessed on 1 November 2014).
24. Saaty, T.L. *The Analytic Hierarchy Process*; McGraw-Hill: New York, NY, USA, 1980.
25. Specht, D.F. Probabilistic Neural Network. *Neural Netw.* **1990**, *3*, 109–118.
26. Jun, H.; Loganathan, G.V.; Kim, J.H.; Park, S. Identifying Pipes and Valves of High Importance for Efficient Operation and Maintenance of Water Distribution Systems. *Water Resour. Manag.* **2008**, *22*, 719–736.
27. Walski, T.M. Water Distribution Valve Topology for Reliability Analysis. *Reliab. Eng. Syst. Saf.* **1993**, *42*, 21–27.
28. Yoo, D.G.; Kang, D.; Kim, J.H. Seismic Reliability Assessment Model of Water Supply Networks. In Proceedings of the World Environmental and Water Resources Congress, Cincinnati, OH, USA, 19–23 May 2013; pp. 955–966.
29. Yoo, D.G.; Kang, D.; Kim, J.H. Applications of Seismic Reliability Assessment Model for Water Distribution Networks. In Proceedings of the World Environmental and Water Resources Congress, Portland, OR, USA, 1–5 June 2014.
30. Cullinane, M.J.; Lansey, K.E.; Mays, L.W. Optimization-Availability-Based Design of Water-Distribution Networks. *J. Hydraul. Eng.* **1992**, *118*, 420–441.
31. Tabucchi, T.; Davidson, R.; Brink, S. Simulation of Post-Earthquake Water Supply System Restoration. *Civil Eng. Environ. Syst.* **2010**, *27*, 263–279.
32. Lansey, K.E. Sustainable, Robust, Resilient, Water Distribution Systems. In Proceedings of the 14th Water Distribution Systems Analysis Symposium, Engineers Australia, Adelaide, Australia, 24–27 September 2012.
33. Knoll, A.L.; Engelberg, A. Weighting Multiple

Objectives—The Churchman-Ackoff Techniques Revisited. *Comput. Oper. Res.* **1978**, *5*, 165–177.

34. Eckenrode, R.T. Weighting Multiple Criteria. *Manag. Sci.* **1965**, *12*, 180–192.
35. Hwang, C.L.; Yoon, K. *Multiple Attribute Decision Making Methods and Applications: A State-of-The-Art Survey*; Springer-Verlag: New York, NY, USA, 1981.
36. Soofi, E.S. Generalized Entropy-Based Weight for Multi-Attribute Models. *Oper. Res.* **1990**, *32*, 362–363.
37. Xanthopulos, Z.; Melachrinoudis, E.; Solomon, M.M. Interactive Multiobjective Group Decision Making with Interval Parameters. *Manag. Sci.* **2000**, *46*, 1721–1732.
38. Yoo, D.G.; Kim, J.H.; Jun, H. *Determination of Rehabilitation Priority Order of Subareas in Water Distribution Systems Considering the Relative Importance of Pipes*; Water Distribution Systems Analysis 2010; Americal Society of Civil Engineers: Reston, VA, USA, 2010; pp. 1045–1052.
39. Korea Water and Wastewater works Association (KWWA). *Water Supply Facility Standard*; Ministry of Environment: Seoul, Korea, 2010.

CHAPTER 7

THE EFFECT OF HYDRAULIC CHARACTERISTICS ON ALGAL BLOOM IN AN ARTIFICIAL SEAWATER CANAL

Seung Oh Lee , Sooyoung Kim, Moonil Kim, Kyoung Jae Lim and Younghun Jung

Keywords: Artificial seawater channel, algal bloom, nutrient (T-N and T-P), hydraulic characteristics, vorticity, FLOW-3D

INTRODUCTION

Algal blooms occur naturally and may occur with regularity depending on weather and water conditions [1,2]. The likelihood of occurrence of an algal bloom depends on local conditions and the hydraulic characteristics of the water [3]. Algal blooms generally occur in places where high levels of nutrients are present, in warm, sunny, and calm conditions [4,5]. At shallow depths in a canal or a lake, during summer and fall, sunlight penetrates the entire depth of the water column at that point and the water temperature increases greatly. Sunlight has a direct effect on the growth rates of phytoplankton. Phytoplankton uses sunlight to create their new cells. In addition, the growth rates of microorganisms increase in proportion to the increase in water temperature. Under sufficient sunlight and nutrient conditions, algal blooms have an increased probability of development. In Korea, these conditions occur during summer and early fall. Natural sources of nutrients, such as phosphorus or nitrogen compounds, can be supplemented by a variety of human activities. For example, in rural areas, agricultural runoff from fields can wash fertilizers into the water. In addition, animal wastes, atmospheric deposition, groundwater inflow, and sedimentation can be the nutrient sources [6,7]. In urban areas, nutrient sources can include treated wastewater from septic systems and sewage treatment plants, as well as urban storm water runoff that carries nonpoint-source pollutants such as lawn fertilizers and septic tank [8,9]. A healthy ecosystem needs to maintain a balance of nutrients that can sustain each organism's life cycle. However, in an unhealthy ecosystem, this balance is destroyed by added nutrients and the associated blooms. Added nutrients are the main cause of organism growth. In the first stage, micro-algae grow rapidly to a critical point (algal blooms). They use the additional materials for the growth of macro-algae or laver. Next, the primary consumers grow by eating the algae.

Natural estuaries maintain a biological balance through buffer reactions and naturally controlled biodiversity. However, this balance could be disturbed by the inflow of additional nutrients, namely total phosphorus (T-P) and total nitrogen (T-N), due to heavy rainfall or human activities [10,11]. One of the results is a bloom identified by pond scums, a bad smell, and micro- and macro-algal bloom. In a natural ecosystem, organics (carbon) and nutrients (nitrogen and phosphorus) are recycled through biochemical reactions. Laver often appears in an adjacent coastal area in the West Sea during the summer season [12]. Generally, laver is found floating or attached onto the surface of rocks. Under the conditions of high nutrient levels and high temperature with strong sunlight, the growth rates of laver increase rapidly and the seaweed covers large areas of the water surface [13,14]. The green laver's bloom directly affects the photosynthesis of other phytoplankton. After the nutrients have been consumed and the temperature has decreased, the growth rate declines and the green laver forms spores for the next generation.

There have been many studies on the control of algal growth, and in particular, on the control of algal growth using hydraulic characteristics [15,16]. Chung *et al.* [17] and

Citation: Seung Oh Lee, Sooyoung Kim, Moonil Kim, Kyoung Jae Lim, and Younghun Jung, The Effect of Hydraulic Characteristics on Algal Bloom in an Artificial Seawater Canal, Water 2014, 6, 399-413; doi:10.3390/w6020399.

Lee *et al.* [12] simulated the effect of a weir on the control of algal blooms using a laterally averaged two-dimensional hydrodynamic and eutrophic model that could accommodate the vertical displacement of the weir following water level fluctuations in the Daecheong reservoir. Morillo *et al.* [18] determined the tracer dilution ratio (the ratio of inflow to outflow concentration) and two residence time scales (arrival time and cleansing time) as a function of lake geometry to verify the effect of impermeable curtains on the Silvan reservoir in Australia. Asaeda *et al.* [19] investigated the effect of impermeable curtains on algal growth to control algal growth in the Terauchi reservoir. Also, Dai *et al.* [20] evaluated the effect of the longitudinal hydrodynamic characteristics on the algal blooms for tributary embayments of a man-made lake in China. These methods are very effective for controlling algal growth by blocking the surface inflow. In this regard, it is necessary to investigate the impact of hydraulic characteristics and the algal blooms. Accordingly, the objectives of this study are to (1) analyze the ecological change in an artificial seawater channel through measurement; and (2) identify the effect of hydraulic characteristics on an algal bloom using a three-dimensional numerical model (FLOW-3D). Due to the temporary emergence of the scums and algal blooms in summer, it is difficult through statistical analysis to find direct correlations between the hydraulic characteristics and algal bloom. For this development, after detecting the spots of scum and algal blooms in summer, 3-D numerical simulations were conducted for the same domain.

MATERIALS AND METHODS

Study Site Description

An artificial canal using seawater as its primary water source was constructed in Central Park, New Songdo City, South Korea (Figure 1). The Central Park has the area of 411,324 m^2, and was constructed at a cost of 190 billion KRW in 2009. As a waterfront area to Incheon citizens, the Central Park was constructed to provide recreation and aesthetic pleasure in an urban region. The canal is a semi-closed water system, and its length is 1.8 km. Water depth is shallow (1.5 m) with variable width (maximum width of 110 m). Filtered seawater by STF (seawater treatment facility) is supplied by two inlets to the canal. Here, discharge conditions are 116 L/s at Inlet #1 and 15 L/s at Inlet #2. For the inflow of seawater from two inlets, the water depth of 1.5 m is regulated by the weir of the outlet. The man-made canal has two major conditions for maintenance: (1) the water source is seawater, filtered by a seawater treatment facility (STF); and (2) the canal has a semi-enclosed flow of very low velocity, equal shallow depth throughout, and a varying width. Seawater taken from an adjacent coastal area in the West Sea is supplied to the canal after filtering with a 5 μm pore size filter from two supply points. The filtered water from the STF flows into two inflow points (the Boat House and Anapji) in the canal and is discharged into the West Sea. It was reported that various species of phytoplankton communities are living in the coastal water of the West Sea [21]. Since the seawater is collected from an adjacent coastal area in the West Sea, invading phytoplankton communities cause the growth of algal blooms when excess nutrients (eutrophication) are introduced into the canal. It is clear that nutrient-related eutrophication is a significant problem with respect to water quality worldwide [22,23,24,25,26,27]. The unique conditions of the canal in Central Park, New Songdo City directly affect the water quality, nutrient loading, and flushing rates. There are several water quality issues including the deterioration of water quality due to the geometric layout of the canal and hydraulic causes of flow, and red tide algal growth caused by nutrients in the artificial seawater canal. Total 54 phytoplankton including Bacillariophyceae, Dinophyceae, Ebryophyceae, and Euglenophyceae were found in the canal. Green laver spores from West Sea might come into canal through a testing booster pump because fully matured Green lavers were unexpectedly found in the canal during summer. Therefore, a proper nutrient balance in the canal is very important for prohibiting and controlling algal blooms, especially for the formation of scum, bad smell, and red tides. Filtration with 5 μm pore size and fewer than 10 days of hydraulic retention time may prevent algal blooms and the deterioration of the water quality.

Figure 1: Study site: Central Park, New Songdo City, South Korea, reprinted with permission from [28]. Copyright 2013 NAVER.

BACKGROUND

Monitoring of Nutrients

There are two main sources of pollution, *i.e.*, point sources and non-point sources. The main point source in the canal is seawater, especially non-filtered seawater, because the water quality of such water changes dynamically, depending on the regional daily weather, seasonal climate, and sea currents. Although the seawater quality changes, the filtered water quality could be kept in a steady state. However, it may change because the minimum filter pore size of the STF was 5 μm and the average removal efficiency was 59.7% in this study. The other main sources of pollution are non-point sources such as soils washed by heavy rainfall, sand winds, and so on. There are several characteristics of non-point source pollution. One is the high pollution rate within a short time, while another is the high fluctuation of the pollutant concentrations. To establish a good and balanced ecosystem in the canal, organics (Chemical Oxygen Demand (COD), Total Organic Carbon (TOC), or Dissolved Organic Carbon (DOC)) and nutrient (T-N and T-P) conditions are very important. Nutrient conditions, in particular, are the most important for algal blooms. Microorganisms grow using organics and nutrients, and hence, under conditions of surplus nutrients, algal blooms could develop in the canal, changing the water color.

Numerical Model Theory (The FLOW-3D Model)

The numerical model employed in this study, Flow-3D, is a commercially available numerical analysis model developed by Flow Science Co. (New Mexico, NM, USA) [29]. The governing equations of FLOW-3D were validated for surface water flow [30,31]. The model uses the continuity and momentum equations as governing differential equations. The general continuity equation (Equation (1)) is as follows:

$$V_F \frac{\partial \rho}{\partial t} + \frac{\partial}{\partial x}\left(\rho u A_x\right) + \frac{\partial}{\partial y}\left(\rho v A_y\right) + \frac{\partial}{\partial z}\left(\rho w A_z\right) = 0 \qquad (1)$$

where, V_F is the fluid elementary volume; ρ is the fluid density; u, v, and w are the velocities of the moving fluid in the x, y, and z directions, respectively; and A_x, A_y, and A_z are the elementary areas in each corresponding direction. The general momentum equation following the Navier-Stokes equation (Equation (2)) is as follows:

$$\frac{\partial u}{\partial t} + \frac{1}{V_F}\left\{uA_x \frac{\partial u}{\partial x} + vA_y \frac{\partial u}{\partial y} + wA_z \frac{\partial u}{\partial z}\right\} = -\frac{1}{\rho}\frac{\partial p}{\partial x} + G_x + f_x$$

$$\frac{\partial v}{\partial t} + \frac{1}{V_F}\left\{uA_x \frac{\partial v}{\partial x} + vA_y \frac{\partial v}{\partial y} + wA_z \frac{\partial v}{\partial z}\right\} = -\frac{1}{\rho}\frac{\partial p}{\partial y} + G_y + f_y$$

$$\frac{\partial w}{\partial t} + \frac{1}{V_F}\left\{uA_x \frac{\partial w}{\partial x} + vA_y \frac{\partial w}{\partial y} + wA_z \frac{\partial w}{\partial z}\right\} = -\frac{1}{\rho}\frac{\partial p}{\partial z} + G_z + f_z \qquad (2)$$

where, G_x, G_y, and G_z denote the body force and f_x, f_y, and f_z denote the viscous force. For the interpretation of the free surface, the fluid volume per unit volume within a grid is expressed as the volume of fluid (VOF) function, $F(x, y, z, t)$. Equation (3) represents the general VOF function.

$$\frac{\partial F}{\partial t} + \left\{\frac{\partial}{\partial x}(Fu) + \frac{\partial}{\partial y}(Fv) + \frac{\partial}{\partial z}(Fw)\right\} = F_{dif} + F_{sour} \qquad (3)$$

where, F_{dif} is the diffusion of fluid fraction and F_{sour} is the fluid source and sink.

Monitoring and Measurement in the Canal

Two types of monitoring activities were performed in the Songdo canal. One was the monitoring of nutrients (T-N and T-P), and the other was the monitoring of hydraulic characteristics.

To monitor nutrients (T-N and T-P), 10 representative sampling points were selected, as shown in Figure 2. The first sampling point was the raw seawater in the storage tank of the STF. In the second sampling point, seawater that had been filtered by the multimedia filter of the STF was collected. The last six sampling points were located inside the canal and included the two inflow areas of the filtered seawater. They are the traffic bridge (SP#9), the two walking bridges across the canal (SP#6 and SP#8), the two input points (SP#5 and SP#7), and an output point of seawater (SP#10). To monitor hydraulic characteristics, five cross sections were selected, as shown in Figure 3. The cross-sectional lines were determined by considering the characteristics, location, and purpose of canal management. In each cross section, three to five measuring points were selected, depending on the width of the cross section, where the maximum number of measurement points in one cross section was seven while the minimum was two.

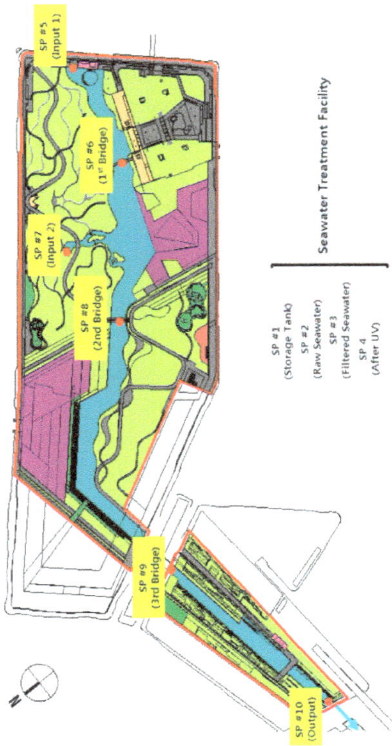

Figure 2. Sampling points (SP) for nutrients monitoring (T-N and T-P).

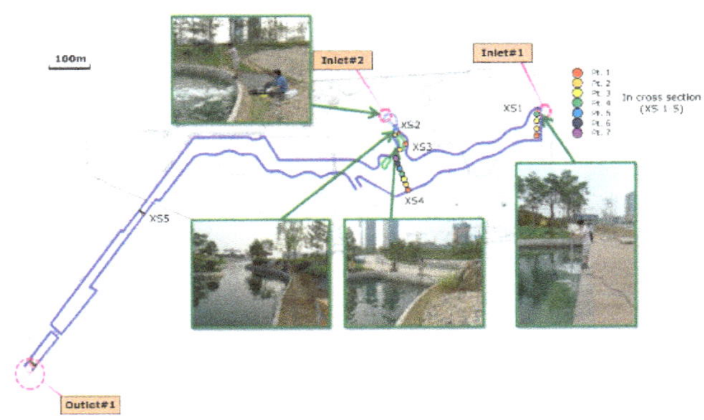

Figure 3: Sampling points (Pt.) for hydraulic characteristics monitoring in the cross section (XS).

Monitoring of Nutrients (T-N and T-P)

In this study, the plastic containers were used to collect water for analysis of the T-N and T-P. Sampling was conducted for each sampling point in the canal and STF (Figure 2). The T-N and T-P determination tests were performed according to Ultraviolet Spectrophotometric Screening Method and Ascorbic Acid Method [32]. For each sampling, the nutrients (T-N and T-P) were estimated three times, and the averaged values were used as representative values of nutrients at each point.

Monitoring of Flow Velocity

Two methods were selected to calculate the mean velocity of the measured data. One is the six-tenths depth method and the other is the three-point method. A velocimeter was used to measure the velocity at normal and low water levels, while a float rod was used to measure velocity at high water levels during flooding. The canal flow velocity measurements were performed under normal conditions (10 days turnover) using the two-dimensional and three-dimensional velocimeters.

The six-tenths depth method (employing the representative mean velocity measured at 0.6 of the total depth measured down from the water surface) uses the observed velocity at this depth. The three-point method (0.2, 0.6, and 0.8 of the total depth measured down from the water surface) uses the observed velocity for the three-dimensional velocity meter.

Velocity measurements at cross sections XS1 to XS5 were carried out using the two-dimensional velocity meter deployed from a boat, and the time series data from the velocity meter were recorded using a laptop computer. Measurements of each of the cross-sectional velocities were carried out onboard the boat for points in the canal, and standing on the canal wall for points near the edge of the canal.

The range of possible velocity measurements was 0.0001–0.001 m/s with an accuracy of ±0.000025 and ±0.005 m/s when using the three-dimensional Acoustic Doppler Velocimeter (3D ADV) and the two-dimensional Electromagnetic Current Meter (2D ECM), respectively. The measuring frequency was 25 Hz for the 3D ADV and 5 Hz for the 2D ECM, correspondingly. The recording time at each point was set to at least 2 min to obtain statistically measureable values.

Numerical Simulation

To conduct a three-dimensional numerical analysis to evaluate the hydraulic characteristics of the canal, a three-dimensional hydraulic numerical model, FLOW-3D, was employed to predict cross-sectional velocities, areas of the 'dead zones', and changes of vorticities inside the canal. The numerical results for the hydraulic characteristics were used to predict the locations of algal blooms based on the monitoring results. The calibrations of the FLOW-3D model for the meandering stream were shown in a previous study [33]. FLOW-3D showed the high agreement between the field data and the simulations in the analysis of the velocity distribution of the open channel flow and the secondary flow. From these reasons, the FLOW-3D model is applied to the canal including an artificial island and a culvert.

In this study, a total of 285,000 meshes were used for the numerical experiment, divided into six blocks to improve calculation efficiency. The large eddy simulation (LES) model, designed for complex geometries, was employed as the turbulence closure model. The retention time of the filtered seawater in the canal was designed with the aim of reducing sedimentation and harmful algal blooms. Furthermore, considering the regional climate and weather changes, the canal operation could be modified. To satisfy the changes, the total water volume in the canal could be changed based on a retention time of 10 days (11,320 m^3/day). When a 10-day retention time was maintained, the flow rate in the canal was 131 L/s. In this study, the simulations were conducted by incompressible flow condition with water density of 1030 kg/m^3, and two boundary conditions were applied to the FOW-3D model. Upstream boundary condition is to supply constant discharge of 131 L/s from two inlets to the canal and downstream condition is to keep the water depth of 1.5 m. From these boundary conditions, the transient flow conditions were set for the numerical simulation, but as time passes the flow status can be steady because inflow and outflow were constantly regulated. The boundary conditions used for the numerical simulation in the canal are 1800 m for canal length, 110 m for maximum width, 1.5 m for initial water depth, and 11,320 m^3/day for flow rate.

RESULTS

Results from Monitoring of Nutrients (T-N and T-P)

Figure 4 shows the monitoring results of the nutrients (T-N and T-P) before, during, and after the canal filling periods with 10 days turnover at Central Park, New Songdo City. During relatively low temperatures (February to May 2009), the level of nutrients in the raw seawater was kept below 0.8 mg/L. However, the T-N concentration increased up to 2.45 mg/L and T-P levels doubled when the temperature increased. On 21 and 23 July 2009, the T-N concentration at sampling point 7 (Anapji) was very high and was around 1.5 mg/L (Figure 5). By comparing the T-P concentration at the same point and time, it was determined that T-N contributes more to the growth of algal blooms and scum in the canal. With increasing temperature, nutrient concentrations in the West Sea increased. On 30 July 2009, the T-N and T-P concentrations were 2.407 and 0.101 mg/L, respectively, and the concentration was over Level III of the guidelines for seawater quality (Korean Environmental Standards). Level III means that the seawater is of bad quality and can be used only as industrial cooling water.

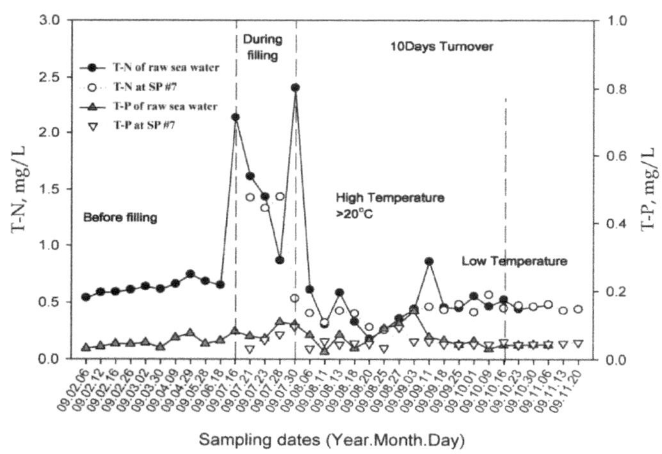

Figure 4: Nutrient variations in the canal within a 10-day turnover.

On the other hand, the T-N concentration decreased to 0.534 mg/L at Anapji (Input 2, SP#7) on 16 and 30 July 2009. It was estimated that phytoplankton and algae in the canal might consume the nutrients because there was no nutrient removal process in the STF. Furthermore, this was supported by the fact that the nutrient levels had decreased in the canal two to three days after the appearance of the periodic scum. During heavy rainfall, the T-N concentrations in Anapji (Input 2, SP#7) were higher than those of the raw seawater (SP#2). However, the T-N concentration decreased to the same level as the raw seawater on 25 August. Another high T-N concentration in the raw seawater was observed on 11 September, but the concentration at Anapji (SP#6, Input 2) was half that of the raw seawater.

The growth rates of all living things are directly affected by temperature. The temperature variation is correlated with the seasonal changes in Korea; therefore, the phytoplankton growth pattern in the canal of Central Park in New Songdo City is mitigated by seasonal changes. Micro- and macro-algal blooms and scum are expected to be produced from July to October 2009. To control and manage algal blooms and scum development, the growth rate must be kept at the same levels as in May and June 2009.

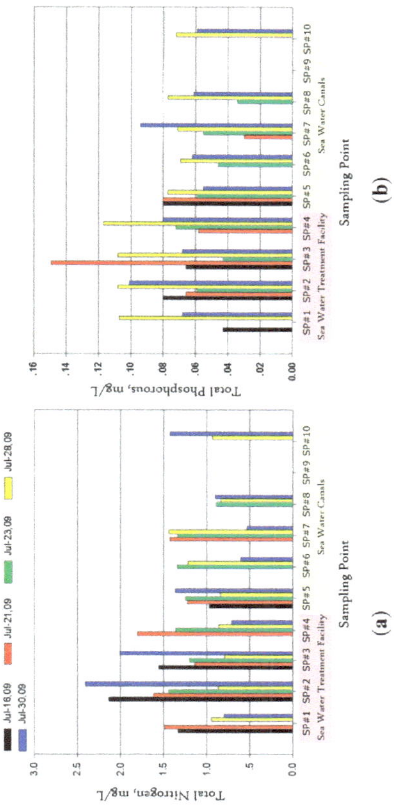

Figure 5: Monitoring results of (**a**) T-N; and (**b**) T-P.

Measurement Results of Flow Velocity

The velocities for each cross-section measurement for this period are presented in Table 1. The overall approximated mean flow velocity was under 0.02 m/s in the canal, and the maximum velocity was 0.0162 m/s at cross section XS1 near the inflow inlet. The cross section XS1 has four measuring points near the inlet of the canal (Figure 6). Due to asymmetric structural inflow, circulation occurs near cross section XS1, as is evident by the negative velocity that was detected on the left side of the bank. Flow velocities measured at cross sections XS1 ranges from −0.0368 to 0.0162 m/s.

For XS2, the velocity is 0.0179 m/s at right side and 0.0181 m/s at left side. Considering geometric effect at XS2 that seawater flowing from the second inlet meets the dramatically narrow cross section, velocity around inlet#2 is expected to be relatively lower than velocity at XS2. After passing XS2, the flow is divided into two directions due to the existence of a small island (Anapji). In this regard, the flows at XS3 are asymmetric velocities at both sides due to the irregular shape of a small island. The velocity on the left side of Anapji with the flow direction was 0.0147 m/s, while it was 0.005 m/s on the right side. The left side flow of Anapji is assumed to be the main flow. The flow from the first and second inlets meets at the central area of the canal (XS4). Although velocities at points of XS1 have big differences in flow direction and magnitude, there is no significant difference in flow velocities at seven measuring points in cross section XS4. Considering the main flow of XS3, the velocity of XS4 decreases to 0.009 m/s again due to the expansion of the canal cross section from XS3 to XS4. The direction and size of the arrow indicate the direction and velocity of flow. In addition, the color of the arrow shows the point number in the cross section.

Table 1: Two-dimensional velocity at measuring point (Pt.) in each cross section (XS) (m/s).

XS	Pt. 1	Pt. 2	Pt. 3	Pt. 4	Pt. 5	Pt. 6	Pt. 7
XS1	−0.0368	−0.0195	0.0091	0.0162		–	
XS2	0.0179	0.0181			–		
XS3	0.0147	0.0150			–		
XS4	0.0092	0.0104	0.0105	0.0092	0.0088	0.0087	0.0093
XS5	0.0175	0.0143	0.0184		–		

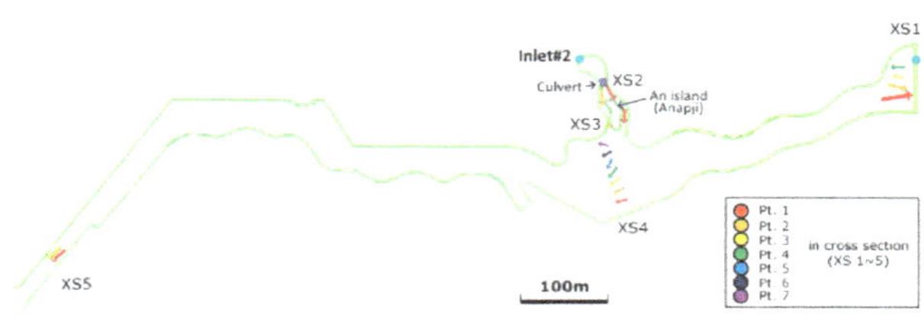

Figure 6: Measured velocity of each cross section (m/s).

Results from Numerical Simulation

The overall approximated mean flow velocity was 0.01 m/s in the canal. This low flow velocity was maintained throughout most of the canal area except around the inflow inlets and outlets because the inflow discharges from the inlet were relatively small with respect to the overall canal volume. Therefore, the influence of the inflow discharge is evident only near the inlet weir. The maximum velocity occurred in front of the inflow weir (Inlet#1). As the seawater flowed through the canal, its velocities gradually decrease due to the widening of the cross section in the pond. With the narrowing of the canal, velocities increase as water passes through the culverts (Figure 7). This velocity distribution is similar to distribution of the measured velocity (Figure 6). The results show a good agreement between the observations and the simulations with a coefficient of determination (R^2) of 0.74 (Figure 8). However, note that the velocity at Pt. 1 of XS1 was excluded to estimate a coefficient of determination because the circulation of the flow near the inlets can produce errors in measuring water velocity. The effect of geometry on the simulated velocity is also shown in the measured velocity around a small island (Anapji), but the velocities near inlets were dissimilar. The flow measured alongside the island gradually slowed down before reaching the Central Park Lake area. Velocity vectors in the middle of the lake area were close to zero due to the widening of the area. The existence of a 'dead zone' (stagnation of flow) should be considered in most of the canal area, which may evoke biological issues including water containing suspended nutrients. Such 'dead zones' can cause serious problems related to the water quality, such as the growth of algal blooms and periphyton communities. It is therefore important to suggest a solution for the reduction of the 'dead zone' in the canal and for maintaining optimum ecological conditions.

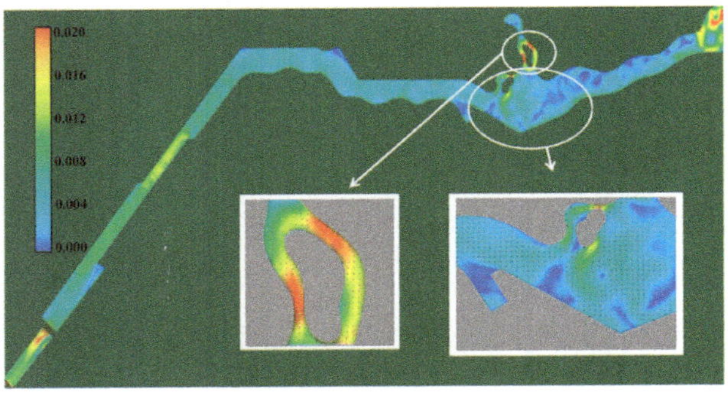

Figure 7: Results of the numerical simulation (velocity distribution (m/s)).

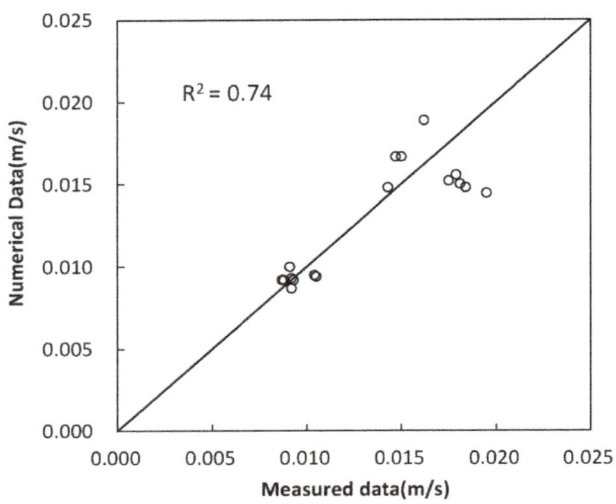

Figure 8: Comparison of the simulations and the observation for velocity.

The Effect of Hydraulic Characteristics (Vorticity) on Algal Blooms

According to the results of the monitoring of nutrients and hydraulic characteristics, high nutrient points and stagnation of flow areas were determined in the canal. Five locations of expected algal blooms were estimated from the monitoring results, as shown Figure 8. However, although the monitored algal blooms were not found in all stagnant zones (blue colors in Figure 7), we could find a cause of algal blooms from the *z*-direction vorticity analysis. The occurrence of the algal blooms was shown in stagnant places with high values of *z*-direction vorticities as shown in Figure 9. From this result, in addition to the effect of velocity of *x* and *y* direction on algal blooms, this study shows that *z*-direction vorticity give a significant contribution to algal blooms. An area with partially high vorticity cannot transfer large volumes of water outside the area when the canal water is in a stagnant condition. High values of *z*-direction vorticities disturb the sinking of nutrients (T-N and T-P), increasing the probability of algal blooms. Agricultural soils that contain high levels of nutrients (organic nitrogen and phosphorus) from around the Central Park area are continuously washed into the canal, accumulating at the bottom of the canal. This increases the amount of nutrients in the water and the probability of growth of algal blooms.

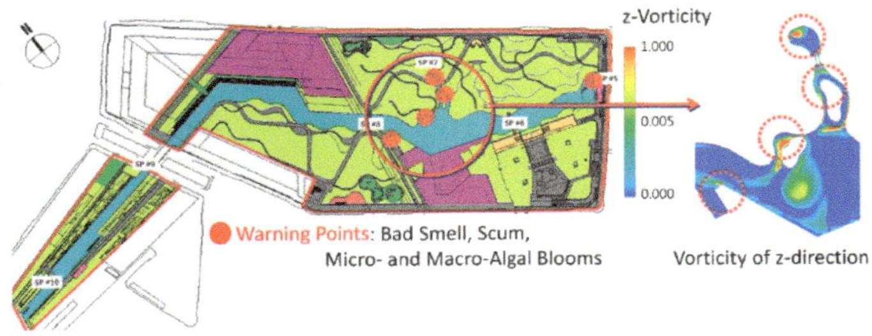

Figure 9: Warning points of algal blooms in the canal and vorticity of the *z*-direction using FLOW-3D.

DISCUSSION AND CONCLUSIONS

Unlike natural streams, water quality in the artificial canal is typically maintained by the related facilities such as waste water treatment and seawater desalination. Nevertheless, there often happen scum and algal blooms in the artificial canal. While hydraulic characteristics are mainly affected by discharge, roughness, and topography, scum and algal blooms occurred by more sources such as climatic and chemical factors with hydraulic characteristics. Therefore, this study attempts to evaluate three dimensional turbulent flow velocities (*x*, *y*, and *z*) under the assumption that hydraulic characteristics are related to occurrence of scam and algal blooms. Also, since the pond scum and algal blooms temporally emerge in summer it is difficult to statistically analyze correlation between the hydraulic characteristics and algal blooms. From these reasons, we first investigated the place where the pond scum and algal blooms easily emerged in summer, and then conducted 3-D numerical simulations in the same domain.

This study was conducted to identify the relevance of the hydraulic characteristics in an artificial seawater channel using FLOW-3D for algal boom. Also, the monitoring results of algal blooms were used to compare with the specific hydraulic characteristic. The meaning of this study could be found in that (1) scum and algal blooms have a high chance to occur in the location with high vertical vorticities; and (2) thus, vorticity among hydraulic characteristics can efficiently be used as an indicator of algal blooms occurrence in artificial seawater channel. In this regard, the algal bloom area in an artificial canal could be predicted throughout 3-D hydrodynamic simulation because the locations of partially high vorticity cannot transfer water flow to the outer area when the canal water is in a stagnant condition. In addition, *z*-direction vorticities with high values disturb the sinking of nutrients which can be feeding sources to harmful planktons, and thus, increase the probability of algal blooms.

When algal blooms occur in a canal, some techniques such as red clay input and the use of equipment such as submersible aerating or mixing systems can be used to control it. The algal blooms are affected by various factors such as climate and nutrient. Also, hydraulic characteristics are another factor dependent on the design of the canal. In this context, analysis of hydraulic characteristics including the *z*-direction vorticity in the canal can help to predict the zone where algae occurs, and prepare precautionary actions for algal blooms. Accordingly, this study will contribute to an economic and reliable design of the artificial seawater canal which will enhance the environmental functions of the waterfront area as well as algal bloom prevention.

ACKNOWLEDGMENTS

This study was supported by Basic Science Research Program through the National Research Foundation of Korea (NRF) funded by the Ministry of Science, ICT and Future Planning (grant Nos. 2013-027882) and a grant (No. 2012000540017, Development of soil erosion reduction techniques with vegetation control and engineering method, and for environment-friendly urban land development) from the GAIA project in Republic of Korea.

REFERENCES

1. Carpenter, E.; Montoya, J.P.; Burns, J.; Mulholland, M.; Subramaniam, A.; Capone, D.G. Extensive bloom of a N_2-fixing diatom/cyanobacterial association in the tropical Atlantic Ocean. *Mar. Ecol. Prog. Ser.* **1999**, *185*, 273–283.
2. Hallegraeff, G.M. Ocean cclimate change, phytoplankton community responses, and harmful algal blooms: A formidable predictive challenges. *J. Phycol.* **2010**, *46*, 220–235.
3. Smayda, T.J. Complexity in the eutrophication–harmful algal bloom relationship, with comment on the importance of grazing. *Harm. Algae* **2008**, *8*, 140–151.
4. Zevenboom, W.; Rademaker, M.; Colijn, F. Exceptional algal blooms in Dutch North Sea waters. *Water Sci. Technol.* **1991**, *24*, 251–260.
5. Murphy, T.P.; Brownlee, B.G. Ammonia volatilization in a hypertrophic prairie lake. *Can. J. Fish. Aquat. Sci.* **1981**, *38*, 1035–1039.
6. Anderson, D.M.; Glibert, P.M.; Burkholder, J.M. Harmful algal blooms and eutrophication: Nutrient sources, composition, and consequences. *Estuaries* **2002**, *25*, 704–726.
7. Beman, J.M.; Arrigo, K.R.; Matson, P.A. Agricultural runoff fuels large phytoplankton blooms in vulnerable areas of the ocean. *Nature* **2005**, *434*, 211–214.
8. Brett, M.T.; Arhonditsis, G.B.; Mueller, S.E.; Hartley, D.M.; Frodge, J.D.; Funke, D.E. Non-point-source impacts on stream nutrient concentrations along a forest to urban gradient. *Environ. Manag.* **2005**, *35*, 330–342.
9. US Environmental Protection Agency (US EPA). *National Water Quality Inventory: 1988 Report to Congress*; EPA Report 440-4-90-003; US EPA: Washington, DC, USA, 1990.
10. Hwang, S.J.; Lee, S.W.; Son, J.Y.; Park, G.; Kim, S.J. Moderating effects of the geometry of reservoirs on the relation between urban land use and water quality. *Landsc. Urban Plann.* **2007**, *82*, 175–183.
11. Somura, H.; Takeda, I.; Arnold, J.G.; Mori, Y.; Jeong, J.; Kannan, N.; Hoffman, D. Impact of suspended sediment and nutrient loading from land uses against water quality in the Hii River basin, Japan. *J. Hydrol.* **2012**, *450–451*, 25–35.
12. Lee, H.S.; Chung, S.W.; Jeong, H.Y.; Min, B.H. Analysis the effects of curtain weir on the control of algal bloom according to installation location in Daecheong reservoir. *J. Korean Soc. Water Environ.* **2010**, *26*, 231–242.
13. Lewis, J.R. *The Ecology of Rocky Shores*; The English Universities Press Ltd.: London, UK, 1964.
14. Hiraoka, M.; Shimada, S. Biology of a special green laver, Ulva prolifera from the Shimanto river. *Aquabiology* **2004**, *26*, 508–515.
15. Fovet, O.; Belaud, G.; Litrico, X.; Charpentier, S.; Bertrand, C.; Dollet, P.; Hugodot, C. A model for fixed algae management in open channels using flushing flows. *River Res. Appl.* **2012**, *28*, 960–972.
16. Lundgren, V.M.; Roelke, D.L.; Grover, J.P.; Brooks, B.W.; Prosser, K.N.; Scott, W.C.; Laws, C.A.; Umphres, G.D. Interplay between ambient surface water mixing and manipulated hydraulic flushing: Implications for harmful algal bloom mitigation. *Ecol. Eng.* **2013**, *60*, 289–298.
17. Chung, S.W.; Park, J.H.; Kim, Y.K.; Yoon, S.W. Application of CE-QUAL-W2 to Daecheong Reservoir for eutrophication simulation. *J. Korean Soc. Water Environ.* **2007**, *23*, 52–63.
18. Morillo, S.; Imberger, J.; Antenucci, J.P. Modifying the residence time and dilution capacity of a reservoir by altering internal flow-paths. *J. River Basin Manag.* **2006**, *4*, 255–271.
19. Asaeda, T.; Priyantha, D.G.; Saitoh, S.; Gotoh, K. A new technique for controlling algal blooms in the withdrawal zone of reservoirs using vertical curtains. *Ecol. Eng.* **1996**, *7*, 95–104.
20. Dai, H.; Mao, J.; Jiang, D.; Wang, L. Longitudinal hydrodynamic characteristics in reservoir tributary embayments and effects on algal blooms. *PLoS One* **2013**, *8*.
21. Yeo, H.G.; Kang, H. Water quality and phytoplankton community in the coastal waters of Inchon. *Korean Environ. Sci.* **1998**, *7*, 321–326.
22. Bonsdorff, E.; Blomqvist, E.M.; Mattila, J.; Norkko, A. Coastal eutrophication: Causes, consequences and perspectives in the Archipelago areas of the northern Baltic Sea. *Estuar. Coast. Shelf Sci.* **1997**, *4*, 63–72.
23. Hodgkin, E.P.; Hamilton, B.H. Fertilizers and eutrophication in southwestern Australia: Setting the scene. *Fertil. Res.* **1993**, *36*, 95–103.
24. Mahlon, G.K.; Naguib, M. *Eutrophication in Coastal Marine Area and Lagoon: A Case Study of Lac de Tunis*; UNESCO Reports in Marine Science; United Nations Educational, Scientific and Cultural Organization: Paris, France, 1984.
25. Gillbricht, M. Phytoplankton and nutrients in the Helgoland region. *Helgol. Meeresunters.* **1988**, *42*, 435–467.
26. Mohamed, M.D.; Nagwa, E.A.; Kamal, Z.E.; Mohamed, A.O. Eutrophication problems in the western harbour of Alexandria, Egypt. *Oceanologia* **2004**, *46*, 25–44.
27. Tomotoshi, O. *Red Tides*; Kluwer Academic: Tokyo, Japan, 2004.

28. Naver Map Home Page. Available online: http://map.naver.com/ (accessed on 20 October 2013).

29. Flow Science. *Flow-3D User's Manual*; Flow Science: Los Alamos, NM, USA, 2003.

30. Spanoudaki, K.; Stamou, A.I.; Nanou-Giannarou, A. Development and verification of a 3-D integrated surface water-groundwater model. *J. Hydrol.* **2009**, *375*, 410–427.

31. Fischer-Antze, T.; Stoesser, T.; Bates, P.; Olsen, N.R.B. 3D numerical modelling of open-channel flow with submerged vegetation. *J. Hydraul. Res.* **2001**, *39*, 303–310.

32. Andrew, D.E.; Clescenri, L.S.; Breenberg, A.E. *Standard Method for the Examination of Water and Wastewater*, 19th ed.; American Public Health Association, American Water Work Association, Water Environment Federation: Washington, DC, USA, 1995.

33. Rodriguez, J.F.; Bombardelli, F.A.; García, M.H.; Frothingham, K.M.; Rhoads, B.L.; Abad, J.D. High-resolution numerical simulation of flow through a highly sinuous river reach. *Water Resourc. Manag.* **2004**, *18*, 177–199.

CHAPTER 8

IMPROVEMENT OF HYDRAULIC AND WATER QUALITY RENOVATION FUNCTIONS BY INTERMITTENT AERATION OF SOIL TREATMENT AREAS IN ONSITE WASTEWATER TREATMENT SYSTEMS

José A. Amador , David A. Potts , George W. Loomis, David V. Kalen, Erika L. Patenaude and Josef H. Görres

Keywords: Septic tank effluent, intermittent aeration, hydraulic function, water quality, SoilAir

INTRODUCTION

Approximately 25% of the population of the USA relies on onsite wastewater treatment systems (OWTS) for dispersal and renovation of domestic wastewater, a number that has remained constant for the past 35 years [1]. Improperly functioning OWTS constitute public health and environmental hazards because of their potential to degrade ground, surface, and coastal water quality. Of special concern are elevated emissions of N, P , biodegradable organic C and pathogens, which can adversely impact the utilization of ground and surface water resources [2,3].

The ability of soil treatment areas (STA) in conventional OWTS to accept and renovate septic tank effluent (STE) is subject to fluctuations in dosing frequency, volume, and strength of wastewater. These systems are prone to hydraulic failure due to a number of factors, including lack of proper maintenance [4]. Improvements in the quality of wastewater as it passes through the STA vary considerably, particularly with respect to N removal, which can range from 0 to 30% [2,5].

Current advanced treatment technologies for improving water quality discharged from OWTS include those aimed specifically at removing nitrogen (e.g., recirculating media filters and activated sludge systems) or biochemical oxygen demand (BOD_5) and pathogens (e.g., single-pass sand and peat filters, UV/ozonation units) [2]. Although effective, these technologies can be expensive, require maintenance for proper functioning, and involve hiring a maintenance specialist, all of which can make them difficult to adopt by homeowners [6]. Alternatives for hydraulically failed systems, such as STA replacement to meet current regulations, can also be costly, and may only postpone subsequent hydraulic failure in the absence of routine inspection and maintenance.

In the present study we tested intermittent aeration of STAs for its ability to improve the hydraulic and water quality functions of existing OWTS. The tests were conducted at three medical group homes within the Narragansett Bay watershed in southern Rhode Island, USA. The technology, SoilAir™, involves intermittent aeration (ISA) of the soil treatment area, rather than aeration of septic tank effluent [7]. It has been shown to enhance infiltration of STE in numerous field tests, and was found to enhance the quality of water draining from STAs in pilot studies [8]. Over 1,000 field applications of this technology in commercial and household OWTS in eleven states in the USA and in Canada have shown that it can rejuvenate septic systems by restoring hydraulic function in a matter of days. In a previous study, Potts et al. [8] evaluated the effects of intermittent soil aeration on removal of total

Citation: JoséA. Amador, David A. Potts, George W. Loomis, David V. Kalen, Erika L. Patenaude, and Josef H. Görres, Improvement of Hydraulic and Water Quality Renovation Functions by Intermittent Aeration of Soil Treatment Areas in Onsite Wastewater Treatment Systems, Water 2010, 2, 886-903; doi:10.3390/w2040886.

N, biochemical oxygen demand (BOD_5) and fecal coliform bacteria from STE in sand-filled leachfield mesocosms relative to unaerated mesocosms. The results of this pilot-scale study indicated that intermittent soil aeration enhances removal of total N, fecal coliform bacteria, and BOD_5, with improvements in water quality generally increasing with hydraulic load.

MATERIALS AND METHODS

Study Sites

Field evaluations of the performance of ISA technology were carried out at three sites in Washington County, Rhode Island, USA. All were medical group homes managed by the Rhode Island Department of Mental Health, Retardation and Hospitals (RIMHRH), with one site located in the town of South Kingstown and two in Charlestown (Table 1). Medical group homes were selected for the study because they typically have sustained wastewater flows that exceed design flow and thus can stress the STA. Site selection was restricted to those that were similar in use, type and structural condition of the OWTS, accessibility to OWTS components, and soil properties (Table 1). Sites that met these criteria were inspected for functionality and identified as failed or not failed, with failure defined as the STA not accepting STE. Managers of the selected group homes had the septic tanks pumped regularly to prevent STE back-up into the homes. All of the systems selected for our study relied on stone-lined trenches for STE dispersion and were experiencing hydraulic failure. For the purposes of our study, the STA was considered to include the trenches receiving STE and the soil surrounding these trenches.

Table 1: Location, soil series, and soil profile descriptions for study sites [9].

Site	Town	Soil series	Soil profile description
1	South Kingstown	Canton-Urban complex	Coarse loamy over sandy skeletal, mixed mesic Typic Dystrudepts. Upper solum fine sandy loam to gravelly fine sandy loam in the lower B horizons; massive, very friable. Substratum gravelly to very gravelly loamy sand to a depth of 60 inches or more.
2	Charlestown	Merrimac	Sandy, mixed, mesic Typic Dystrudepts. Weakly structured fine sandy loam in the A and upper B horizons and gravelly analogs in lower B. Substratum very gravelly sand; single grain; loose stratified; 40% gravel.
3	Charlestown	Canton-Charlton complex	Coarse loamy over sandy skeletal, mixed, mesic Typic Distrudepts. Upper solum fine sandy loam to gravelly fine sandy loam; massive, very friable. Substratum gravelly to very gravelly loamy sand extending to a depth of 60 inches or more; single grain friable to very friable.

Each septic tank was fitted with a plastic sampling pipe placed near the tank outlet and 45–60 cm below the water surface for sampling of STE. Three nests of ceramic suction cup lysimeters were installed at each site in February of 2006 (Figure 1). Cluster locations were as follows:

Cluster 1 (C1):

Upgradient from STA

Cluster 2 (C2):

Between leachfield trenches, within the STA

Cluster 3 (C3):

Downgradient from STA (within 15–20 cm of outer wall of most downgradient trench)

Within each cluster, two lysimeters were installed with the bottom of the cups at a depth of 30 and 90 cm below infiltrative surface (designated C1-30, C1-90; C2-30, C2-90; C3-30, C3-90).

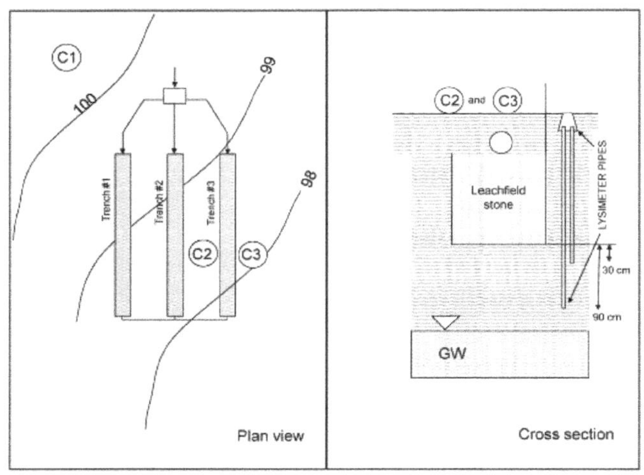

Figure 1: Plan view (left) and cross section (right) of placement of sampling clusters at the study sites. Drawings are not to scale.

Treatments

Two treatments are distinguished in the study: Phase I represents the period prior to operation of the ISA system, between March and August 2006; Phase II represents the period during which the ISA system was in operation, between September 2006 and April 2007.

Installation of Intermittent Soil Aeration System

ISA system components, time dosing pumps and associated controls were retrofitted onto Sites 1, 2 and 3 in mid-September, 2006 (Figure 2). In accordance with established installation protocols developed by Geomatrix, LLC (Old Saybrook, CT), the septic tank at every site was pumped prior to installation of the ISA system components and septic tank effluent pump (STEP) components. In addition, approximately 50%, 50% and 68% of the leachfield system capacity was isolated from the wastewater flow at Sites 1, 2 and 3, respectively, as part of the installation protocols by blocking the flow of STE to particular trenches at the distribution box (D-box) (Figure 2b).

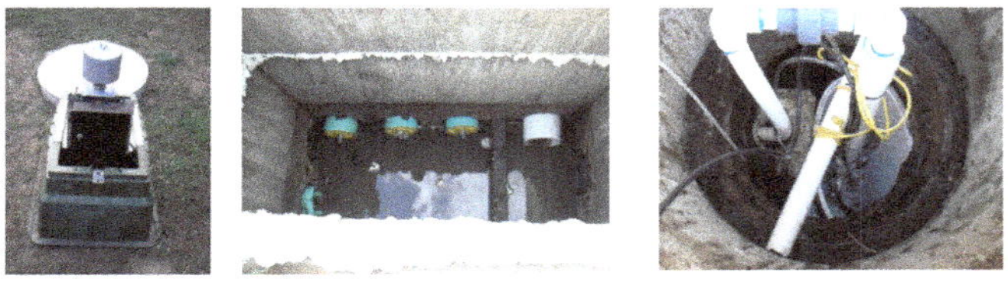

a b c

Figure 2: (a) Air blower, filter and controller components of the intermittent soil aeration system; (b) septic tank effluent flow to a portion of the leachfield trenches was blocked at the distribution box to test the robustness of the technology; (c) septic tank effluent pump (STEP) system used in the study.

At each site a STEP system was installed in the second compartment of the most downstream septic tank (Figure 2c). Float switches were set to drop the effluent levels in the tank by no more than ~15 cm to provide for flow equalization capacity and to quantify flow. Air lines, consisting of 5 cm Schedule 40 polyvinyl chloride (PVC) pipe, were run from the blower (placed within a protective enclosure) and connected to the pipe between the septic tank and the STA (Figure 2a). Distribution boxes and inspection ports were sealed to prevent the short-circuiting of air. Electrical power and control cables were connected to the ISA system, STEP system and associated float switches. A telecommunications cable was connected between the device controls and the building to facilitate modem communication for remote telemetry of the systems. The microprocessor-based controllers were configured to operate both the blower and the timed-dose STEP system. The STEP system was programmed to be on for a set period of time, which correlated to the desired dose. The STEP dose interval was programmed to run every 6 h, unless a high or low-level float switch was triggered. In addition to controlling dosing, this allowed us to quantify the hydraulic load to the STA. The blower was programmed to remain inactive for approximately 1 h after a wastewater dose. After this time interval, the blower was programmed to run intermittently for a set period of time. The blower and STEP system were interlocked to prevent simultaneous operation, with priority given to the STEP system.

Sampling and Analyses

A hand pump was used to apply a vacuum (~80 kPa) to the suction cup lysimeters 24 h prior to sampling. Samples of soil pore water from suction cup lysimeters and STE samples were collected using a peristaltic pump fitted with silicon tubing, placed in autoclaved polyethylene screw-cap bottles, and stored in a cooler filled with ice packs immediately after collection.

STE samples were analyzed for dissolved oxygen (DO) immediately after collection in the field. The temperature and concentration of ferrous iron (Fe^{2+}) of samples of STE and of soil pore water samples from cup lysimeters were also determined immediately after collection. Prior to filtering, all samples were analyzed for pH immediately upon arrival at the laboratory. Unfiltered STE samples were analyzed for biochemical oxygen demand (BOD_5). A portion of all unfiltered samples was frozen for subsequent determination of total N (TN) and total P (TP) content. The remaining sample was passed through a nylon membrane filter (0.45-μm pore-size, 47-mm dia.; Osmonics, Watertown, MA) and the filtrate stored in plastic, screw-cap vials at 4 °C.

DO was measured using the azide modification of the Winkler titration method [10]. The concentration of Fe^{2+} in water was determined using EM Quant ® Iron (Fe^{2+}) test strips (EM Industries, Inc., Gibbstown, NJ). The pH of water samples was determined using a combination pH electrode and a model UB-10 pH meter (Denver Instruments, Denver, CO). The concentration of sulfate was measured using the barium chloride turbidimetric method [10]. Nitrate, ammonium, and phosphate concentrations of water samples were determined colorimetrically using an automated nutrient analyzer (model Flow Solution IV, Alpkem, College Station, TX). The total N and total P content of water samples were determined using the persulfate digestion method [10] followed by colorimetric analysis of nitrate and phosphate, respectively. Chloride was determined using the argentometric titration method [10]. BOD_5 was measured on undiluted, unamended samples by manometric respirometry using an OxiTop® BOD system (WTW, Fort Myers, FL) at 21 ± 1 °C. The dissolved organic carbon (DOC) content of filtered samples was determined using a TOC-5000A Total Organic Carbon Analyzer (Shimadzu Instruments, Inc., Laurel, MD).

Calculation of Apparent and Actual Constituent Removal Rates

The mean concentration of N, P, chloride and DOC in samples from C1 (upgradient, or background) suction cup lysimeters (Figure 1) was subtracted from the mean concentration for samples from either C2 (within the STA) or C3 (downgradient from the STA) lysimeters at the same depth to correct for background effects. The apparent removal rate, R, for constituent X (%) was calculated using the equation:

$$R_X = 100 \times [(C_{X\text{-}STE} - C_{X\text{-}LYS})/C_{X\text{-}STE}] \quad (1)$$

where $C_{X\text{-}STE}$ is the concentration of constituent X in STE and $C_{X\text{-}LYS}$ is the background-corrected concentration of X in a soil pore water sample from a suction cup lysimeter, both expressed in mg/L.

The actual removal rate, Φ, for a constituent X ($mg/m^2/d$) in the STA was calculated using the equation:

$$\Phi_X = [((R_X - R_{Cl})/100) \times (V_{STE} \times C_{X\text{-}STE})]/A/t \quad (2)$$

where R_X is the apparent mean removal rate for constituent X (%), R_{Cl} is the observed mean removal rate for Cl^- (%), V_{STE} is the mean volume of septic tank effluent applied to the leachfield, $C_{X\text{-}STE}$ is the mean concentration of constituent X in septic tank effluent (mg/L), A is the estimated basal area of leachfield (m^2) (Table 2), and t is time (d). This analysis assumes that Cl^- acts as a conservative tracer of the movement of STE through the soil, with reduction in its concentration attributed exclusively to dilution. An apparent removal rate larger than that observed for chloride is assumed to represent loss processes other than dilution (e.g., biological uptake, abiotic sorption), whereas an apparent removal rate lower than observed for chloride is assumed to represent production processes (e.g., mineralization, desorption). Values of A prior to and after installation of the ISA system at Sites 1, 2 and 3 (Table 2) were estimated from design plans submitted to Rhode Island Department of Environmental Management as part of the original permitting process for septic system installation, and from knowledge of which trenches were isolated when the intermittent aeration system was installed. Values of Vrepresent the mean of automated measurements of the volume of STE dosed daily to the leachfield and were assumed to be the same prior to and during operation of the ISA system, since group home occupancies remained the same (Table 2).

Table 2: Estimates of bottom leachfield area (A), mean volume of septic tank effluent dosed daily to a leachfield (V), and hydraulic load in Sites 1, 2, and 3.

Site	Phase	A (m^2)	V (L)	Hydraulic load (L/m^2)
1	I	76.6	1440	18.8
	II	38.3	1440	37.6
2	I	24.5	1876	76.6
	II	12.3	1876	152.5
3	I	99.8	4686	47.0
	II	32.1	4686	146.0

Statistical Analyses

Student's t test ($P < 0.05$) was used to evaluate differences in water quality parameters between Phase I and Phase II.

RESULTS AND DISCUSSION

Hydraulic Function

All three systems were in hydraulic failure during Phase I, as indicated by STE levels that were considerably above the outlet pipe invert in the D-box and/or the STA not accepting effluent from the septic tank. Once ISA was in operation, measurements of depth to the water surface within the D-box suggest that the hydraulic load at all three sites infiltrated readily, as indicated by the relatively constant values of depth to water surface (Figure 3). Levels of STE were at the D-box outlet pipe inverts during Phase II in all cases. Depth values for Site 3 during the early part of Phase II were affected by an increased hydraulic load from a leaking toilet. Variations in depth to the water surface within a site are the result of differences in timing of measurements relative to dosing events.

Hydraulic failure, which was experienced at all three sites prior to implementation of ISA, results from a combination of factors, including hydraulic overload and/or longer periods of saturation resulting from overloading, which results in a biomat that is increasingly restrictive to STE infiltration. The biomat is thought to form from accumulation of suspended solids and organic debris released by microorganisms under anaerobic conditions, which results in a shift towards smaller pore sizes, restricting infiltration [2]. Fine mineral particles present in STE are trapped within this organic matrix, further restricting infiltration. The thickness of the biomat, and thus its infiltrative capacity, is the net result of microbial processes that produce organic polymers and those that consume it, with production favored under saturated, anaerobic conditions, and consumption favored under unsaturated, aerobic conditions. As such, intermittent aeration of the STA would be expected to enhance infiltration by promoting aerobic conditions that support microbial oxidation of organic polymers in the biomat.

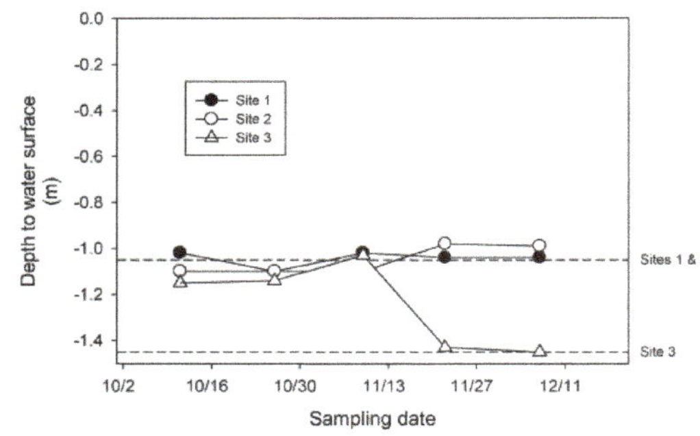

Figure 3: Depth to water surface in the distribution box (D-box) at Sites 1, 2 and 3 during operation of the intermittent soil aeration system. Values indicate distance below the ground surface. Dashed line indicates depth to invert at each site. Depth values for Site 3 during the early part of Phase II were affected by an increased hydraulic load from a leaking toilet.

Water Quality

Levels of DO in STE were below detection limits at all three sites. Mean ($n = 27$) values of BOD_5 were 309, 365 and 202 mg/L for Sites 1, 2 and 3. These and other water quality parameter values for STE (Table 3) were in the range of those reported by others [11].

The pH of water below the infiltrative surface was not significantly different between Phase I and Phase II regardless of sampling depth or location at any of the test sites (Table 3). In those instances where Fe^{2+} was present prior to intermittent soil aeration (Phase I), the levels decreased while the ISA was operational, with statistically significant differences observed in most cases (Table 3). Levels of sulfate were not significantly different between Phase I and Phase II at any of the test sites, regardless of sampling depth or position (data not shown).

There was no statistically significant difference in concentration of total P (Table 3) and PO_4 (data not shown) in soil pore water before and during intermittent soil aeration. No significant differences in DOC concentration were observed between Phase I and Phase II at Site 1 or Site 3 (Table 3). By contrast, the concentration of DOC was significantly lower in water from 30 and 90 cm below the infiltrative surface at Site 2 during Phase II (Table 3).

The concentration of total N responded differently to intermittent soil aeration at the three sites. At Site 1 there was a significant increase in total N in soil pore water at 90 cm between Phase I and Phase II, from 12 to 24 and 9 to 20 mg N/L for C2 and C3, respectively (Table 3). These increases were observed despite the fact that the concentration of total N in STE inputs did not change (Table 3), although the load of N per unit area was higher as a result of increase hydraulic load. By contrast, there were no significant differences between Phase I and II in total N levels at 30 or 90 cm in either Site 2 or Site 3 (Table 3). Levels of NO_3 and NH_4 in soil pore water below the infiltrative surface were not significantly different between Phase I and Phase II regardless of sampling depth, position or test site. The relative distribution of inorganic N species differed among study sites during both phases of the experiment. Nitrate generally accounted for >90% of inorganic N at Sites 1 and 2, whereas at Site 3, NH_4 accounted for the bulk of the inorganic N in soil pore water. Previous studies of the composition of soil pore water below the infiltrative surface of conventional OWTS report that NO_3 accounts for 80 to 90% of inorganic N [12,13].

Table 3: Values (mean ± S.D.) of Cl^-, pH, Fe(II), total N, NO_3, total P and dissolved organic carbon (DOC) in STE and in soil pore water at Site 1, 2 and 3 prior to (Phase I) and during (Phase II) operation of intermittent aeration. n ≤ 12 for Phase I; n ≤ 15 for Phase II. Units are mg/L except for pH. Values in bold indicate significant difference between Phase I and Phase II.

Site	Paramete	Phase	STE	C2-30	C2-90	C3-30	C3-90
1	Cl^-	I	341 ± 129	27 ± 35	46 ± 31	27 ± 23	19 ± 13
		II	323 ± 128	39 ± 45	135 ± 40	96 ± 93	115 ± 77
	pH	I	6.7 ± 0.2	6.2 ± 0.5	6.2 ± 0.5	6.2 ± 0.6	6.2 ± 0.6
		II	6.7 ± 0.3	6.1 ± 0.6	5.8 ± 0.5	6.0 ± 0.5	5.9 ± 0.5
	Fe^{2+}	I	0 ± 0	**1 ± 1**	**1 ± 0**	0 ± 0	**1 ± 0**
		II	0 ± 0	**0 ± 0**	**0 ± 0**	0 ± 0	**0 ± 0**
	Total N	I	41 ± 8	14 ± 3	**12 ± 7**	15 ± 6	**9 ± 1**
		II	47 ± 7	14 ± 9	**24 ± 9**	20 ± 13	**20 ± 7**
	NH_4	I	23 ± 7	3 ± 3	1 ± 1	1 ± 1	1 ± 1
		II	26 ± 5	0 ± 0	0 ± 0	1 ± 1	1 ± 3
	NO_3	I	0 ± 0	8 ± 6	8 ± 6	12 ± 5	7 ± 1
		II	1 ± 1	8 ± 7	12 ± 6	9 ± 8	11 ± 7
	Total P	I	12 ± 5	2 ± 1	2 ± 1	2 ± 1	2 ± 1
		II	13 ± 2	3 ± 2	3 ± 1	3 ± 2	4 ± 2
	DOC	I	106 ± 24	12 ± 11	4 ± 2	5 ± 3	2 ± 3
		II	118 ± 20	3 ± 3	2 ± 5	2 ± 3	6 ± 5
2	Cl^-	I	49 ± 20	42 ± 6	41 ± 5	2 ± 2	14 ± 14
		II	45 ± 5	42 ± 16	41 ± 13	5 ± 7	15 ± 9
	pH	I	6.6 ± 0.1	6.7 ± 0.2	6.5 ± 0.1	6.7 ± 0.2	6.5 ± 0.3
		II	6.6 ± 0.2	6.5 ± 0.2	6.5 ± 0.2	6.9 ± 0.1	6.5 ± 0.5
	Fe^{2+}	I	0 ± 0	9 ± 1	0 ± 0	0 ± 0	0 ± 0
		II	0 ± 0	2 ± 1	0 ± 0	0 ± 0	0 ± 0
	Total N	I	41 ± 7	37 ± 8	30 ± 6	14 ± 9	21 ± 9
		II	40 ± 7	32 ± 11	33 ± 4	18 ± 15	14 ± 9
	NH_4	I	21 ± 4	20 ± 9	1 ± 1	0 ± 0	0 ± 0
		II	22 ± 4	20 ± 8	1 ± 1	0 ± 0	0 ± 0
	NO_3	I	0 ± 0	9 ± 4	22 ± 7	9 ± 7	12 ± 6
		II	0 ± 0	1 ± 1	19 ± 6	5 ± 4	5 ± 4
	Total P	I	9 ± 4	7 ± 4	2 ± 1	2 ± 1	2 ± 1
		II	12 ± 3	9 ± 2	3 ± 2	3 ± 1	4 ± 3
	DOC	I	105 ± 23	63 ± 11	**37 ± 9**	**36 ± 3**	**6 ± 3**
		II	100 ± 16	67 ± 6	**16 ± 4**	**16 ± 3**	**1 ± 2**
3	Cl^-	I	42 ± 11	27 ± 13	44 ± 10	41 ± 6	44 ± 11
		II	115 ± 88	93 ± 77	88 ± 79	102 ± 88	117 ± 75
	pH	I	6.7 ± 0.1	6.5 ± 0.2	6.5 ± 0.1	6.8 ± 0.1	6.7 ± 0.1
		II	6.8 ± 0.3	6.6 ± 0.2	6.6 ± 0.2	6.8 ± 0.2	6.7 ± 0.2
	Fe^{2+}	I	0 ± 0	15 ± 7	17 ± 9	4 ± 2	7 ± 5
		II	0 ± 0	5 ± 6	8 ± 6	0 ± 0	0 ± 0
	Total N	I	38 ± 7	20 ± 3	26 ± 4	32 ± 3	30 ± 5
		II	36 ± 6	27 ± 6	29 ± 9	27 ± 11	31 ± 7
	NH_4	I	24 ± 11	15 ± 5	19 ± 6	22 ± 7	21 ± 8
		II	19 ± 6	18 ± 5	20 ± 8	18 ± 8	19 ± 9
	NO_3	I	0 ± 0	0 ± 0	0 ± 0	0 ± 1	0 ± 0
		II	0 ± 0	1 ± 3	2 ± 1	1 ± 2	1 ± 2
	Total P	I	6 ± 2	3 ± 1	4 ± 1	5 ± 1	5 ± 2
		II	9 ± 2	6 ± 2	6 ± 1	7 ± 2	7 ± 1
	DOC	I	77 ± 13	42 ± 6	56 ± 10	58 ± 6	57 ± 7
		II	91 ± 25	58 ± 16	68 ± 16	67 ± 23	70 ± 15

Previous mesocosm-scale evaluation of the effects of intermittent soil aeration of STAs on nitrogen transformations have shown high levels of NO_3 and nearly complete absence of NH_4 in water draining below 30 cm of sand or soil, in contrast to unaerated mesocosms [8,14,15]. Furthermore, drainage water from intermittently aerated mesocosms had a considerably lower pH than unaerated treatments [8,14,15]. We did not observe these effects in the present study. Mesocosms provide a much greater degree of control over variables such as penetration of air, homogeneity of soil properties, movement of STE through soil—and the reproducibility of these conditions—all of which can affect the dynamics of NH_4 and NO_3 production and consumption. The effects of ISA on nitrate dynamics may have taken place at shallow depths, such that the concentrations of NH_4 and NO_3 measured at 30 or 90 cm do not provide an accurate representation of inorganic N dynamics as a result of other production and consumption processes that can take place deeper within the soil profile, such as plant uptake, and microbial mineralization and/or immobilization. Using experimental mesocosms we have shown that total N removal appears to take place within the top 7.5 cm of aerated sand [15]. In addition, spatial and temporal variability may mask the effects of aeration on speciation of inorganic N below the STAs.

Apparent and Actual Removal of N, P and DOC

Intermittent soil aeration generally enhanced the apparent removal rates (R_x) of N, P and DOC at all three sites (Figure 4) despite marked reductions in nominal infiltration area (Table 2). In most instances, mean values of R_X for total N at 90 cm below the infiltrative surface were negligible or negative (higher N levels in soil pore water relative to STE inputs) during Phase I, and became positive (lower N levels in soil pore water relative to STE inputs) during Phase II (Figure 4). Intermittent soil aeration also increased apparent removal rates for total P, as indicated by higher values of R_X during Phase II (Figure 4). The exception was observed at Site 2, C2, where apparent removal rates were negative during the ISA phase. Intermittent soil aeration had mixed effects on apparent rates of DOC removal (Figure 4). During Phase I there was a higher concentration of DOC in soil pore water relative to STE inputs, whereas during intermittent soil aeration apparent removal of DOC was observed. By contrast, at Sites 2 and 3 ISA resulted in lower apparent removal rates for DOC, stemming from a higher concentration of DOC in soil pore water than in STE inputs.

Apparent removal rates do not account for two potentially important factors that affect contaminant removal in the STA: (i) differences in size of STA utilized, and (ii) dilution by ground water, precipitation or snowmelt. In the present study, reductions in treatment area (Table 2), which would be expected to influence the effectiveness of these systems, were implemented to test the robustness of the ISA system with respect to its ability to restore hydraulic flow. Dilution of contaminants results in lower apparent concentrations, which overestimates the effectiveness of contaminant removal in the STA. Differences in Cl^- concentrations between STE and soil pore water suggest that dilution needs to be accounted for (Table 3). To address these two issues, we calculated an actual removal rate (mass removed/area/time) using Equation 2.

Actual removal rates for N during Phase I were relatively small, with both net production (negative values) and net removal (positive values) observed at a depth of 90 cm under the infiltrative surface (Figure 5). By contrast, actual removal rates for N at 90 cm below the leachfield during Phase II were consistently positive and up to two orders of magnitude higher at all sites (Figure 5). Intermittent soil aeration also resulted in higher actual P removal rates at all three sites, with values that were between 2 and 15 times higher than those observed during Phase I (Figure 5). Similarly, actual rates of DOC removal were 4 to 10 times higher during the ISA phase than during Phase I at all three sites (Figure 5).

Removal of total N in STA of conventional OWTS is generally low [16,17], owing to poor denitrification [17]. The results of our study show that apparent N removal rates in our systems prior to intermittent aeration ranged widely, from removal of up to 60% to increases in N concentration in soil pore water of 5 to 30% above the levels in STE inputs. Literature reports on apparent N removal rates for conventional systems vary considerably, from 21% [18], to 35% [12], to 50% [13]. These values are for systems receiving normal hydraulic loads, and are within the range of those observed by us prior to intermittent aeration in some instances. By contrast, the net increases in total N levels below the infiltrative surface relative to STE inputs during Phase I may be due to mineralization of nitrogenous exopolymers and dead microbial biomass accumulated within the STA, likely due to an increased hydraulic load.

The intermittent aeration of the STA (Phase II) resulted in an increase in both apparent and actual removal rates for N, P and C (Figure 4 and Figure 5). Actual removal values for N, P, and C under field conditions during intermittent aeration (Phase II) were comparable to those observed for mesocosm scale experiments [8,14]. Furthermore, values of actual N removal rates during Phase II were comparable to those observed for wood-based denitrification reactors used to treat drainage water (950 to 2500 mg N/m²/day) [19]. The positive effects of intermittent aeration are unlikely to be due to differences in temperature between Phase I and Phase II, inasmuch as mean temperature (as reflected in values for STE) was higher during Phase I (20.1 to 22.7 °C)) than during Phase II (16.8–17.4 °C). The rate of microbial processes, which are presumed to be involved in removal of N, P and C in these systems, is expected to decrease as temperature decreases [20].

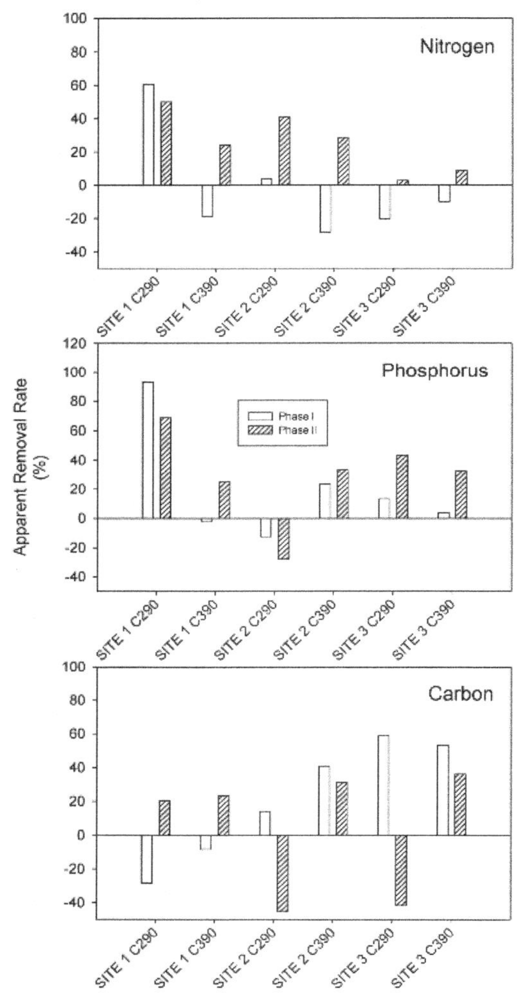

Figure 4: Apparent removal rate for nitrogen, phosphorus and dissolved organic carbon at the three study sites prior to (Phase I) and during (Phase II) operation of intermittent soil aeration. Values are for lysimeters at 90 cm below the infiltrative surface placed between (C2) and downstream from (C3) leachfield trenches. Apparent removal rates were calculated according to Equation (1).

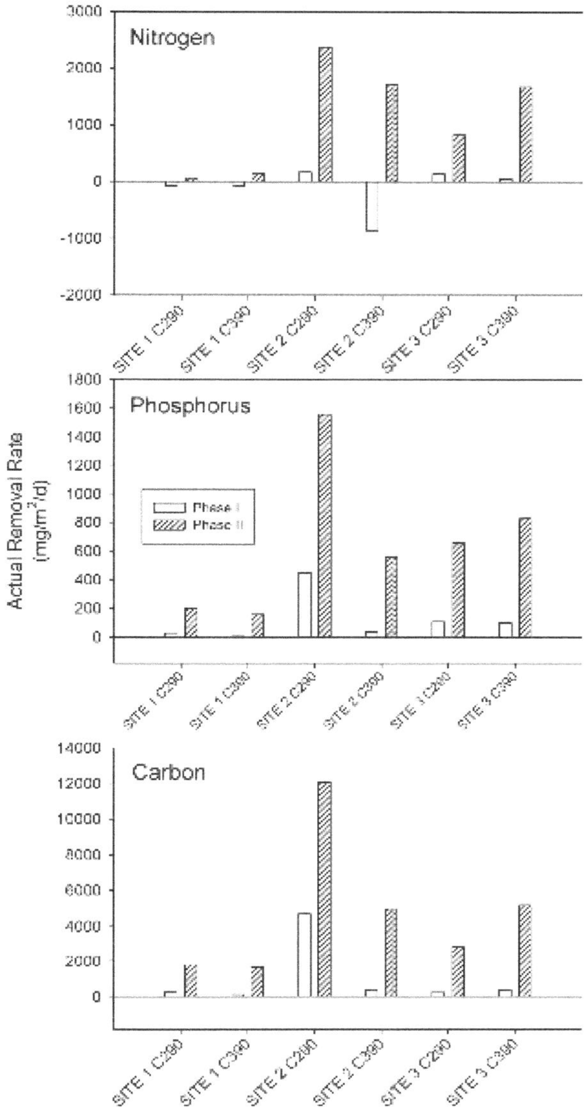

Figure 5: Actual removal rates for nitrogen, phosphorus and dissolved organic carbon at the three study sites prior to (Phase I) and during (Phase II) operation of intermittent soil aeration. Values are for lysimeters at 90 cm below the infiltrative surface placed between (C2) and downstream from (C3) leachfield trenches. Actual removal rates were calculated using Equation (2). Negative values indicate net constituent production.

Relationship between Hydraulic Load and Removal of N, P and DOC

Differences in the effects of ISA on actual removal rates of N, DOC and P among test sites (Figure 5) led us to examine the possibility that these effects are related to hydraulic load. In previous studies we have observed that actual nitrogen removal from STE in intermittently aerated mesocosms increased with increasing hydraulic load [8,14]. The differences in hydraulic load at the three sites in the present study provided an opportunity to examine the applicability of pilot-scale results to field systems.

A plot of hydraulic load vs. actual N removal rate shows that prior to intermittent aeration, actual removal rates for N, P and C remained the same or decreased as a function of hydraulic load (Figure 6). By contrast, actual N removal rates increased with increasing hydraulic load at the field sites. Furthermore, the actual removal rates for N were similar to those for sand-filled mesocosms. Actual removal rates for DOC for the field sites exhibited a similar relationship with hydraulic load to that observed for nitrogen, with DOC removal rates increasing with hydraulic load, and field removal rates values that were close to those reported for sand mesocosms (Figure 6). Actual removal rates for P at the field sites also increased with hydraulic load; however, unlike N and C, these values were closer to those for soil-filled mesocosms (Figure 6). By contrast, actual removal rates for C and P during Phase I changed little with increasing hydraulic load, and rates of N removal appeared to decrease as hydraulic load increased (Figure 6).

In general, actual removal rates for N, P and C are expected to be inversely proportional to hydraulic loading rate because frequent saturation of the STA for long periods of time is expected to lead to anaerobic processes that are less efficient at C and nutrient removal [21]. In our study, actual removal rates under field conditions increased with increasing hydraulic load at all sites. The response of actual removal rates to hydraulic load in the present study suggests that removal mechanisms may be similar under field and laboratory conditions. We have previously suggested that enhanced N removal with intermittent aeration of the STA stems from a sequence in which aeration promotes ammonia oxidation to nitrate in soil, with the subsequent addition of STE to soils promoting removal of N via denitrification, by providing anaerobic conditions as well as organic C sources [8,14]. This is similar to the processes involved in N removal from wastewater using sequencing batch reactors (SBR) [22]. Recent studies of SBR have shown that simultaneous nitrification/denitrification can also take place in SBR [23]. Increases in removal of N with increased hydraulic load during the ISA phase in the present study suggest that N removal is enhanced by a longer period of anaerobic conditions and/or a larger volume of anaerobic soil. Because denitrification consumes organic C, intermittent aeration of STAs should also increase C removal, and this removal should increase with increasing hydraulic load, as was the case under both field and mesocosm conditions. Phosphate removal is also enhanced in SBR, apparently as a result of phosphate uptake by polyphosphate-accumulating microorganisms during the aerobic phase [24], which may explain enhanced P removal with increased hydraulic load.

Both C and N removal under field conditions were more similar to values observed for sand-filled mesocosms, whereas values for P removal were closer to those observed for soil-filled mesocosms (Figure 6). This difference may be explained by differences in removal mechanisms for these three constituents. Both C and N removal—assuming it takes place primarily via denitrification—involve gaseous losses (as CO_2, N_2, and N_2O) and do not require reaction with soil minerals. By contrast, net removal of P from STE likely involves the reaction of PO_4 (released during the anaerobic phase of enhanced biological phosphorus removal) with soil minerals, which were absent in the synthetic sand used in mesocosm studies [8].

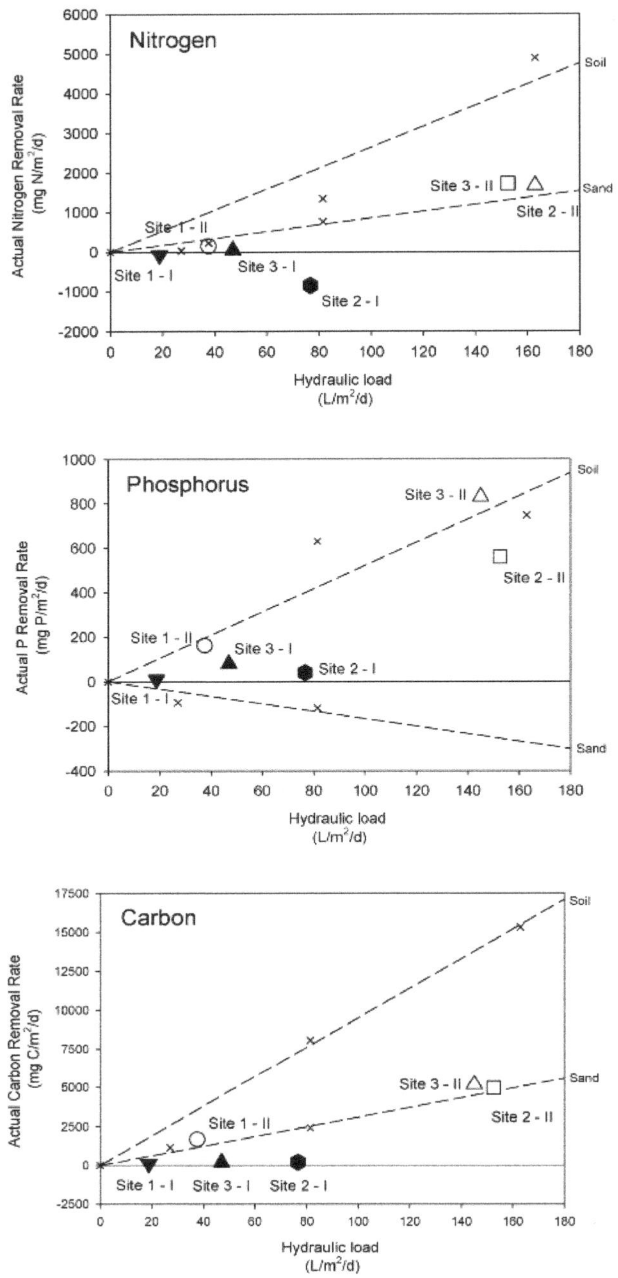

Figure 6: Relationship between hydraulic load and actual removal rates for nitrogen, carbon and phosphorus at position C3 (downstream from leachfield trench; Figure 1) for the three field sites evaluated in the present study and for mesocosms containing 30 cm of sand [8] or soil [14]. Dashed lines represent linear regressions for soil and sand mesocosms data. Solid symbols represent values for Phase I (prior to intermittent soil aeration); open symbols represent values for Phase II (during intermittent soil aeration).

CONCLUSIONS

Our results suggest that intermittent aeration of STAs can enhance the hydraulic and water quality improvement functions of hydraulically-failed conventional OWTS. Intermittent soil aeration improves the infiltrative capacity of these systems, with normal hydraulic function observed despite reduction in infiltrative surface area and concomitant higher hydraulic load. Apparent removal rates for N, P, and DOC were comparable to or higher than those observed prior to intermittent aeration. Furthermore, these rates were achieved even though the potential infiltrative surface area was reduced between 33 and 50%. When differences in infiltrative surface area and dilution are taken into account, actual rates of removal for N, P, and DOC increase markedly in conventional STAs.

ACKNOWLEDGEMENTS

This study was funded in part by a grant from the NOAA/UNH Cooperative Institute for Coastal and Estuarine Environmental Technology (CICEET) and by the Rhode Island Agricultural Experiment Station. We thank Tim Leviness for field assistance. The cooperation of Steven Denoyelle at the Rhode Island Department of Mental Health, Retardation and Hospitals and Brian Moore at the Rhode Island Department of Environmental Management is gratefully acknowledged. This is contribution no. 5421 of the Rhode Island Agricultural Experiment Station.

REFERENCES

1. US Census Bureau. American Housing Survey for the United States: 1999, Available online: www.census.gov/jjes/www/housing/ahs/ahs99/tab1a4.html (accessed on 29 August 2010).
2. Onsite Wastewater Treatment Systems Manual; USEPA (US Environmental Protection Agency): Washington, DC, USA, 2002.
3. Reneau, R.B., Jr.; Hagedorn, C.; Degen, M.J. Fate and transport of biological and inorganic contaminants from on-site disposal of domestic wastewater. J. Environ. Qual. **1989**, 18, 135–144.
4. Analyzing Wastewater Treatment Systems; CIDWT (Consortium of Institutes for Decentralized Wastewater Treatment): North Carolina State University: Raleigh, NC, USA, 2008; p. 228.
5. Kaplan, O.B. Septic Systems Handbook; Lewis Publishers: Chelsea, MI, USA, 1987.
6. Residential Onsite Wastewater Treatment Systems: An Operation and Maintenance Service Provider Program; CIDWT: North Carolina State University: Raleigh, NC, USA, 2005; p. 272.
7. Converse, J.C.; Tyler, E.J. Renovating Failed Septic Tank-soil Absorption Systems Using Aerated Pretreated Effluent. In Proceedings of the Seventh International Symposium on Individual and Small Community Sewage Systems; ASABE (Amercian Society of Agricultural and Biological Engineers): St. Joseph, MI, USA, 1994; pp. 416–423.
8. Potts, D.A.; Görres, J.H.; Nicosia, E.L.; Amador, J.A. Effects of aeration on water quality from septic system leachfields. J. Environ. Qual. **2004**, 33, 1828–1838.
9. Soil Survey Staff. Soil Survey of Rhode Island; US Government Printing Office: Washington, DC, USA, 1981.
10. American Public Health Association. Standard Methods for the Examination of Water and Wastewater, 20th ed.; APHA (American Public Health Association): Washington, DC, USA, 1998.
11. Whelan, B.R.; Titamnis, Z.V. Daily chemical variability

of domestic septic tank effluent. Water Air Soil Pollut. **1982**, 17, 131–139.

12. Andreoli, A.; Bartilucci, N.; Forgione, R.; Reynolds, R.L. Nitrogen removal in a subsurface disposal system. J. Water Pollut. Contr. Fed. **1979**, 51, 841–854.

13. Bunnell, J.F.; Zampella, R.A.; Morgan, M.D.; Gray, D.M. A comparison of nitrogen removal by subsurface pressure dosing and standard septic systems in sandy soils. J. Environ. Manage. **1999**, 56, 209–219.

14. Amador, J.A.; Loomis, G.W.; Kalen, D.V.; Patenaude, E.L.; Gorres, J.H.; Potts, D.A. Evaluation of Leachfield Aeration Technology for Improvement of Water Quality and Hydraulic Functions in Onsite Wastewater Treatment Systems. Report to the Cooperative Institute for Coastal and Estuarine Environmental Technology (CICEET); NOAA (National Oceanic and Atmopsheric Administration)/University of New Hampshire: Durham, NH, USA, 2007.

15. Amador, J.A.; Potts, D.A.; Patenaude, E.L.; Gorres, J.H. Effects of depth on domestic wastewater renovation in intermittently aerated leachfield mesocosms. ASCE J. Hydrol. Eng. **2008**, 13, 729–734.

16. Gardner, T.; Geary, P.; Gordon, I. Ecological sustainability and on-site effluent treatment systems. Aust. J. Environ. Manage. **1997**, 4, 144–156.

17. Beal, C.; Gardner, E.; Menzies, N. Process, performance, and pollution potential: A review of septic tank-soil absorption systems. Aust. J. Soil Res. **2005**, 43, 781–802.

18. Costa, J.E.; Heufelder, G.; Foss, S.; Milham, N.P.; Howes, B. Nitrogen removal efficiencies of three alternative septic system technologies and a conventional septic system. Environment Cape Cod **2002**, 5, 15–24.

19. van Driel, P.W.; Robertson, W.D.; Merkley, L.C. Denitrification of agricultural drainage using wood-based reactors.Trans. ASABE **2006**, 49, 565–573.

20. Hartel, P.G. The Soil Habitat. In Principles and Applications of Soil Microbiology; Sylvia, D.M., Fuhrmann, J.J., Hartel, P.G., Zuberer, D.A., Eds.; Pearson/Prentice Hall: Upper Saddle River, NJ, USA, 2005; pp. 26–53.

21. Fuhrmann, J.J. Microbial Metabolism. In Principles and Applications of Soil Microbiology; Sylvia, D.M., Fuhrmann, J.J., Hartel, P.G., Zuberer, D.A., Eds.; Pearson/Prentice Hall: Upper Saddle River, NJ, USA, 2005.

22. Irvine, R.L.; Ketchum, L.H.; Asano, T. Sequencing batch reactors for biological wastewater treatment. Crit. Rev. Environ. Contr. **1989**, 18, 255–294.

23. Münch, E.V.; Lant, P.; Keller, J. Simultaneous nitrification and denitrification in bench-scale sequencing batch reactors.Water Res. **1996**, 30, 277–284.

24. Manning, J.F.; Irvine, R.L. The biological removal of phosphorus in a sequencing batch reactor. J. Water Poll. Contr. Fed. **1985**, 57, 87–94.

CHAPTER 9

DETERMINATION OF WATERSHED INFILTRATION AND EROSION PARAMETERS FROM FIELD RAINFALL SIMULATION ANALYSES

Mark E. Grismer

Keywords: Rainfall-runoff modeling, rainfall simulations, effective hydraulic conductivity, infiltration, laminar or turbulent flows, erodibility

INTRODUCTION

Modeling watershed runoff and erosion processes realistically requires field determination or model calibration estimation of surface soil infiltration rates or effective hydraulic conductivities (K_m) and erosion rates because they vary with soil cover/tilth/slope conditions, and seasonally with changing water contents. The field methods often used to measure in situ saturated or effective hydraulic conductivities (K_s or K_m, respectively, where K_m is some fraction of K_s) include surface (e.g., disk, single/double-ring infiltrometers, and rainfall simulators), or subsurface techniques (e.g., bore-hole methods). Indirect estimates of K_m and erodibilities available from NRCS soil survey information that are typically derived from soil texture information are also more often used in modeling than field measured values due to the difficulty associated with the field measurements and their variability. While each measurement method may have particular advantages depending on the intended use of the data, surface methods enable measurement of actual conditions as affected by soil tilth and surface cover while also providing insights into the effectiveness of soil restoration methods in the field. However, they are not without complications associated with surface disturbance/roughness, type and configuration of surface cover, steep slope (double-ring methods require mild slopes), disk plate hydraulic contact and estimation from rainfall minus runoff rates using rainfall simulators. Similarly, subsurface measurement techniques or K_m estimation from soil texture information may miss the effect of surface layers on infiltration storage or excess leading to runoff–erosion prediction failure.

Here, we focus on determination of K_m and erodibilities from rainfall simulation (RS) methods in the field as these are not limited by surface conditions, slope or soil type. Following development of an integrated infiltration and time-to-ponding/runoff equation, we use infiltration and runoff data from over 425 RS plots in the Tahoe Basin, together with simultaneous measurements of saturated hydraulic conductivity, K_s, using bore-hole or mini-disc infiltrometers at selected sites to determine what estimation method of K_m yields the most consistent results. Then we consider the erosion process and soil detachment data from disturbed skirun soils to investigate the relationship between soil detachment and stream power to determine what form (laminar or turbulent) of the stream power function is most appropriate to describe overland sheet flows typical in RS test and forest soils. Finally, we consider test plot results from three different rainfall simulators used at similar locations to demonstrate application of the equations for determination of K_m and erodibilities associated with RS

Citation: Mark E. Grismer, Determination of Watershed Infiltration and Erosion Parameters from Field Rainfall Simulation Analyses, Hydrology 2016, 3(3), 23; doi:10.3390/hydrology3030023.

tests having a range of raindrop impact energies.

Brief Literature Review

Rainfall simulators (RSs) are essential tools for investigating the dynamic processes of infiltration, runoff and erosion under a variety of field conditions [1,2,3,4], and information from RS test plots can provide the infiltration/runoff parameterization required for watershed modeling. Though multiple RS designs for field application exist, no single RS design (including plot runoff frame installation) has emerged as a standard. Similarly, while in RS plot tests, typically, data about rainfall, runoff and erosion rates and time to runoff are collected from which infiltration rates, or K_m and erodibilities are estimated, no standard RS data analysis methodology has evolved. Despite a number of years of research into the plant/cover effects on soil erosion [5,6,7,8,9,10,11,12], it remains difficult to understand erosion processes and mechanics due to the lack of sufficiently comparable data or results [8,13]. Moreover, the differences in soil properties, slope surface conditions and vegetation types in field experiments have tended to complicate interpretations of field measurements. It is difficult, therefore, to draw meaningful comparisons between RS plot data reported in different studies [2,3,4,14,15]. In addition, there are few, if any, actual comparisons of RS performance with respect to infiltration, erosion, or soil detachment rates. For example, Lascelles et al. [16] only considered the variability in drop sizes, distributions and energies from two different RSs and speculated on implications with respect to erosion, but did not offer comparison RS plot studies. Similarly, Kinnell [17,18] completed thorough reviews of the processes associated with raindrop impacted erosion and noted that both conceptual models and measurements fail in various respects to adequately characterize field observed erosion processes from bare soils. Concerns such as these have also arisen in the Tahoe Basin, because a variety of methods for measurement of infiltration and erosion rates have been deployed, but comparisons between results of different studies remain less than definitive.

Project Hypothesis & Objectives

Data from RS field plots can be used to assess the initiation time of runoff, an indication of surface conditions (though, admittedly, from perhaps unusually large rainfall rates, or associated durations), the subsurface infiltration characteristics through determination of the infiltrated volume after a RS test time, as well as soil detachment rates or "erodibilities". In order to develop more widely usable information/data, standardized approaches to the RS plot setups and the subsequent data analyses are needed [2]; here we focus on the latter, using RS test data from 425 plots across the Lake Tahoe Basin. We hypothesize that there exists a physically consistent set of equations that relate infiltration and erosion data collected from different RS test plots to surface soil effective hydraulic conductivity and erodibility. The primary project objectives include: (a) develop the appropriate infiltration and erosion equations applicable to field RS plots, (b) evaluate the applicability of these equations to a large RS data set, and (c) use these equations to compare RS test plot results from three different field rainfall simulators having a range of rain drop energies and rainfall configurations to both demonstrate their applicability and determine whether differences in raindrop impact energy result in significantly different K_m values or erodibilities for the Tahoe Basin soils.

Theory—Infiltration Equation Development

Several infiltration equations have been derived during the past few decades and the most often used in watershed modeling revolve around time-to-ponding estimates and the ponded infiltration Green–Ampt type square-wave wetting-front formulation, among others. In RS plot studies, the applied rainfall rates are generally large and for durations that exceed that of natural rainfall so as to achieve runoff after some acceptable elapsed time after rainfall initiation. As such, there are typically a few minutes during which the plot infiltration capacity exceeds that of the applied rainfall followed by shallow ponded infiltration. Ideally, the effective hydraulic conductivity that satisfies both the time-to-ponding and ponded infiltration equations would represent the self-consistent field value that then should apply in watershed modeling. We briefly define terms and develop these equations here, eventually coupling the ponding time and Green–Ampt–type equations to allow implicit determination of K_m from the RS data outlined above.

Basic definitions from Grismer [19] include:

$$h_e = \frac{h_d}{h_c} \tag{1}$$

where h_d = displacement pressure head (mm), and h_c = capillary pressure head (mm)

$$S_e = \frac{S - S_r}{S_m - S_r} \cong h_e^{\lambda} \tag{2}$$

where S = saturation, S_r = residual saturation, S_m = maximum saturation associated with infiltration, and λ = pore-size distribution index.

And the Green–Ampt infiltration equation takes the form

$$I = K_m(H + h_f + z_f)/z_f \tag{3}$$

where I = infiltration rate (mm/hr),

- K_m = "natural saturated hydraulic conductivity" (mm/hr) ≈ $0.5K_s$,
- H= ponding depth (mm),
- h_f = wetting front capillary pressure head (mm), and
- z_f = wetting front depth (mm).

Combining the Green–Ampt Equation (3) with the continuity equation yields

$$t = \frac{1}{K_m}\left\{V - \phi(S_m - S_i)\left(H + h_f\right)\ln\left[1 + \frac{V}{\varphi(S_m - S_i)\left(H + h_f\right)}\right]\right\} \tag{4}$$

where V = volume infiltrated per unit area (mm).

The Green–Ampt equation above is readily modified to account for air resistance to wetting flow by using β to replace the 1 at the beginning of the RHS of Equation (4) and h_f within the parentheses with H_c [19]. β and H_c are defined as follows

$$\beta = \frac{1}{\mu_w}\int_1^{S_e} \frac{d^2 f_w/dS_e^2}{\left[(k_{rg}/\mu_g) + (k_{rw}/\mu_w)\right]}\, dS_e \tag{5}$$

where μ_w and μ_g are the water and air viscosities, respectively, similarly k_{rw} and k_{rg} are the relative permeabilities of the porous media to water and air, respectively. S_e is the effective saturation and f_w is the fractional flow function of water, or the ratio of water flow to total (air and water) flowrate. For all practical purposes, β takes on a value between 1.2 and 1.3 for typical λ values between 2 and 3, an index range that covers loamy to sandy soils [19]. The wetting front driving head, H_c, can be determined from

$$H_c = h_d \int_1^{h_c} \frac{f_w}{h_c^2}\, dh_c \tag{6}$$

where h_e is defined in Equation (1). Again for all practical purposes, H_c takes on a value between 1.12 h_d – 1.08 h_d for λ between 2 and 3. Substituting Equations (5) and (6) in the Green–Ampt Equation (4) to correct for air resistance or counterflow during infiltration results in [19]

$$t = \frac{\beta}{K_m}\left\{V - \varphi(S_m - S_i)(H + H_c)\ln\left[1 + \frac{V}{\varphi(S_m - S_i)(H + H_c)}\right]\right\} \tag{7}$$

Finally, the corresponding 'time–to-ponding' infiltration equation that accounts for air resistance or counterflow as well is given by

$$t_p = \frac{\varphi(S_m - S_i)}{(1 - f_w)q_o} H_c \exp\left[\left(\frac{q_o\beta}{K_m} - 1\right)^{-1} - 1\right]\Big\} \tag{8}$$

where q_o is the rainfall rate and f_w is practically zero for ponded infiltration (see Morel-Seytoux [20] discussion linking the transition of rainfall rate controlled infiltration conditions to that for ponded infiltration).

With a relationship between h_d and K_m together with the time-to-ponding, infiltrated depth and times from RS plot data, Equations (7) and (8) can be solved simultaneously for K_m (assuming that K_m = 0.5 K_s) by solving Equation (8) for $\varphi(S_m - S_i)$ and letting f_{wi} = 0. Using lab column data collected to date, Grismer [19] determined that for h_d in meters, the semi-empirical theoretical relationship for permeability was k (μm²) ≈ $0.84/h_d^2$. Using only that lab data for sands that more closely replicate the Tahoe soils and changing to more directly applicable units, we found that K_m (mm/hr) = $19.9/h_d^{2.4}$ for h_d (mm), as shown in Figure 1. And as noted above, taking $H_c \cong 1.1\, h_d$, β ≅ 1.3 for λ = 3, and the ponded depth H of Equations (4) or (7) as typically assumed to be small (e.g., 1 mm as in watershed modeling, [21]), we can implicitly solve for K_m from Equations (7) and (8) by re-arranging Equation (8) to solve for $\Delta\theta = \varphi(S_m - S_i)$, as shown below as Equation (9) and substituted into Equation (7).

$$\Delta\theta = \frac{(q_o t_p/H_c)}{\exp\left[\left(\frac{\beta q_o}{K_m} - 1\right)^{-1} - 1\right]} \tag{9}$$

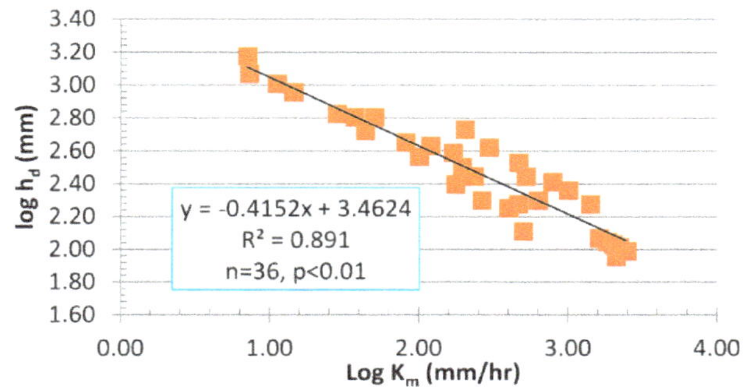

Figure 1: Relationship between field-saturated hydraulic conductivity, K_m, and displacement pressure head, h_d, for fine sands (data taken from Grismer, [19]).

Alternatively, we can use the simpler Main–Larson equations analogous to Equations (7) and (9) below, but still accounting for air-resistance to infiltration

$$t_p = \frac{\varphi(S_m - S_i)H_c}{q_o\left(\frac{q_o}{K_m} - 1\right)} \tag{10}$$

and

$$\Delta\theta = \left(\frac{q_o t_p}{H_c}\right)\left(\frac{q_o}{K_m} - 1\right) \tag{11}$$

Finally, we can use the estimated effective Δθ, taken as the measured infiltrated depth compared to the visual wetting front depth, from RS plot data when no runoff occurs, combined with the infiltrated depth and RS test duration to determine K_m from Equation (7). For example, Figure 2 shows the relationship between visual wetting and infiltrated depths from the long-term monitoring site at Heavenly ski area on the coarser-textured granitic soils where the average regression slope suggests an effective Δθ ≅ 45%.

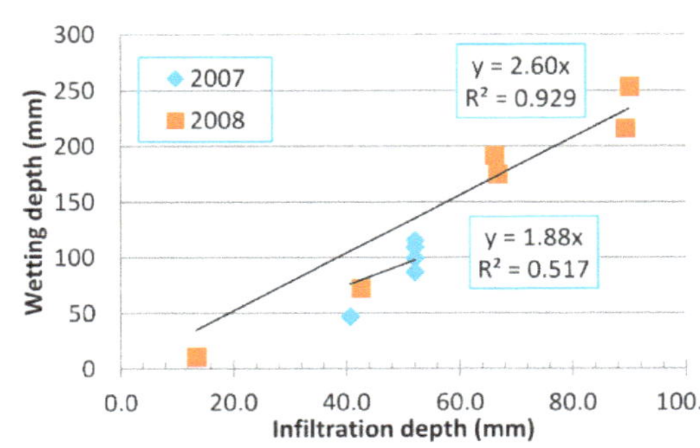

Figure 2: RS (rainfall simulation) plot infiltrated vs. visible wetting depths at Heavenly LT site (granitic soils).

Theory—Erosion Equation Development

Most watershed modeling efforts and associated estimation of sediment detachment or erosion rates employ the well-known Manning's equation to relate overland or channel flowrates and velocity to flow depth and hillslope, or channel gradient. Of course, use of Manning's equation implies that an appropriate surface roughness value, "n", can be identified, that the driving force, or slope angle <10% and that flows are fully turbulent. The assumption of turbulent flow conditions during sheet flow across the RS plot or the landscape is questionable when flow depths are quite small and laminar flows are more likely present [21]. The general derivation of the laminar flow equation for thin films on inclined planes at any angle, α, to the horizontal requires only the assumptions of the "no-slip" boundary condition together with constant fluid properties [22]. Considering two-dimensional steady laminar flow of depth, y, down an inclined plane at angle, α, to the horizontal, the shear force as given by the fluid viscosity, μ, and the parabolic velocity function is balanced by the gravitational force (unit weight, ρg) on the fluid body in the direction of flow. That is, the flowrate Q per unit width is given by

$$Q = u_m y = (\rho g y^3/3\mu)\sin\alpha \tag{12}$$

where u_m is the mean velocity taken as 2/3rds of the maximum velocity assuming a parabolic velocity distribution perpendicular to the inclined plane. Of course, the equivalent flowrate equation assuming turbulent flow and mild slopes (<10%) is taken from the approximation for the Chezy–Mannings equation assuming flow depths are very small compared to flow width, B,

$$Q = (1/n)\, y^{1.67} S^{0.5} \tag{13}$$

The flow mean velocity needed to determine shear stresses on soil particles is given by Q/y from Equations (12) or (13), depending on whether laminar or turbulent flow is assumed. Replacing the sine function with a power function approximation in Equation (14), it is apparent that under steady laminar flow conditions the mean surface flow velocity is practically proportional to the slope, S, rather than the square root of the slope, as in the mean velocity determined from Manning's Equation (15); moreover, there is no slope limitation or need to identify the roughness value, "n".

$$u_m = (\rho g y^2/3\mu)\sin\alpha \cong (0.7524\,\rho g y^2/3\mu) S^{0.983} \tag{14}$$

$$u_m = (1/n)\, y^{0.67} S^{0.5} \tag{15}$$

In RS plot studies, the runoff rate, q, is often expressed as the steady flowrate collected at the runoff lip frame divided by the area of the plot, as this allows convenient comparison to the infiltration and rainfall rates, and because rarely is the actual sheet flow depth measured or estimated so as to enable calculation of the mean, or average, overland flow velocity, u_m. The relationship between q and u_m is simply given by the mass balance

$$q = Byu_m/A, \text{ or } u_m = qA/(By) \tag{16}$$

where A = runoff frame area (m²), B = the runoff lip width (m) and the other parameters are as previously defined. Solving Equation (16) for the mean velocity then requires substitution for the flow depth based on either Equation (15) for turbulent flow, or Equation (14) for laminar flow, as shown below, respectively, in Equations (17) and (18).

$$u_m = (qA/B)^{0.4} S^{0.3}/n^{0.6} = \left(\frac{A}{Bn^{1.5}}\right) q^{0.4} s^{0.3} \tag{17}$$

$$u_m = (qA/B)^{0.67} (0.7524\rho g/3\mu)^{0.33} S^{0.328} = \left(\frac{A}{B}\right)^{0.67} \left(\frac{0.25\rho g}{\mu}\right)^{0.33} q^{0.67} S^{0.328} \tag{18}$$

Quantitative description of soil particle detachment or erosion processes perhaps originated with Ellison's [23] observation that "erosion is a process of detachment and transport of soil materials by erosive agents". These "erosive agents", of course, include raindrop impact and overland flow. Of course, rain drop impact energy diminishes with increasing flow depth and presence of soil cover that both "absorb" the drop impact energy. Raindrop impact energy, while an erosive agent on bare soils should probably be related to the aggregate strength, that is, the energy needed to break up and mobilize surface aggregates [24,25]. Subsequent research, more or less, begins with Ellison's paradigm of sorts that continues in concept through the soil-detachment equation review by Owoputi and Stolte [26]. Similarly, Foster and Meyer [27] interpreted results of several experiments in terms of Yalin's equation that assumes "sediment motion begins when the lift force of flow exceeds a critical force … necessary to … carry the particle downstream until the particle weight forces it out the flow and back to the bed." Bridge and Dominic [28] built on this concept and described the critical velocities and shears needed for single particle transport over fixed rough planar beds. Gilley et al. [29,30,31] included the Darcy–Weisbach friction factor as a measure of the resistance to flow that was eventually adopted in the WEPP model. Moore and Birch [32] combined slope and velocity and suggested that particle transport and transport capacity for both sheet (interrill) and rill flows is best derived from the unit stream power. In all these descriptions, the erodibility process essentially involves momentum transfer with an energy loss to friction. More recently, the concept of stream power as the primary factor controlling soil detachment rates has been adopted in several reviews e.g., [33,34,35,36]. Assuming turbulent and laminar flow conditions, respectively, from Equations (17) and (18), stream power, ***P***, can be expressed as

$$P = \rho g u_m S = \left(\frac{A}{Bn^{1.5}}\right)^{0.4} \rho g q^{0.4} S^{1.30} \tag{19}$$

and

$$P = \left(\frac{A}{B}\right)^{0.67} \left(\frac{0.251}{\mu}\right)^{0.33} (\rho g)^{1.33} q^{0.67} S^{1.328} \tag{20}$$

Note that in both Equations (19) and (20), slope, S, has a practically the same effect on stream power, hence detachment rate, whether the flow is laminar or turbulent; that is, ***P*** is proportional to $S^{\sim 1.3}$. However, the runoff rate under laminar flow conditions has a much greater affect than that under turbulent flow conditions (i.e., ***P*** is proportional to $q^{0.67}$ as compared to $q^{0.4}$), as do the unit weight and plot dimensions (though offset to some degree by 'n'). Nonetheless, in terms of practical analyses of RS plot runoff and erosion data, when rain splash impacts are negligible, it is apparent that the soil particle detachment rate is proportional to $\sim S^{1.3}$ and qa, where 'a' takes on a value between 0.4 and 0.7.

Experimentally, the dependence of stream power on slope between laminar and turbulent flow is not well articulated. In fact, at slopes of 4%–12%, McCool et al. [37] found soil loss rates dependent on $S^{1.37}$–$S^{1.5}$, rather than $S^{\sim 1.3}$. In flume studies,

Zhang et al. [38] found across a slope range of 3%–47% their detachment data was proportional to $q^{1.04}$ $S^{1.27}$ suggesting that both Equations (19) and (20) may underestimate the effects of runoff rate. At small slopes, detachment rate was more sensitive to q than S, however, as S increased, its influence on detachment rate increased. Later, Zhang et al. [39] found that for undisturbed "natural" soils across a similar slope range (9%–47%), detachment rates were most proportional to $q^{0.89}S^{1.02}$. In contrast, on nearly flat slopes (1%–2%) with deep flow depths (~10 mm), Nearing and Parker [40] found that turbulent flow resulted in far greater soil detachment rates than did laminar flow, in part as a result of greater shear stresses. Following Gilley and Finkner [29], Guy et al. [41] examined the effects of raindrop impact on interrill sediment transport capacity in flume studies at 9%–20% slopes. Assuming a laminar flow regime, they found that raindrop splash accounted for ~85% of the transport capacity, in some contrast to earlier studies indicating that raindrop impact had little or no effect on slopes greater than about 10%. Sharma et al. [42,43,44] systematically examined rain splash effects on aggregate breakdown and particle transport in the laboratory but did not relate their results to stream power. At larger slopes, Lei et al. [45] found that both slope and runoff rate were important towards transport capacity on slopes up to about 44%, but that transport capacity increased only slightly at still steeper slopes. Zhang et al. [39] found the best linear regression quantifying soil detachment rates occurred for equations that included the square of the rainfall rate (I^2) times the WEPP slope factor, or I times the runoff rate and slope, S, as compared to I times the square root of the runoff rate times $S^{0.67}$. This observation of a better fit with the first two equations suggested that detachment rate is proportional to stream power. Similarly, considering soil detachment from overland flow only across a range of burned, disturbed and relatively undisturbed rangeland soils with slopes of 6%–57%, Al-Hamdan et al. [35] found that detachment rates were proportional to $\boldsymbol{P}^{1.18}$, but that the exponent of 1.18 did not differ significantly from unity. Gabriels [34] found that detachment rates were related to $\boldsymbol{P}^{1.3}$ for a range of clay fractions of 7%–41%, an exponent value consistent with that for the plot slope in Equations (19) and (20). In the RS plot studies considered here, we assume that detachment rates are proportional to $S^{1.3}$ and then determine the optimal exponent applicable to the runoff rate.

EXPERIMENTAL METHODS—INFILTRATION-RUNOFF AND HYDRAULIC CONDUCTIVITY MEASUREMENTS IN THE FIELD

In the analyses here, we use field data collected from 423 RS plots and associated measurements of hydraulic conductivity for which overviews of the methodologies and some results were described elsewhere [46,47,48,49,50,51]. While most of the Lake Tahoe Basin (See Figure 3) soils are of andesitic or granitic origin and of sandy textures from soil surveys (Table 1), we have found that the surface soils can be broadly classified as coarser-textured "granitics" and finer-textured "volcanics" (Table 2). In addition to the 1 m and 3 m tall drop-former (DF) type RSs used here, we also include information from the USDA-FS sprinkler RS described by Foltz et al. [52]. Table 3 summarizes the field methods and associated references used here to develop the data set considered in the infiltration and soil detachment analyses. The MDI and Precision permeameter methods both involved repeated measurements (5–10 times per hole, or MDI test) and the bore-holes used varied from 150 to 300 mm deep and 70 mm in diameter, with a static water depth of 100–120 mm.

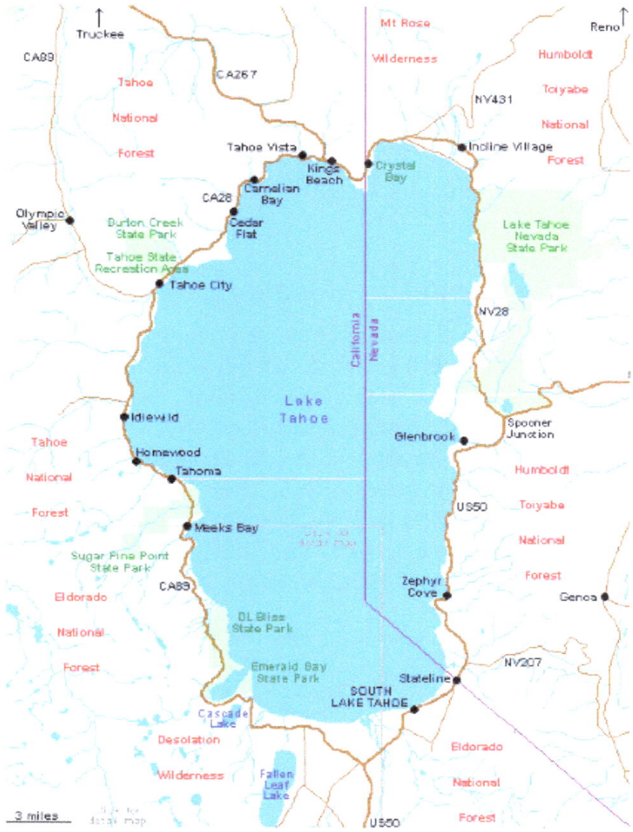

Figure 3: Lake Tahoe basin map (http://www.americansouthwest.net).

Table 1: Typical Tahoe soils encountered in the field measurements [53].

RS Sites	Soil Series	Taxonomic Classification	Surface Texture	pH	Conductivity (mm/h)
Bliss SP & Rubicon—granitic	Meeks	Sandy-skeletal, mixed, frigid Humic Dystroxerepts	very stony loamy coarse sand	6.1–6.5	16–51
Blackwood Cyn—volcanic	Waca	Medial-skeletal, amorphic, frigid Humic Vitrixerands	cobbly coarse sandy loam	5.6–6.5	5.1–16
Brockway & Dollar Hill—volcanic	Jorge-Tahoma	Fine-loamy, isotic, frigid Ultic or amorphic, frigid Ultic Haploxeralfs	very stony sandy loam	5.1–6.0	5.1–16

Table 2: Summary of sieved particle-size distribution measurements for the Tahoe Basin disturbed soils (>63 μm size fraction) and estimated K_{sat} values. From Grismer and Hogan [46].

Soil Type	n	D_{10} (μm)	D_{30} (μm)	D_{60} (μm)	[a] C_u	[b] K_{sat} (mm/hr)
Granitic—mean	33	117.1	322.5	946.4	8.23	332
Std. Deviation ([c] CV %)	33	20.4 (17.4)	73.9 (22.9)	208 (22.0)	1.96 (23.8)	116 (34.8)
Volcanic—mean	56	100	278	1320	13.6	248
Std. Deviation (CV %)	56	23.2 (23.2)	120 (43.3)	568 (48.5)	6.58 (53.1)	125 (50.4)

[a] C_u is the Coefficient of Uniformity = D_{60}/D_{10}; in geotechnical engineering, $C_u > 4$ indicates "well-graded" (i.e., broad range) of particles sizes; [b] K_{sat} = (Constant)xD_{10}^2 from Harleman et al. [53]; [c] *CV* = Standard Deviation/Mean.

Table 3: Summary of field measurement methodologies used to develop the datasets considered in the infiltration, soil detachment and RS comparisons analyses.

Device/Method	Description	Parameters Measured	Reference
1 m Tall Drop-Former RS (~21% of terminal velocity raindrops)	1 m^2 Needle tank RS raining on 0.8 × 0.8 m^2 plot	Rain rate, wetting depth, soil moisture, time-to-ponding, runoff rate, runoff particle size distributions and sediment & OM concentrations	Grismer & Hogan [46]
3 m Tall Drop-Former RS (~70% of terminal velocity raindrops)	1 m^2 Needle tank RS raining on 0.8 × 0.8 m^2 plot	Rain rate, surface roughness, time-to-ponding, runoff rate, runoff sediment concentrations	Battany & Grismer [14,15] Grismer & Hogan [46]
Modified Purdue sprinkler RS (USDA-FS)	Single oscillating VeeJet nozzle RS raining on 1 m^2 plot	Rain rate, time-to-ponding, runoff rate, runoff sediment concentrations	Foltz et al. [52]
Decagon Mini-Disk Infiltrometer (MDI)	Surface infiltrometer with ceramic disk	Saturated conductivity (K_s) with & without surfactants	Robichaud et al. [54] Rice & Grismer [51]
Precision Permeameter	Bore-hole method using mariotte siphon on graduated cylinder	Saturated conductivity (K_s) of near surface soils	Johnson [55] Glover solution in Zanger [56]

RESULTS

Infiltration–Runoff Field Data

Table 4 and Table 5 summarize the average RS plot characteristics for the granitic plots that both resulted in runoff, or not during the 30–60 min RS tests in the Tahoe Basin, respectively. Similarly, Table 6 and Table 7 summarize the average RS plot characteristics for the volcanic plots that both resulted in runoff, or not during the RS test, respectively. Table 4 and Table 5also distinguish between non-hydrophobic and hydrophobic RS plot tests within the granitic soils. These data provide a unique opportunity to evaluate application of the combined Infiltration Equations (7) and (8) across a broad range of plot characteristics, such as initial soil moistures ranging from 2% to 12%, rainfall rates from 60–120 mm/h, RS test durations of 10–70 min, plot slopes from 6% to 72% and ponding times from 1–20 min. Also included in this analysis, were RS test plots having clearly identified hydrophobic surface soils; these plots always resulted in quite large Δθ values. In most cases, the infiltrated depths were obtained at the end of the RS test associated with the infiltrated time to be used in Equation (7).

Table 4: Summary of RS plot setting data averages for granitic soil plots generating runoff.

Granite Soil Site Descriptions—Non-Hydrophobic Plots	n	Rain Rates (mm/h)	Initial Moisture (%)	Average Slope (%)	Mean Time to Runoff (min)	Infilt Depth (mm)	Infilt Time (min)	Mean Δθ	St.Dev. Δθ
Bliss SP RC—bare	4	60	ND	64.3	3.5	13.68	15	ND	NA
Bliss SP RC—treated	3	60	ND	57.5	14	14.8	15	ND	NA
CalTrans BS Plots	3	72–120	ND	30.2	10.9	28.9	18.0	ND	NA
Cave Rock/Incline Village RC	10	60	2.3	59.2	7.0	35.7	37.8	0.20	0.13
Heavenly CAN LT—2005	4	72–120	2.3	16.9	5.3	45.3	30.0	ND	NA
Heavenly CAN LT—2006	1	120	ND	17.1	5.7	25.8	30.0	0.49	NA
Heavenly CAN LT—2007	6	72–120	ND	18.1	2.9	61.3	30.0	0.42	0.10
Heavenly GB MSR & OLB	17	120	4.0	25.3	8.1	69.9	36.2	0.36	0.16
Heavenly LRR	6	120	4.2	25.8	6.6	90.7	46.5	ND	NA
Heavenly LT—2006	3	72–90	2.4	22.9	3.2	37.5	28.3	ND	NA
Heavenly PON/MCD	5	72–120	4.2	23.5	4.6	27.3	23.0	0.41	0.16
Heavenly STC	5	120	4.4	35.8	3.4	39.5	21.5	0.93	0.23
Luther Pass RC	12	60	ND	51.5	2.64	12.4	15	ND	NA
Mammoth	7	120	4.8	34.9	7.1	50.7	27.7	0.30	0.14
Meyers RC	7	72–120	2.3	30.0	3.9	81.8	22.0	ND	NA
Roundhill landing	5	65–70	1.0	6.2	6.9	33.6	52.9	0.38	0.14
Roundhill landing—USFS	4	86	1.0	15.6	3.0	26.2	60.0	ND	NA
Rubicon RC, treated & native	12	60	ND	49.3	10.3	14.2	15.0	ND	NA
Truckee Bypass	6	72–120	1.0	26.2	16.3	44.5	31.1	0.37	0.09
Upper Cutthroat 1D & 2T	7	72–90	2.7	6.0	6.5	30.8	22.6	0.28	0.04
Upper Cutthroat 3T	4	120	2.7	5.1	17.7	89.9	46.1	0.54	0.33
Upper Cutthroat 4T & 6T	4	120	6.6	9.8	19.4	78.7	43.7	0.63	0.34
Upper Cutthroat 7T	3	120	2.0	7.3	5.1	36.9	25.9	0.37	0.09
Upper Cutthroat 8T	6	120	6.3	12.7	4.8	33.8	21.5	0.23	0.09
Angora Fire	8	72	ND	17.6	8.2	25.5	20.3	1.19	0.35
Bliss SP & Meyers RC—water	6	120	2.4	12.7	2.5	91.0	60.0	6.23	4.32
Heavenly LT—Natives—2006	3	72	2.2	27.3	1.4	6.8	8.6	ND	NA
Upper Cut. 10T & 11T	3	120	4.4	6.4	14.3	110.0	55.8	1.28	0.69
Upper Cut. 3N—natives	4	72–120	4.8	4.8	1.3	9.9	22.2	1.12	0.70

ND = No Data and NA = Not Applicable.

Table 5: Summary of RS plot setting data averages for granitic soil plots not generating runoff.

	n	Rain Rates (mm/h)	Initial Moisture (%)	Avg. Slope (%)	Infilt Depth (mm)	Infilt Time (min)	Mean Δθ	St.Dev. Δθ
Granite Soil Site Descriptions—Non-Hydrophobic Plots								
CalTrans OP Plots	3	72–120	ND	37.4	52.3	30.0	ND	NA
Cave Rock–Inc. Village RCs	4	60	1.29	53.1	30.0	30.0	0.59	NA
Heavenly CAN LT	10	72–120	1.70	14.9	82.4	39.3	ND	NA
Heavenly CAN LT	13	72–120	ND	17.9	102.9	51.4	0.43	0.11
Heavenly GB MSR & OLB	1	120	3.56	34.7	92.0	46.0	ND	NA
Heavenly STC	1	120	4.0	27.8	90.0	45.0	0.39	NA
Mammoth	2	120	4.39	38.4	90.0	45.0	0.59	0.08
Meyers RC	4	90–120	2.5	28.7	48.8	30.0	ND	NA
Roundhill landing	1	55	1.0	6.7	50.4	55.0	0.25	NA
Truckee Bypass	9	72–120	2.7	28.5	57.3	30.0	0.27	0.07
Upper Cutthroat 3T	2	72	5.38	6.8	69.8	58.2	0.47	0.05
Upper Cutthroat 4T & 6T	1	120	5.6	5.9	90.0	45.0	0.80	NA
Upper Cutthroat 8T	4	120	5.77	9.6	105.6	58.6	0.53	0.28
Granite Soil Site Descriptions—Hydrophobic plots								
Bliss SP & Meyers RC—surfactant	6	120	2.60	12.6	120.0	60.0	0.82	0.30
Angora Fire	1	72	ND	21.3	72.0	60.0	1.29	NA
Upper Cut. 10T & 11T	6	120	5.05	7.5	76.8	38.4	1.50	0.59

ND = No Data and NA = Not Applicable.

Table 6: Summary of RS plot setting data averages for volcanic soil plots generating runoff.

Volcanic Soil Site Descriptions	*n*	Rain Rates (mm/h)	Initial Moisture (%)	Average Slope (%)	Mean Time to Runoff (min)	Infilt Depth (mm)	Infilt Time (min)	Mean Δθ	St.Dev. Δθ
Blackwood Canyon–Truckee (hydrophobic)	8	120	1.89	12.5	18.2	105.9	60.0	0.68	0.16
Brockway Summit—2006	1	72	9.50	42.1	18.2	35.0	30.8	0.40	NA
Brockway Summit—2003	6	60	ND	51.0	3.5	12.3	15	ND	NA
Dollar Hill—Tall RS	4	61	11.46	52.2	4.15	13.83	15	ND	NA
Homewood #31 Rd	4	72	5.5	18.5	4.8	25.3	22.7	0.24	0.03
Homewood LL A&B Rd	6	120	5.0	37.5	5.1	43.0	22.8	0.26	0.02
Homewood MAR Rd	9	120	3.00	30.7	3.6	33.5	18.8	0.28	0.13
Homewood SCD RD&BK Rd	9	120	5.3	28.5	5.0	43.8	20.4	0.32	0.09
Homewood WDG Rd	8	120	9.9	22.5	3.2	17.5	9.8	0.40	0.19
Northstar—small RS	2	60	4.75	49.2	8.0	24.2	26.5	0.25	NA
Juniper Mtn Skirun	5	60–75	4.20	36.4	4.4	15.4	15.8	0.47	0.07
Juniper Mtn Skirun—WCs	3	100–125	8.67	35.4	9.3	41.0	22.8	1.58	0.40
Northstar—Tall RS	7	60	7.6	53.3	5.6	21.1	21.6	0.20	0.04
Northstar—bare ski Tall RS	13	60	NA	35.8	5.2	14.1	16.8	0.23	0.06
Homewood—Tall RS	18	60	3.7	42.5	9.6	19.5	22.4	0.14	0.13
Northstar BP CON plots	2	120	3.95	19.4	17.4	79.6	40.8	0.36	NA
Northstar LT	12	72–120	2.9	34.4	8.5	35.6	23.8	ND	NA
Northstar STA plots	3	120	5.3	42.6	2.7	23.1	12.6	0.48	0.06
Northstar Unit 7	2	120	3.75	39.7	4.4	44.8	23.8	ND	NA
Ward Cr. Landing—IERS	4	70	10.0	7.4	6.4	38.3	44.2	0.49	0.15
Ward Cr. Landing—USFS	3	91	10.0	18.1	2.1	26.1	60.0	ND	NA

ND = No Data and NA = Not Applicable.

Table 7: Summary of RS plot setting data averages for volcanic soil plots not generating runoff.

Volcanic Soil Site Descriptions–	*n*	Rain Rates (mm/h)	Initial Moisture (%)	Mean Slope (%)	Infilt Depth (mm)	Infilt Time (min)	Mean Δθ	St.Dev. Δθ
Blackwood–Truckee (Surf)	4	120	1.70	12.5	120.0	60.0	0.62	0.21
Brockway Summit plots	5	120	2.21	45.6	51.2	26.7	0.34	0.01
Northstar—Short RS	5	60	6.9	53.2	30.8	29.5	0.48	0.18
Northstar—Tall RS	7	60	8.9	55.7	39.6	38.2	0.30	0.08
Homewood—Tall RS	2	60	ND	20.9	41.0	41.0	0.12	NA
Juniper Mtn Skirun	16	60–125	7.46	38.7	42.4	30.0	ND	NA
Northstar BTD & BP CON	4	120	2.61	19.6	90.0	45.0	0.34	0.03
Northstar LT	6	72–120	2.3	32.4	44.0	30.0	ND	NA
Northstar Unit 7—Short RS	9	75–120	3.81	40.6	54.3	30.0	ND	NA

ND = No Data and NA = Not Applicable.

Finally, Table 8 lists the results of the saturated hydraulic conductivity, K_s measurements completed using the mini-disk infiltrometer (MDI) and Precision Permeameter (PP) devices at the various RS sites around the Tahoe Basin. The two K_s measurement methods yielded very similar results, in general, the average non-hydrophobic K_s values were 970 and 870 mm/h, respectively, for the granitic and volcanic soils, while the hydrophobic MDI K_s values were roughly 10%–40% of the non-hydrophobic values. The measured K_s values are 3–4 times greater than that estimated from particle-size distributions (Table 2), nearly an order of magnitude larger than the NRCS estimates (Table 1) and, as will be discussed later, an order of magnitude larger than the RS derived estimates of K_m; however, they are consistent with the measured values of ~900 mm/h for a fine sand surrogate for Tahoe basin soils [21].

Table 8. Summary of hydraulic conductivity measurements associated with RS plots.

	Device	RS Test Runoff?	Mean K_s (mm/h)	*CoV* K_s (%)
RS Sites—Granitic Soil				
Bliss SP	PP	Yes	896	*8.7*
Luther Pass	PP	Yes	828	*12.1*
Rubicon	PP	Yes	627	*27.5*
Bliss/Meyers—w/surfactant	MDI	No	1386	*44.5*
Angora Fire—w/ surfactant	MDI	No	1093	*42.9*
Bliss/Meyers hydrophobic	MDI	Yes	376	*118*
Heavenly LT—hydrophobic	MDI	Yes	143	*43.4*
RS Sites—Volcanic Soil				
Blackwood Cyn -Truckee -hydrophobic plots	MDI	Yes	412	*125*
Brockway Summit—2006	PP	Yes	729	*43.1*
Brockway Summit—2003	PP	Yes	558	*24.1*
Dollar Hill—Tall RS	PP	Yes	883	*9.64*
Northstar—Tall RS	PP	Yes	907	*32.3*
Homewood—Tall RS	PP	Yes	803	*27.7*
Northstar—Tall RS	PP	No	978	*23.1*
Homewood—Tall RS	PP	No	762	*22.0*
Blackwood-Truckee—w/surf.	MDI	No	871	*34.3*
Brockway—w/surfactant	MDI	No	1311	*42.4*

DISCUSSION

Estimating Effective Hydraulic Conductivity K_m *from RS Test Plot Data*

Modeling hillslope infiltration and runoff rates requires measurement or estimation of the field effective hydraulic conductivity, K_m, some measure of the antecedent soil moisture and moisture retention conditions of the hillslope soils. Data collected from RS plot studies uniquely provides this information from a field-based assessment, though the data collected is used to infer both water retention and hydraulic conductivity parameters. Typically, the K_m value is estimated directly from the difference between the measured steady rain and runoff rates, while the infiltrated depth, V, is related to the visual wetting depth at the end of the RS test to estimate the soil water storage capacity. Alternatively, in WEPP modeling efforts, the K_m value is determined from that value which by trial and error yields the visual best-fit between the modeled runoff hydrograph assuming a single hillslope flow element and that measured during the RS test. Such a modeling fit to estimate K_m that relies on the Green–Ampt formulation of the wetting process, as outlined above in Equations (1)–(4), implicitly uses the infiltrated depth values from the RS test, but not the time-to-ponding/runoff directly, as described in Equation (8). Here, we use the individual RS plot data summarized in Table 4 and Table 6 to simultaneously solve both Equations (7) and (8) using the K_m approximation for h_d from Figure 1 for sandy soils to determine the K_m value that meets both the infiltrated depth and time-to-ponding, t_p, requirements. We use time to runoff from the RS test as a conservative estimate of t_p. For the data summarized in Table 5 and Table 7 for the plots not generating runoff, hence no t_p values, we use the $\Delta\theta$ value estimated from the visual wetting depth measurements and Equation (7), assuming a wetting h_d = 50 mm to determine K_m.

Table 9 and Table 10 summarize the steady rain–runoff and infiltration-equation based K_m values from the RS runoff plots on granitic and volcanic soils, respectively, while Table 11 summarizes those K_m values for the non-runoff RS plots on both soils. Mean K_m values determined from the steady rain–runoff and infiltration equation calculations were tested for significant difference based on a two-tailed test at $p < 0.01$ and $p < 0.05$, as indicated in the tables. Mean K_m value ranges were similar for both soils (10–100 mm/h in the volcanic soils and 20–120 mm/hr in granitic soils), and overall average values were several times greater than the NRCS estimates in Table 1, though 3–4 times less than the particle-size estimated K values (Table 2) and roughly an order of magnitude less than the average permeameter and MDI estimates of K (Table 8). Overall, data collected from 20% to 25% of the RS plots resulted in undeterminable K_m values in the simultaneous solution of Equations (7) and (8). As the simultaneous infiltration equation solution is highly sensitive to the time-to-ponding/runoff value, this result suggests that those times estimated during about ¼ of the RS tests may be problematic.

Table 9: Summary of steady Rain–Runoff and Infiltration Eq. K_m values for granitic RS plots having runoff.

Non-hydrophobic Plots	Rain—Runoff K_m (mm/h)	*CoV* K_m (%)	Infilt-Eq. K_m (mm/hr)	*CoV* K_m (%)	# of UD [a] K_m
Bliss SP RC—bare	47.1 [c]	7.21	40.2	*85.1*	0
Bliss SP RC—treated	55.4	8.70	54.6	*NA* [b]	2
CalTrans BS Plots	89.9	*21.5*	UD	*NA*	3
Cave Rock/Incline Village RC	49.9	*17.0*	56.4	*30.9*	0
Heavenly CAN LT—2005	83.7 [d]	*27.2*	101	*36.7*	0
Heavenly CAN LT—2006	95.9	*NA*	UD	*NA*	1
Heavenly CAN LT—2007	96.9	*23.8*	97.6	*NA*	5
Heavenly GB MSR & OLB	97.0	*21.9*	99.2	*62.7*	7
Heavenly LRR	110	*5.0*	108	*55.1*	1
Heavenly LT—2006	79.9	*10.8*	97.0	*9.2*	0
Heavenly PON/MCD	60.4	*45.7*	57.2	*44.8*	0
Heavenly STC	99.3	*9.5*	104	*33.3*	0
Luther Pass RC	43.2	*16.7*	43.5	*19.2*	0
Mammoth	108	*9.7*	108	*44.1*	1
Meyers RC	81.8	*18.1*	82.9	*33.7*	0
Roundhill landing-	26.5	*39.0*	19.1	*83.2*	1
Roundhill landing—USFS	20.2	*47.2*	27.2	*NA*	3
Rubicon RC, treated & native	48.4	*17.7*	52.8	*10.1*	1
Truckee Bypass	72.5	*16.8*	48.5	*NA*	4
Upper Cutthroat 1D & 2T	67.3	*42.1*	90.6	*24.8*	3
Upper Cutthroat 3T	107	*7.6*	85.2	*12.7*	0
Upper Cutthroat 4T & 6T	106	*22.7*	110	*51.4*	2
Upper Cutthroat 7T	81.7	*7.6*	87.3	*24.9*	0
Upper Cutthroat 8T	98.7	*22.5*	110	*25.6*	2
Angora Fire	62.5	*30.6*	35.3	*82.6*	3
Hydrophobic plots					
Bliss SP & Meyers RC	91.0	18.72	107	24.6	0
Heavenly LT'06—Natives	37.0	*38.7*	17.7	*95.5*	0
Upper Cut. 10T & 11T	119	*1.1*	149	*1.2*	1
Upper Cut. 3N—Natives	54.0	*13.1*	41.4	*55.4*	0

[a]: UD = UnDefined values; no real solution for K_m; [b]: NA = Not Applicable usually because $n < 3$; [c]: Mean values highlighted in *RED* do not differ significantly ($p < 0.01$); [d]: Mean values highlighted in *BLUE* do not differ significantly ($p < 0.05$).

Table 10: Summary of the steady Rain–Runoff and Infiltration Equation based K_m values associated with volcanic RS plots having runoff.

Site Descriptions	Rain–Runoff K_m (mm/h)	CoV K_m (%)	Infilt. Equation K_m (mm/h)	CoV K_m (%)	# of UD [a] K_m
Blackwood Cyn–Truckee	106 [c]	12.6	117	24.9	2
Brockway Summit—2006	59.3	NA [b]	62.0	NA	0
Brockway Summit—2003	36.2	25.3	37.4	42.8	0
Dollar Hill—Tall RS	45.0	2.2	39.6	47.5	0
Homewood #31 Rd	61.8	1.7	62.1	53.0	0
Homewood LL A&B Rd	102	8.9	101	42.8	0
Homewood MAR Rd	86.5	10.6	104	25.4	2
Homewood SCD RD&BK Rd	85.7	15.9	72.9 [d]	67.5	5
Homewood WDG Rd	85.4	17.2	88.1	16.1	4
Northstar—Short RS	48.5	NA	35.8	NA	0
Juniper Mtn Skirun	40.5	19.2	26.00	73.1	0
Juniper Mtn Skirun—WCs	103	17.0	95.7	37.1	0
Northstar—Tall RS	52.5	44.8	42.2	75.8	0
Northstar—Bare ski Tall RS	37.7	16.9	28.1	48.3	5
Homewood—Tall RS	41.0	24.6	25.7	48.3	11
Northstar BP CON plots	97.6	NA	87.0	NA	1
Northstar LT	87.3	26.9	99.5	51.4	3
Northstar STA plots	102	6.8	99.0	27.5	0
Northstar Unit 7	102	NA	77.0	NA	0
Ward Cr. Landing	43.4	27.2	47.1	46.4	0
Ward Cr. Landing—USFS	23.1	13.7	9.2	NA	1

[a]: UD = UnDefined values; no real solution for K_m; [b]: NA = Not Applicable usually because $n < 3$; [c]: Mean values highlighted in *RED* do not differ significantly ($p < 0.01$); [d]: Mean values highlighted in *BLUE* do not differ significantly ($p < 0.05$).

Table 11: Summary of Infiltration Equation K_m values associated with RS plots not having runoff.

	Infilt. Equation K_m (mm/h)	CoV K_m (%)
Granite Soil Sites—Non-Hydrophobic Plots		
CalTrans OP Plots	79.7	38.6
Cave Rock—Inc. Village RCs	36.3	NA [a]
Heavenly CAN LT	42.4	29.9
Heavenly CAN LT	63.6	32.6
Heavenly GB MSR & OLB	97.3	NA
Heavenly STC	89.6	NA
Mammoth	96.7	NA
Meyers RC	58.4	23.0
Roundhill Landing- IERS	35.7	NA
Truckee Bypass	84.9	18.1
Upper Cutthroat 3T	51.9	0.1
Upper Cutthroat 4T & 6T	86.2	NA
Upper Cutthroat 8T	91.7 [b]	10.5
Granite Soil Sites—Hydrophobic plots		
Bliss SP & Meyers RC—surf	115	NA
Angora Fire	50.8	NA
Upper Cut. 10T & 11T	83.3	7.0
Volcanic Soil Sites		
Blackwood-Truckee (Surf)	101	NA
Brockway Summit plots	97.5	NA
Northstar—Short RS	40.5	13.6
Northstar—Tall RS	37.2	12.0
Homewood—Tall RS	50.7	NA
Juniper Mtn Skirun	42.9	50.4
Northstar ETD & BP CON	97.5	NA
Northstar LT	55.2	41.3
Northstar Unit 7—Short RS	75.2	24.9

[a]: NA = Not Applicable usually because $n < 3$; [b]: Mean values highlighted in *RED* between sister runoff plots do not differ significantly ($p < 0.01$).

For roughly 80% of the RS sites on both soils, the simpler rain–runoff rate estimate of K_m values was essentially equivalent to that obtained from the infiltration equation solution, lending credence to use of the rain–runoff rate estimates of K_m for watershed modeling purposes. Considering K_m values estimated from Equation (7) for the RS plots lacking runoff, there was some agreement between the rain–runoff rate K_m values from the sister plots as indicated in Table 11, but few direct comparisons were available. On the other hand, the overall soil average rain–runoff rate K_m values of 73.2 and 68.9 mm/h from the granitic and volcanic runoff RS plots, respectively, were essentially equivalent to the Equation (7) estimates from the non-runoff RS plots of 72.7 and 66.4 mm/h, respectively. Thus, use of the Equation (7), combined with visual estimation of the $\Delta\theta$ values (e.g., Figure 2), appears to also be a reasonable estimate of K_m values for modeling purposes. Finally, to illustrate some of the variability in the comparison between both estimates of K_m values as well as determining any possible bias, Figure 4 and Figure 5 show comparisons of the K_m values summarized in Table 9 and Table 10, respectively. Interestingly, for the granitic soils, infiltration equation K_m values appear to be a slightly greater on average (~3%) as compared to the rain–runoff rate estimates for the granitic soils, while for the volcanic soils, they are on average ~4% less, such that, overall across all RS plots, there appears to be no systematic bias or preference between the two methods of estimating K_m values.

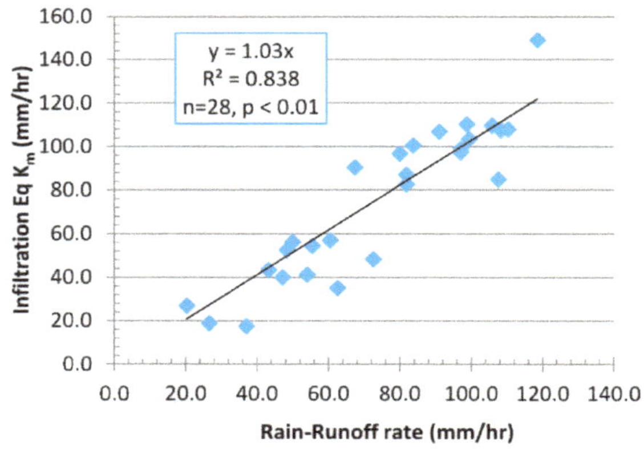

Figure 4: Dependence of Infiltration Equation K_m values on Rain–Runoff rates for granitic soil RS plots having runoff.

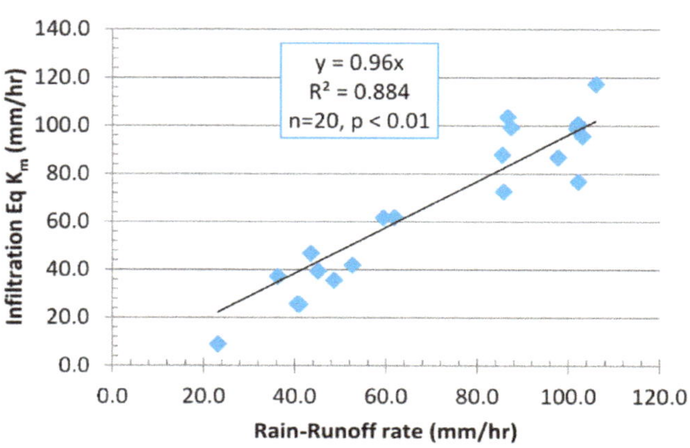

Figure 5: Dependence of Infiltration Equation K_m values on Rain–Runoff rates for volcanic soil RS plots having runoff.

Runoff Erosion Field Data—How Are Sediment Detachment Rates Related to Stream Power?

Perhaps one of the primary purposes of RS plot studies is to estimate surface runoff transport rates of sediment and other contaminants so as to develop the required watershed parameters, or evaluate the relative success of soil restoration efforts. Using sediment detachment rates measured from the short DF RS plots on granitic roadcuts on the south west shore and volcanic soil skiruns of the north shore of Lake Tahoe, we evaluate whether turbulent or laminar flow stream power (Equations (19) or (20)) are appropriate to describe these rates. As summarized in Table 4 and Table 6, these RS test plots were on a broad range of slopes and used rainfall rates from 60 to 120 mm/h, with test durations of 30–60 min. While occasionally there was no runoff from the treated or grassed RS plots, these data were not included in the comparisons; however, multiple non-runoff plots in the restored and native soils were represented by two zero values in the analyses. The soil surface conditions included no cover (bare) to light grasses, treated soils (usually pine-needle or woodchip mulch covers), restored soils (incorporated mulches and tillage) and "native" (relatively undisturbed forest soils adjacent to RS test sites). We graph average RS plot soil detachment rates ($\times 10^5$ kg/m²/s) per condition (usually $n = 3$) as a function of $q^{0.67}S^{1.328}$ from Equation (20) to estimate the effective "erodibility" as it depends on stream power in Figure 6 for the granitic soils, and in Figure 7 and Figure 8 for the volcanic soils. Note that the soil detachment rates from the finer-textured volcanic soils are much greater than those from the granitic soils in general. Similarly, considering the linear slopes of the regressions to represent soil "erodibility", erodibilities from the bare granitic and grassed volcanic soils were roughly equivalent, while those from restored and native (undisturbed) volcanic soils were effectively identical, though about three times greater than equivalently treated or native granitic soils. Lastly, comparing sediment detachment rates as a function of the product of the rainfall rate and stream power dramatically lowered the R^2 values in all of the regressions, suggesting that at least for the rainfall rates/energies, infiltration rates and soil cover conditions considered here, rainfall rates had little impact on soil erodibility.

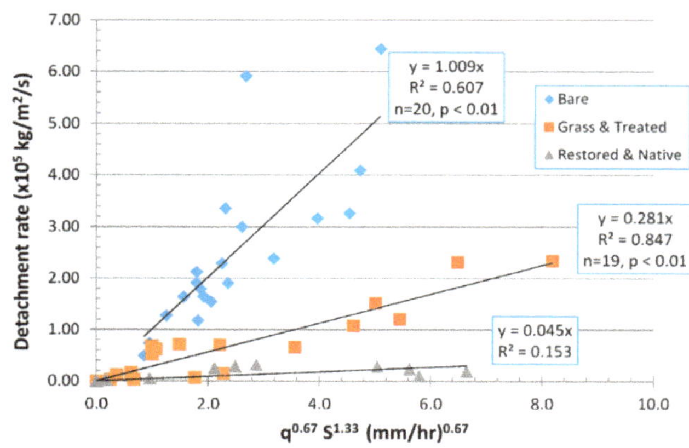

Figure 6: Comparison between and stream power variables for granitic RS plots on the southwest shore of Lake Tahoe.

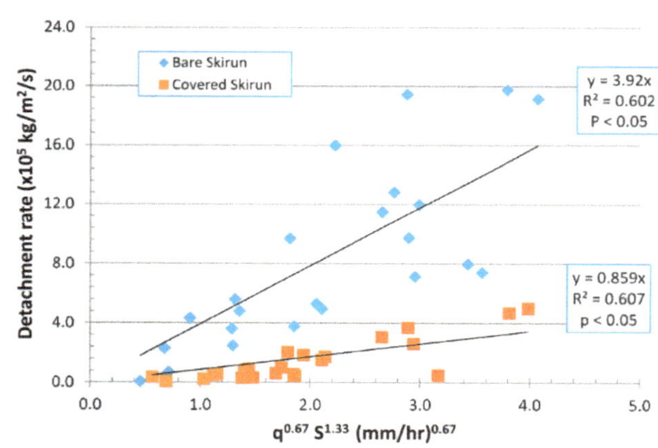

Figure 7: Comparison between and stream power variables for bare and grass-covered volcanic RS plots on the northwest shore of Lake Tahoe.

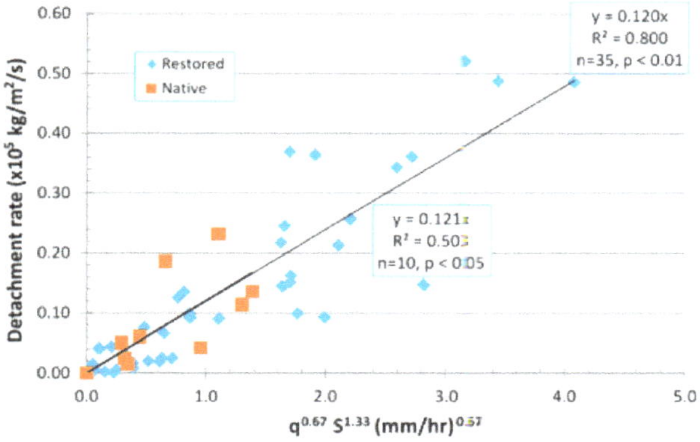

Figure 8: Comparison between and stream power variables for bare and grass-covered volcanic RS plots on the northwest shore of Lake Tahoe.

As a means of assessing the hypothesis that laminar flow definition of stream power by Equation (20) better describes the sediment detachment rates than that from turbulent flow assumptions using Equation (19), we examined the regression coefficient variation as it depends on the exponent of the surface runoff rate for the data shown in Figure 6, Figure 7 andFigure 8. It should be noted, of course, that soil detachment rates as determined from the RS plot runoff sediment concentrations result in a degree of self-correlation with runoff rates as both are calculated from the runoff sampling data. Considering the regression coefficient variation provides some insight into the range of stream power exponent values reported in the literature and perhaps why they range from theoretical values. Figure 9 illustrates the dependence of the linear regression R^2 value on the exponent of q in Equation (20) for the granitic soils of Figure 6, while Figure 10 illustrates this same dependence for the volcanic RS test plot data shown in Figure 7 and Figure 8. For the granitic RS test plots, the greatest R^2 value occurs as predicted by Equation (20) at $q^{0.67}$ for the grassed and treated RS plots, and this is also near the maximum R^2 value for the bare plots. In both cases, however, the R^2 values at $q^{0.67}$ readily exceed those for $q^{0.4}$ from Equation (19), but the relative dependence of the R^2 value is "flat" for exponents ranging from 0.8 to 1.3, as found by Al-Hamdan et al. [35]. The sediment detachment rates from the restored and native granitic RS test plots are so small and variable as to have little discernible dependence on stream power. In the volcanic RS test plots, the greatest R^2 value again occurs as predicted by Equation (20) at $q^{0.67}$ for the grassed and treated RS plots, and this is also near the maximum R^2value for the bare plots. As with the bare granitic RS plot data, the R^2 value for the volcanic RS plots has little dependence on the runoff rate exponent at values greater than about 0.8, but these R^2 values differ little from that associated with an exponent of 0.67. For the bare and native volcanic RS test plots, the greatest R^2 values occur at an exponent of 1.1. In all volcanic RS plots, however, the R^2 values at $q^{0.67}$ readily exceed those for $q^{0.4}$ from Equation (19), suggesting that, as with the granitic soils, the soil detachment rate is better described by laminar as compared to turbulent flows for the Tahoe soils.

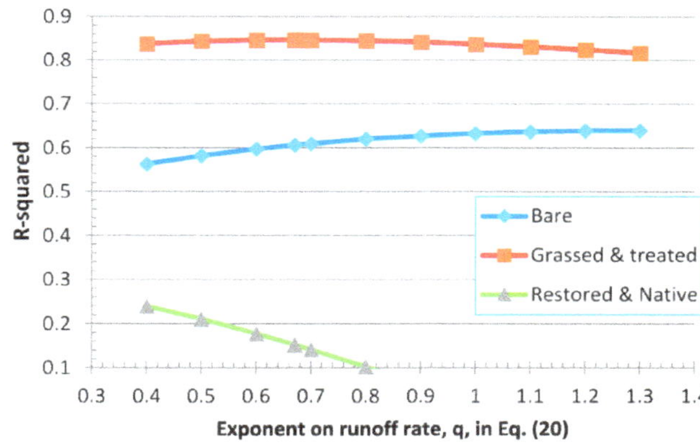

Figure 9: Dependence of linear regression R^2 values on exponent of runoff rate in stream power function when related to soil detachment rates from granitic RS plots.

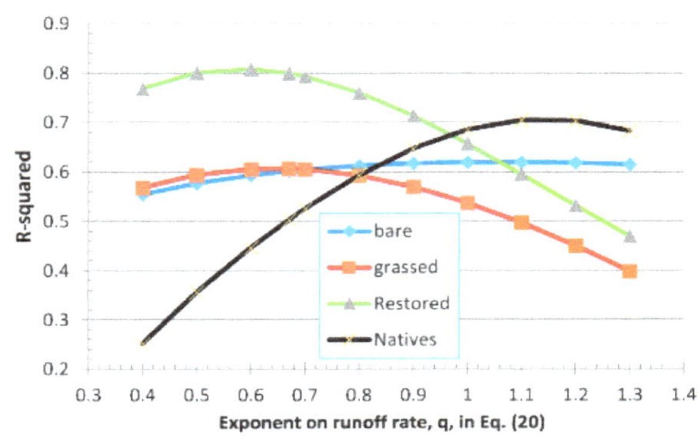

Figure 10: Dependence of linear regression R^2 values on exponent of runoff rate in stream power function when related to soil detachment rates from volcanic RS plots.

Finally, defining RS plot erodibility as the linear regression slope between soil detachment rates, D_i, and laminar stream power, ***P***, or average ratio thereof, the erodibilities likely useful for watershed modeling in the Tahoe basin are summarized in Table 12 from the results shown in Figure 6, Figure 7 and Figure 8. Of course, as the values in Table 12 were derived from the runoff plots from the restored and less-disturbed forest (native) soil conditions, they are likely conservative estimates as there were numerous non-runoff RS plots for these conditions.

Table 12: Summary of stream power (Equation (20)) derived soil erodibility for granitic and volcanic soil RS plots from the southwest and north shores of Lake Tahoe, respectively.

<table>
<tr><th rowspan="2">Soil Type</th><th colspan="4">Erodibilities for Soil Condition = D_i/P (s^2/m)</th></tr>
<tr><th>Bare</th><th>Grass Cover</th><th>Treated</th><th>Restored/Native</th></tr>
<tr><td>Granitic</td><td>3.35×10^{-5}</td><td colspan="2">0.93×10^{-5}</td><td>0.15×10^{-5}</td></tr>
<tr><td>Volcanic</td><td>16.0×10^{-5}</td><td>2.20×10^{-5}</td><td colspan="2">0.40×10^{-5}</td></tr>
</table>

Results & Discussion of Rainfall Simulator Comparisons—Infiltration & Erosion Processes

With the several types of "portable" RS devices used in field plot studies that have a variety of rainfall energies applied to plot areas in the order of one square meter, comparing plot test results (infiltration, runoff and erosion rates) from these different RSs has been a recurring challenge in the literature. Having outlined above possible interpretations methods for determination of infiltration (K_m) and erosion parameter averages from the entire range of RS test plots across the Tahoe Basin, here, we consider the more specific site evaluation of RS test plots using different RSs to determine how simulators with different rain energies may affect these parameters. We focus on a direct comparison of the infiltration and erosion results obtained from three RSs used in the basin that have rain drop energies of < ¼ and about ¾ of maximum values (i.e., short and tall drop-former (DF) simulators, respectively) and the greater energy sprinkler RS used by the USFS. We use RS plot data for the short and tall DF RSs at similar Northstar disturbed and treated volcanic soil sites, and likewise, we use plot data for the short DF and USFS RSs at soil forest landings on the west and east shores of the lake. First, we compare soil K_m values estimated from steady rain–runoff rates and simultaneous solutions to the infiltration equations, as described above, then we consider the estimated erodibilities ($D_i/\boldsymbol{P}$) from the different RSs, as in Table 12 for the same sites.

In the Northstar area, there were several skirun, or adjacent sites, that included RS tests with both the original tall DF RS and the short DF RS on bare and treated soils from different years. The average plot information for the Northstar sites is inTable 6 for the runoff plots, while the plot-specific erosion information is summarized in Table 13. Overall, the average plot slopes for both DF RS plots were similar, however, the short DF RS plots were subject to rain rates twice as great (120 *vs.* 60 mm/h) as that for the tall DF RS plots, though with about 30% less impact energy. The average of the 14 K_m values estimated from the bare and treated soil plot data for the tall DF RS was, surprisingly, roughly half that for the 11 short DF RS test plots subjected to the much greater rain rate. However, the comparison between infiltration equation and rain–runoff rate estimates are functionally the same, as shown in Figure 11.

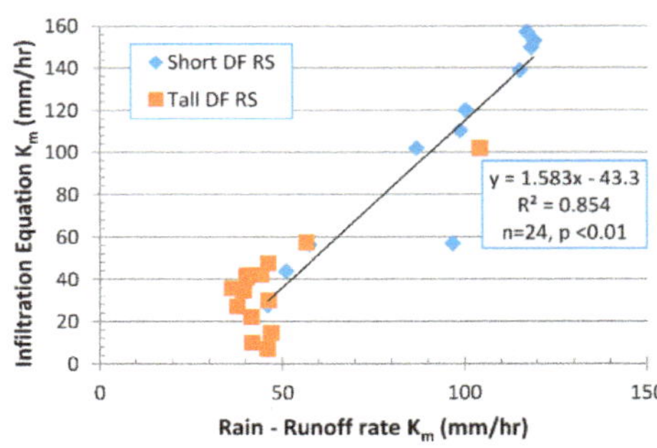

Figure 11: Dependence of Infiltration-equation K_m values on Rainfall–Runoff rates for the short and tall DF type RS at Northstar ski area (volcanic soils).

Table 13: Summary of plot data for stream power derived soil erodibility for volcanic bare and treated soil RS plots using the tall and short DF type RSs at Northstar ski area.

Soil Condition (Short or Tall DF RS)	Slope (%)	Steady q (mm/h)	SY [a] (gm/mm)	$D_i \times 10^5$ (kg/m²/s)	D_i/P (s²/m)
	60.7	19.6	23.2	19.8	1.73×10^{-4}
	42.1	26.9	16.6	19.5	2.24×10^{-4}
Bare soils (Tall DF)	37.0	23.6	15.6	16.0	2.39×10^{-4}
	31.7	23.7	9.4	9.71	1.78×10^{-4}
	25.6	22.4	5.8	5.64	1.42×10^{-4}
Averages	39.4	23.3	14.1	14.1	19×10^{-5}
	27.6	10.9	0.35	0.16	6.07×10^{-6}
	27.3	16.2	0.20	0.14	4.04×10^{-6}
Treated [b] soils (Tall DF)	35.0	10.9	0.60	0.28	7.64×10^{-6}
	35.0	5.7	0.30	0.08	3.12×10^{-6}
	36.4	16.7	0.40	0.29	5.55×10^{-6}
Averages	32.3	12.1	0.37	0.19	0.53×10^{-5}
	47.7	24.4	3.62	3.82	3.99×10^{-5}
Light grass soil covers (Short DF)	51.0	15.9	6.25	4.33	5.48×10^{-5}
	47.9	20.1	2.89	2.52	2.97×10^{-5}
Averages	48.9	20.1	4.25	3.56	4.15×10^{-5}
	34.5	24.6	0.16	0.17	2.72×10^{-6}
	31.1	5.3	0.16	0.04	1.88×10^{-6}
	32.9	19.1	0.26	0.22	4.38×10^{-6}
Treated soils (Short DF)	31.9	22.6	0.15	0.15	2.84×10^{-6}
	35.5	6.3	0.14	0.04	1.44×10^{-6}
	35.9	12.6	0.16	0.09	2.09×10^{-6}
	33.9	10.0	0.26	0.11	3.42×10^{-6}
Averages	33.7	14.4	0.19	0.12	0.27×10^{-5}

[a]: SY = Sediment yield expressed as mass of sediment per mm of runoff determined from slope of accumulated sediment *versus* accumulated runoff from the RS plot (Grismer & Hogan, 2004); [b]: Various combinations of tilled compost, pine needle mulch and wood chips plot covers.

Despite the much greater rain impact energies and greater runoff rates from the tall *versus* short DF RSs, erodibilities expressed as $D_i/\boldsymbol{P}$ were only slightly greater for the tall RS plots, though practically the same for these bare volcanic skirun soils (19×10^{-5} *vs.* 16×10^{-5} s²/m). Similarly, for the lightly covered plots, the short DF RS plots at the Northstar specific site were greater than at the skirun average (~4×10^{-5} *vs.* ~2×10^{-5} s²/m), perhaps due to varying coverages. While mean erodibility values from the tall DF RS for the "treated" plots were twice as great as those from the short DF RS plots (0.53×10^{-5} *vs.* 0.27×10^{-5} s²/m), these values ranged on either side of that mean in Table 12 (0.4×10^{-5} s²/m) for the short DF RS plots. Moreover, the difference between the means was similar to the variability in values between individual RS plots from either RS, and may have reflected the different years of RS tests or variability in treatments. We note that the average runoff organic matter fraction (OM%) from the treated RS plots was ~25%, suggesting presence of some non-mineral soil cover.

Table 14: Summary of plot data for stream power derived soil erodibility for granitic and volcanic soil forest landing RS plots for short DF RS and USFS sprinkler RS.

Site (DF or USFS RS)	Soil Condition	Slope (%)	Steady q (mm/h)	SY [a] (gm/mm)	$D_i \times 10^5$ (kg/m²/s)	D_i/P (s²/m)
Round Hill (DF)	Bare	3.4	51.1	0.41	0.91	19.3×10^{-5}
	Light PN mulch cover	6.5	23.8	0.98	1.01	15.1×10^{-5}
	PN mulch cover, grass	10.4	38.1	0.20	0.33	1.93×10^{-5}
	PN Mulch & WC cover	4.7	54.5	0.06	0.14	1.87×10^{-5}
	Tilled PNM [b] & WC cover	6.1	45.2	0.02	0.04	0.42×10^{-5}
Round Hill (USFS)	Bare	17.1	69.3	4.53	8.72	20.5×10^{-5}
	Light PN mulch cover	14.2	69.6	2.93	5.66	17.0×10^{-5}
	Light PN mulch cover	22.1	71.9	3.51	7.01	11.4×10^{-5}
Ward Cr. (DF)	Bare ~5% mulch cover	6.7	41.2	1.93	3.45	34.3×10^{-5}
	PN Mulch & WC cover	7.0	19.5	0.17	0.14	2.23×10^{-5}
	PN Mulch & WC cover	7.0	14.9	0.19	0.12	2.28×10^{-5}
	PN Mulch & WC cover	8.7	30.7	0.18	0.24	2.05×10^{-5}
Ward Cr. (USFS)	Bare ~5% mulch cover	10.3	62.3	2.86	1.040	24.4×10^{-5}
	Light PN mulch cover	21.2	71.0	2.37	4.674	8.11×10^{-5}
	Light PN mulch cover	22.8	71.6	1.67	3.321	5.20×10^{-5}

[a]: SY = Sediment yield expressed as mass of sediment per mm of runoff determined from slope of accumulated sediment *versus* accumulated runoff from RS plot [46]; [b]: PNM = Pine needle mulch, WC = wood chips plot covers.

In a more direct comparison of results from different RSs, the short DF RS and the USFS sprinkler RS were used on compacted soil forest landing sites near Ward Creek (volcanic soil) and at Round Hill (granitic soil) using randomly selected plots from within the same area and conducting the RS tests on the same days. The average plot information for these sites is in Table 4 and Table 6 for the runoff plots while the plot-specific erosion information is summarized in Table 14. Overall, the RS plot slopes for the DF RS were about half that for the USFS RS, and the DF RS rain rates were also 10–20 mm/h less, as well as having much smaller rain impact energies as compared to the sprinkler RS. As a result, times-to-runoff and runoff rates from the DF RS plots were smaller than those from the sprinkler RS. Finally, one USFS plot at Round Hill responded completely differently than its sister plots and was not included in the erosion analysis.

When considering K_m estimates developed from all RS plot data collected from the two forest landing sites, we found that of the seven USFS RS plots, only three yielded data that enabled computation of K_m from the combined infiltration equations, while eight of the nine short DF RS plots yielded calculable K_m values. At the Round Hill site, the average K_m values from both calculation methods were the same, at about 23 mm/hr. At the volcanic soil Ward Creek site, the DF RS plot average K_m values were more than twice that obtained from the neighboring USFS RS plots. Following the same format as in Figure 4 and Figure 5, the forest landings K_m values are graphed in Figure 12. Both types of RSs resulted in infiltration equation *versus* rain–runoff rate estimated K_m values fitting the same line as well as being consistent with the WEPP based K_m values. Similarly, despite the differing rain impact energies from the two RSs, erodibilities expressed as $D_i/\boldsymbol{P}$ were practically the same for the bare granitic soils (~20×10^{-5} s²/m) and the lightly covered plots (~15×10^{-5} s²/m). Erodibility values for the better covered RS plots were approximately 1.9×10^{-5} s²/m, consistent with the "grass/treated soils" value inTable 12. On the volcanic soils, the bare plot erodibilities differed between the two RSs (~24×10^{-5} *vs.* 34×10^{-5} s²/m), but well within the variability range encountered in replicate plots results. These erodibility values were roughly two times greater than that from the bare skiruns in Table 12, likely reflecting the greater soil compaction at the forest landings. The one tilled-in mulch plot yielded an erodibility value equal to that average found for the "treated/restored" volcanic skirun soils (Table 12). This treated soil plot suggests the possibility that even the compacted forest landing soils may be "treatable" to the point of returning them to something like native forest soil functionality with respect to infiltration and erosion. Finally, we note that the OM% of the runoff from the DF RS plots were quite high (30%–50%), suggesting that visual assessment of the surface conditions as "bare" or "light coverage" likely overlooked the possibility of an apparent fine mulch/duff organic layer integrated with the mineral soil surface.

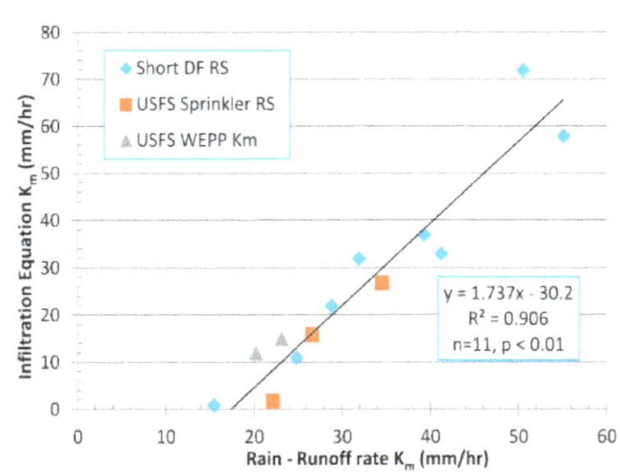

Figure 12: Dependence of Infiltration equation K_m values on Rainfall-Runoff rates for the short DF and Sprinkler type RSs from tests plots at Ward Creek and Round Hill forest landings.

SUMMARY & CONCLUSIONS

A key component of modeling watershed runoff and erosion processes is estimation of surface soil infiltration rates or effective hydraulic conductivities (K_m) and erosion rates, as they vary with soil cover/tilth/slope conditions, and seasonally with changing water contents. Portable rainfall simulators (RSs) are essential tools for measuring infiltration, runoff and erosion under a variety of field conditions, and information from RS test plots can provide the infiltration/runoff parameterization required for watershed modeling. Though multiple RS designs for field application exist, no single RS design (including plot runoff frame installation) has emerged as a standard. Here, we develop a simultaneous solution of time-to-ponding/runoff and Green–Ampt type infiltration equations to determine K_m, as well as developing a laminar flow-based description of stream power that can be used to determine erodibilities and then apply these analyses to data from 423 RS plots across the Tahoe Basin. Finally, we provide direct infiltration and erosion analysis comparisons of results from three RSs that have different rainfall energies. The overall goal of the analyses was to develop a common assessment method, or approach, to evaluating RS plot data for use in watershed modeling efforts.

With respect to estimation of K_m values, the simpler-to-calculate steady rain–runoff rate value was equivalent for all practical purposes to that estimated from infiltration equations. It was possible to calculate infiltration equation based K_m values from ~80% of the RS plot data that spanned a wide range of rain rates (60–120 mm/h), runoff rates (2–70 mm/h) and times-to-ponding (1–20 min). However, this effective conductivity, though assumed to be half of the saturated value, was as much as an order of magnitude less than K_s values derived from other field test methods. In terms of developing comparable "erodibilities", defined here as simply the ratio of sediment detachment rates to stream power, it appears that the laminar flow derived stream power results in a better fit between detachment rate and stream power than that derived from turbulent flow assumptions, and eliminates the need to define 'n' and the restriction to <10% slopes associated with the Mannings equation. Applying the laminar flow stream power derivation to the determination of erodibilities enabled comparison of data from RS test plots having a wide range of slopes and runoff rates, as well as from simulators having different rainfall energies. While rain drop impact energy did not appear to be a factor in these analyses, this may stem from the presence of some cover materials, the very high infiltration rates and the lack of aggregation of the Tahoe surface soils. We suggest, as Grismer [50] did, that RS plot studies include measurements of particle sizes and OM% in the runoff collected in order to provide more useful data from which to understand erosional processes and soil restoration efforts.

REFERENCES

1. Amerman, C.R. Rainfall simulation as a research tool in infiltration. In Proceedings of the Rainfall Simulator Workshop, Tuscon, AZ, USA, 7–9 March 1979.
2. Grismer, M.E. Standards vary in studies using rainfall simulators to evaluate erosion. *Calif. Agric.* **2012**, *66*, 102–107.
3. Grismer, M.E. Rainfall Simulation Studies—A Review of Designs, Performance and Erosion Measurement Variability. Review for the Tahoe Science Consortium. 2011, 110. Available online: http://ucanr.org/sites/californiaagriculture/files/145682.pdf (accessed on 1 June 2016).
4. Iserloh, T.; Ries, J.B.; Arnáez, J.; Boix-Fayos, C.; Butzen, V.; Cerdà, A.; Echeverría, M.T.; Fernández-Gálvez, J.; Fister, W.; Geißler, C.; et al. European small portable rainfall simulators: A comparison of rainfall characteristics. *Catena***2013**, *110*, 100–112.
5. Gallart, F.; Llorens, P.; Latron, J.; Regüés, D. Hydrological processes and their seasonal controls in a small Mediterranean mountain catchment in the Pyrenees. *Hydrol. Earth Syst. Sci.* **2002**, *6*, 527–537.
6. Hessel, R. Modelling soil erosion in a small catchment on the Chinese Loess Plateau: applying LISEM to extreme conditions. *Neth. Geogr. Stud.* **2002**, *307*, 1–21.
7. García-Ruiz, J.M.; Lana-Renault, N.; Beguería, S.; Lasanta, T.; Regüés, D.; Nadal-Romero, E.; Serrano-Muela, P.; López-Moreno, J.I.; Alvera, B.; Martí-Bono, C.; et al. From plot to regional scales: Interactions of slope and catchment hydrological and geomorphic processes in the Spanish Pyrenees. *Geomorphology* **2010**, *120*, 248–257.
8. Pan, C.; Shangguan, Z. Runoff hydraulic characteristics and sediment generation in sloped grassplots under simulated rainfall conditions. *J. Hydrol.* **2006**, *331*, 178–185.
9. Pardini, G.; Gispert, M.; Dunjo, G. Relative influence of wildfire on soil properties and erosion processes in different Mediterranean environments in NE Spain. *Sci. Total Environ.* **2004**, *328*, 237–246.
10. Seeger, M. Uncertainty of factors determining runoff and erosion processes as quantified by rainfall simulations. *Catena* **2007**, *71*, 56–67.
11. Boer, M.; Puigdefábregas, J. Effects of spatially structured vegetation patterns on hillslope erosion in a semiarid Mediterranean environment: a simulation study. *Earth Surf. Process. Landf.* **2005**, *30*, 149–167.
12. Nadal-Romero, E.; Lasanta, T.; Regüés, D.; Lana Renault, N.; Cerdá, A. Hydrological response and sediment production under different land cover in abandoned farmland fields in a Mediterranean mountain environment. *Bull. Assoc. Spanish Geogr.* **2011**, *55*, 303–323.
13. Stoffel, M.; Wilford, D.J. Hydrogeomorphic processes and vegetation: Disturbance, process histories, dependencies and interactions. *Earth Surf. Process. Landf.* **2012**, *37*, 9–22.
14. Battany, M.C.; Grismer, M.E. Development of a portable field rainfall simulator system for use in hillside vineyard runoff and erosion studies. *Hydrol. Proc.* **2000**, *14*, 1119–1129.
15. Battany, M.C.; Grismer, M.E. Rainfall runoff, infiltration and erosion in hillside vineyards: Effects of slope, cover and surface roughness. *Hydrol. Proc.* **2000**, *14*, 1289–1304.
16. Lascelles, B.; Favis-Mortlock, D.T.; Parsons, A.J.; Guerra, A.J.T. Spatial and temporal variation in two

rainfall simulators: Implications for spatially explicit rainfall simulation experiments. *Earth Surf. Process. Landf.* **2000**, *25*, 709–721.

17. Kinnell, P.I.A. Raindrop-impact-induced erosion process and prediction: a review. *Hydrol. Proc.* **2005**, *19*, 2815–2844.
18. Kinnell, P.I.A. Simulations demonstrating interaction coarse and fine sediment loads in rain-impacted flow. *Earth Surf. Process. Landf.* **2006**, *31*, 355–367.
19. Grismer, M.E. Pore-size distribution and infiltration. *Soil Science* **1986**, *141*, 249–260.
20. Morel-Seytoux, H.J. Application of infiltration theory for determination of excess rainfall hyetograph. Water Resources Bulletin. *Am. Water Resour. Assoc.* **1981**, *17*, 1012–1022.
21. Grismer, M.E. Runoff simulations and watershed modeling. *Hydrology* **2016**, *3*, 18–30.
22. Benjamin, T.B. Wave formation in laminar flow down an inclined plane. *J. Fluid Mech.* **1957**, *2*, 554–574.
23. Ellison, W.D. Soil erosion studies. Part I. *Agric. Eng.* **1947**, *28*, 145–146.
24. Fristensky, A.; Grismer, M.E. A modeling approach for ultrasonic soil aggregate stability assessment. *Catena* **2008**, *74*, 153–164.
25. Fristensky, A.; Grismer, M.E. Evaluation of ultrasonic aggregate stability and rainfall erosion resistance of disturbed and amended soils in the Lake Tahoe Basin, USA. *Catena* **2009**, *79*, 93–102.
26. Owoputi, L.O.; Stolte, W.J. Soil detachment in the physically based soil erosion process: A review. *ASAE Trans.* **1995**,*38*, 1099–1110.
27. Foster, G.R.; Meyer, L.D. Transport of soil particles by shallow flow. *ASAE Trans.* **1972** *15*, 99–102.
28. Bridge, J.S.; Dominic, D.F. Bed load grain velocities and sediment transport rates. *Water Resour. Res.* **1984**, *20*, 476–490.
29. Gilley, J.E.; Finkner, S.C. Estimating soil detachment caused by raindrop impact. *ASAE Trans.* **1985**, *26*, 140–146.
30. Gilley, J.E.; Woolhiser, D.A.; McWhorter, D.B. Interrill soil erosion—Part I: Development of model equations. *ASAE Trans.* **1985**, *26*, 147–153.
31. Gilley, J.E.; Woolhiser, D.A.; McWhorter, D.B. Interrill soil erosion—Part II: Testing and use of model equations. *ASAE Trans.* **1985**, *26*, 154–159.
32. Moore, I.D.; Burch, G.J. Sediment transport capacity of sheet and rill flow: Application of unit stream power theory.*Water Resour. Res.* **1986**, *22*, 1091–1107.
33. Grismer, M.E. Soil Restoration and Erosion Control: Quantitative Assessment in Rangeland and Forested areas.*ASABE Trans. Soil Water Centen. Collect.* **2007**, *50*, 1619–1626.
34. Gabriels, D.M. The Stream Power Concept in Estimating Sediment Transport. Available online: http://indico.ictp.it/event/a12165/session/37/contribution/30/material/0/0.pdf (accessed on 1 June 2016).
35. Al-Hamdan, O.Z.; Pierson, F.B.; Nearing, M.A.; Williams, C.J.; Stone, J.J.; Kormos, P.R.; Boll, J.; Weltz, M.A. Concentrated flow erodibility for physically based erosion models: Temporal variability in disturbed and undisturbed rangelands. *Water Resour. Res.* **2012**.
36. Zhang, F.-B.; Wang, Z.-L.; Yang, M.-Y. Validating and Improving Interrill Erosion Equations. *PLoS ONE* **2014**.
37. McCool, D.K.; Brown, L.C.; Foster, G.R.; Mutchler, C.K.; Meyer, L.D. Revised slope steepness factor for the universal soil loss equation. *ASAE Trans.* **1987**, *30*, 1387–1396.
38. Zhang, G.H.; Liu, B.Y.; Nearing, M.; Zhang, K.L. Soil detachment by shallow flow. *ASAE Trans.* **2002**, *45*, 351–357.
39. Zhang, G.H.; Liu, B.Y.; Liu, G.B.; He, X.; Nearing, M. Detachment of undisturbed soil by shallow flow. *Soil Sci. Soc. Am. J.* **2003**, *67*, 713–719.
40. Nearing, M.A.; Parker, S.C. Detachment of soil by flowing water under turbulent and laminar conditions. *Soil Sci. Soc. Am. J.* **1994**, *58*, 1612–1614.
41. Guy, B.T.; Dickinson, W.T.; Rudra, R.P. The roles of rainfall and runoff in the sediment transport capacity of interrill flow. *ASAE Trans.* **1987**, *30*, 1378–1386.
42. Sharma, P.P.; Gupta, S.C.; Foster, G.R. Predicting soil detachment by raindrops. *Soil Sci. Soc. Am. J.* **1993**, *57*, 674–680.
43. Sharma, P.P.; Gupta, S.C.; Foster, G.R. Raindrop-induced soil detachment and sediment transport from interrill areas.*Soil Sci. Soc. Am. J.* **1995**, *59*, 727–734.
44. Sharma, P.P.; Gupta, S.C.; Rawls, W.J. Soil detachment by single raindrops of varying kinetic energy. *Soil Sci. Soc. Am. J.* **1991**, *55*, 301–307.
45. Lei, T.W.; Zhang, Q.W.; Zhao, J. A laboratory study on sediment transport capacity in the dynamic process of rill erosion. *Trans. ASAE* **2001**, *44*, 1537–1542.
46. Grismer, M.E.; Hogan, M.P. Evaluation of Revegetation/Mulch Erosion Control Using Simulated Rainfall in the Lake Tahoe Basin: 1. Method Assessment. *Land Degrad. Dev.* **2004**, *13*, 573–588.
47. Grismer, M.E.; Hogan, M.P. Evaluation of Revegetation/Mulch Erosion Control Using Simulated Rainfall in the Lake Tahoe Basin: 2. Bare Soil Assessment. *Land Degrad. Dev.* **2005**, *16*, 397–404.
48. Grismer, M.E.; Hogan, M.P. Evaluation of Revegetation/Mulch Erosion Control Using Simulated Rainfall in the Lake Tahoe Basin: 3. Treatment Assessment. *Land Degrad. Dev.* **2005**, *16*, 489–501.
49. Grismer, M.E.; Ellis, A.L.; Fristensky, A. Runoff Sediment Particle-sizes associated with Soil Erosion in the Lake Tahoe Basin, USA. *Land Degrad. Dev.* **2008**, *19*, 331–350.
50. Grismer, M.E.; Shnurrenberger, C.; Arst, R.; Hogan, M.P. Integrated Monitoring and Assessment of Soil Restoration Treatments in the Lake Tahoe Basin. *Environ. Monit. Assess.* **2009**, *150*, 365–383.
51. Rice, E.C.; Grismer, M.E. Dry-season soil-water repellency effects on infiltration rates in the Tahoe Basin. *Calif. Agric.***2010**, *64*, 141–148.

52. Foltz, R.B.; Luce, C.H.; Stockton, P. The Kinetic Energy Field Under a Rainfall Simulator. In *Watershed Management Planning for the 21st Century*; Committee on the Water Resources Engineering Division/ASCE: San Antonio, TX, USA, 1995.

53. NRCS, Soil Survey Staff, USDA. Soil Survey of the Tahoe Basin Area, California and Nevada, 2007. Available online: www.websoilsurvey.nrcs.usda.gov (accessed on 7 June 2016).

54. Robichaud, P.R.; Lewis, S.A.; Ashmun, L.E. *New Procedure for sampling infiltration to assess post-fire soil water repellency*; U.S. Department of Agriculture, Forest Service, Rocky Mountain Research Station: ReFort Collins, CO, USA, 2008; p. 14.

55. Johnson, L.K. Precision Permeameter Manual; Measuring saturated hydraulic conductivity of earthen materials in the vadose zone. Unpublished material. 2001.

56. Zangar, C.N. *Theory and Problems of Water Percolation*; United States Bureau of Reclamation: Washington, DC, USA, 1953.

CHAPTER 10

THE INFLUENCE OF STATISTICAL UNCERTAINTY IN THE HYDRAULIC BOUNDARY CONDITIONS ON THE PROBABILISTICALLY COMPUTED HIGH WATER LEVEL FREQUENCY

Hua Zhong , Pieter van Gelder , Wen Wang, Gaoxu Wang, Yongzhi Liu and Shuai Niu

Keywords: Statistical uncertainty, the bootstrap method, high water level frequency, Lower Rhine Delta

INTRODUCTION

The hydrodynamic characteristics of a delta or estuary are mainly governed by discharges of rivers and water levels at sea (or lake) boundaries. The joint probability approach is widely used to quantify the high water level frequency in deltas [1,2,3,4,5,6,7]. In the Rhine Delta, the relevant hydrodynamic loading variables at boundaries, namely the astronomical tides, the wind induced storm surges and the river flows, are jointly investigated and their joint probability distribution is estimated from the historical flood events [7]).

The high water level frequency assessment consists of two steps: first, a large number of stochastic scenarios of extreme boundary conditions are generated from the joint probability distribution; second, these generated scenarios are used as inputs to drive a deterministic hydrodynamic model to result in the peak water levels at locations of interest in the delta. The resulting peak water levels can be converted to the high water level frequency curve. The probabilistically computed high water level frequency is critically influenced by the joint probability distribution, time evolution of the relevant loading variables and accuracy of the hydrodynamic model.

In the joint probability approach, to get the marginal distributions (Rhine flood, wind induced storm surges), the annual maximal value during the observed year is determined and then a distribution to this series of annual maxima is fitted. From the obtained distribution, a flood event with a certain return period can be estimated. In general two different sources of uncertainty exist in this process [8]. Natural uncertainty refers to the variability of the flood events, mostly due to the limited number of flood events. For example, to estimate a 1000-year flood event based on a few dozen years flood records. Epistemic uncertainty consists of two parts: the uncertainty in distribution type chosen and in the parameter estimation. In this article, we focus on this natural uncertainty.

Statistical uncertainty exists in the marginal distributions of the joint probability distribution, which refers to the uncertainty in the parameters of the marginal distributions caused by estimating them from a limited number of flood events.And this is what we have to work with. Hence, better assessment of the statistical uncertainty due to insufficient data would be helpful.

The nonparametric bootstrap method is that from the given sample, new samples are generated by using resampling with replacement [9,10,11]. It relies on re-sampling with replacements from the given observations and providing estimates of uncertainty of distribution variables and quantiles.

Citation: Hua Zhong, Pieter van Gelder, Wen Wang, Gaoxu Wang, Yongzhi Liu, and Shuai Niu, The Influence of Statistical Uncertainty in the Hydraulic Boundary Conditions on the Probabilistically Computed High Water Level Frequency, Water 2016, 8(4), 147; doi:10.3390/w8040147.

It is employed to quantify the natural uncertainty in the distributions as it is simple to present and easy to implement [12,13,14]

The statistical uncertainty can be incorporated in the marginal distributions to form new marginal distributions by a Monte Carlo integration method. It is expected that the new distributions will increase the probability of extreme load values because of the uncertainty in the low quantiles of the marginal distributions. In the joint probability distribution, the marginal distributions with/without incorporating the statistical uncertainty are applied to probabilistically compute the high water level frequency curve. Then the impact of the statistical uncertainty of the marginal distributions on the high water level frequency curve can be investigated and quantified.

This study briefly introduces the joint probability approach to assess the high water level frequency in the Lower Rhine Delta, and aims to quantify the statistical uncertainty in the marginal distributions, and further evaluate the impact on the high water level frequency. This article will specifically focus on the case of the Lower Rhine Delta. It is organized as follows: Section 2 presents the method; the case is shown in Section 3; followed by the conclusion in Section 4.

METHODS

The outline of the method is illustrated in Figure 1. First, the bootstrap method is applied to estimate statistical uncertainty in the marginal distributions. Second, the statistical uncertainty is incorporated into the marginal distributions to form into new marginal distributions. Third, the new marginal distributions are applied to compute the high water level frequency in the Lower Rhine Delta, and to assess the impact of statistical uncertainty on the high water level frequency.

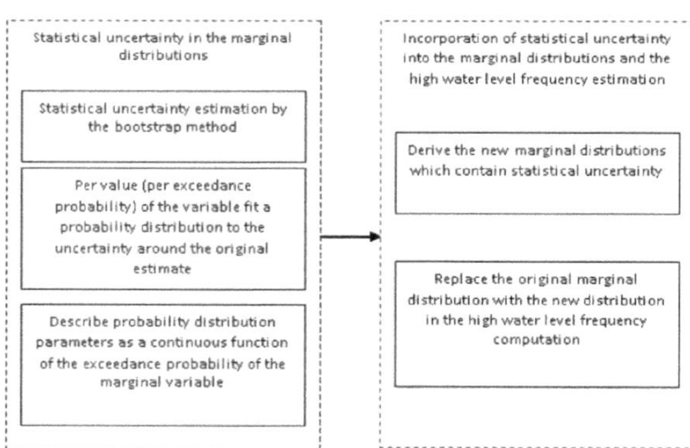

Figure 1: Outline of the method.

Statistical Uncertainty in a Distribution

The statistical uncertainty in a distribution can be estimated using the nonparametric bootstrap method, and subsequently parameterized as a probability distribution function of the exceedance probability of the variable.

The nonparametric bootstrap method generates a predetermined N samples of the variable by randomly re-sampling with replacement from the original sample. Each generated sample has the same size of the original sample. Each sample fits to a same probability distribution, and then there are *N* distributions. For a given exceedance probability of the variable, around the original estimate there are *N* values derived from the *N* distributions. These *N* values can fit to a distribution. For example, the log-normal distribution, a simple skewed distribution can be chosen.

The probability density function of a log-normal distribution is:

$$f_X(x;u,\sigma) = \frac{1}{x\sigma\sqrt{2\pi}}e^{-\frac{(\ln x-u)^2}{2\sigma^2}} \qquad (1)$$

u and σ are the location parameter and the scale parameter on a logarithmic scale respectively.

The log-normal parameters can be made continuous functions of the exceedance probability of the variable. The parameter σ can be modeled as a function of the exceedance probability P by fitting a polynomial.

The mode of the log-normal distribution is the value with maximum probability density. The original estimate of the variable x is assumed to serve as the mode of the log-normal distribution, see Equation (2), and therefore the other log-normal parameter u can be estimated from Equation (3):

$$x = mode(x) = \exp(u - \sigma^2) \qquad (2)$$

$$u(P) = \ln(x(P)) + \sigma(P)^2 \qquad (3)$$

here P is the exceedance probability of the marginal variable.

The statistical uncertainty in the distribution can be estimated by the Log-normal distribution: given a value of x, the corresponding Log-normal distribution, in terms of the parameters u and σ, around that x is known.

Uncertainty-Incorporated Distribution

The statistical uncertainty can be incorporated into the distribution according to Equation (4) by the Monte Carlo Integration method. In Equation (4), x is the original variable, ε is the uncertainty value and x_{un} is the uncertainty incorporated variable, where $x_{un}=x+\varepsilon$.

$$F_{un}(x_{un}) = \int_x F_{logn}((x_{un}); u(x), \sigma(x)) f_X(x)dx \qquad (4)$$

here $F_{un}(x_{un})$ is the uncertainty incorporated distribution of the variable; f_x is the original distribution of the variable; F_{logn} is the Log-normal (statistical uncertainty) distribution conditioned on x.

Impact of the Uncertainty on the High Water Level Frequency in the Rhine Delta

The hydrodynamic characteristics of the Rhine Delta are mainly governed by discharges of the Rhine and Meuse and the water level at the North Sea boundaries, as can be seen in Figure 2 and Figure 3. A joint probability approach using a 1-D hydrodynamic model was applied to assess the high water level frequency in the Lower Rhine Delta [7]. The operational control of the existing flexible hydraulic structures (red points in Figure 4) was taken into account.

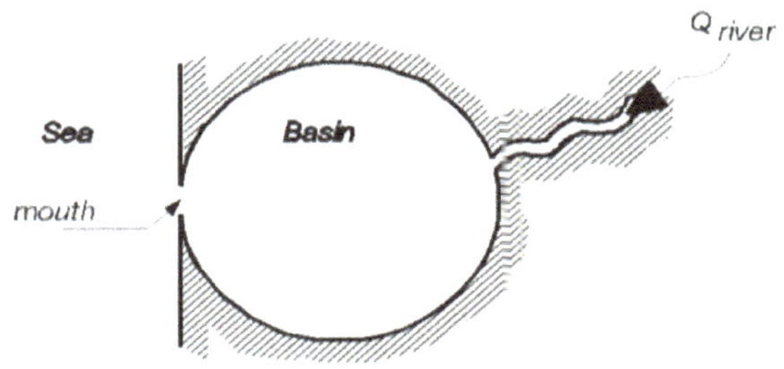

Figure 2: The conceptual model of the open delta.

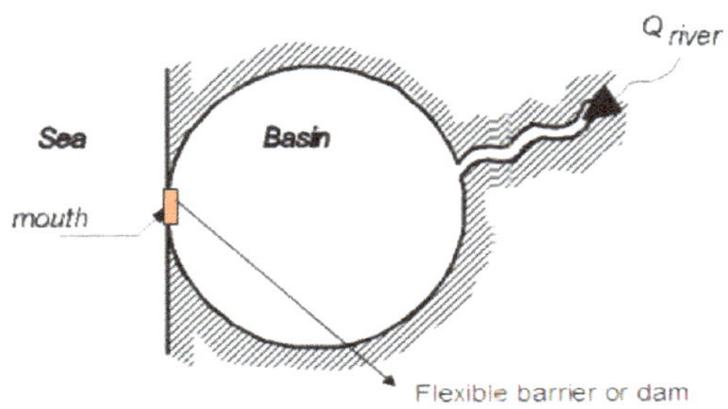

Figure 3: The conceptual model of the closable delta.

Figure 4: The Rhine Delta with operated hydraulic structures [19,20].

To reduce the computation burden, A conceptual model, the "Equal Level Curves" [15,16] is preferred to be applied to examine the interaction of sea level, fluvial flows and infrastructure operations to produce water levels at locations of interest in the Rhine Delta. Equal Level Curves are a simple steady state function, which can simulate the highest water level at Rotterdam by boundary conditions during one tidal period.

Two states of Equal Level Curves are introduced, see Figure 2 and Figure 3: one for the open delta where all dams and barriers are open, and the other for the closable delta where all dams and barriers along the coast are closed. This model has advantages: first, less information needed and a strong reduction in computation time; second, convenient to combine with the present operation control of the Maeslant storm surge barrier at the mouth of the delta. More information is in Appendix A.

According to previous studies [7,16,17], particular attention is paid to the simultaneous occurrence of storm surges and Rhine floods. Facing this kind of flood event, the Haringvliet dam gates and the Maeslant Storm Surge Barrier with the Hartel Storm Surge Barrier should be closed to protect high sea levels from propagating into the delta [18,19,20], but unfortunately the simultaneous high Rhine flow accumulating in a long closure duration can result in extreme high water level behind the dams and barriers.

In the joint probability of this kind of flood events, it is commonly assumed that the magnitude of the Rhine flow is independent of the magnitude of the storm surge [7,21,22,23] found no significant dependence of simultaneous occurrence of storm surges and Rhine floods. The marginal distributions were estimated from the selected flood events (Peak over Threshold values) with the parametric distributions (for example, generalized Pareto distribution). All parameters in the marginal distributions were estimated by the Maximum likelihood method. The joint probability distribution of this kind of event and its marginal distributions are shown in the Appendix B. The statistical uncertainty of three critical marginal distributions in terms of the wind induced storm surge peak level h_{smax}, the wind induced storm surge duration T_s and the high Rhine flow Q_r, are discussed.

The flood risk map of the Netherlands indicated the urbanized areas Rotterdam and Dordrecht are more hazardous and vulnerable than other places in the Netherlands, and with the higher fatalities [24]. As a result, Rotterdam is taken as the study areas of interest for the high water level frequency estimation.

In the joint probability distribution, three marginal distributions with/without incorporating the statistical uncertainties are applied to compute the high water level frequency in Rotterdam. Thus, the impact of the statistical uncertainty on the high water level frequency can be investigated and quantified.

RESULTS

Statistical Uncertainty

For each marginal variable, 1000 bootstrap samples are generated and fitted to 1000 bootstrap curves. Of which 41 curves (by order 25th, 50th ..., 475th, 525th ..., 950th, 975th, as well as 5th and 995th) are selected, shown as blue lines in Figure 5. The exceedance probability curve estimated by the original samples is shown in the red line in Figure 5. For a given exceedance probability of a marginal variable, 41 bootstrap estimates around the original estimate are fitted to a Log-normal distribution, as can be seen in Figure 6. Here only the figures with regard to the wind induce surge peak (h_{smax}) are given.

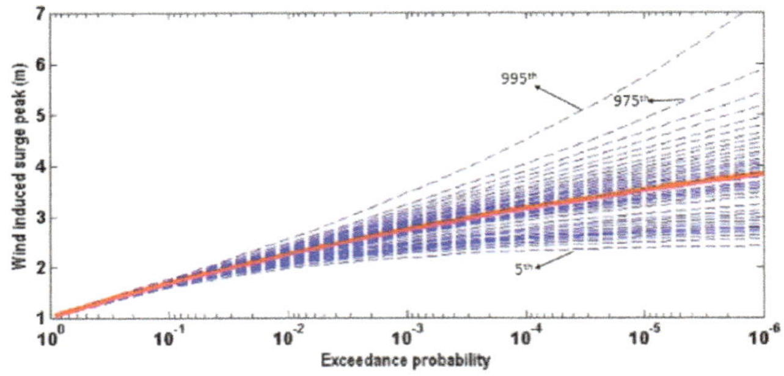

Figure 5: Selected bootstrap quantiles for wind induced surge h_{smax}.

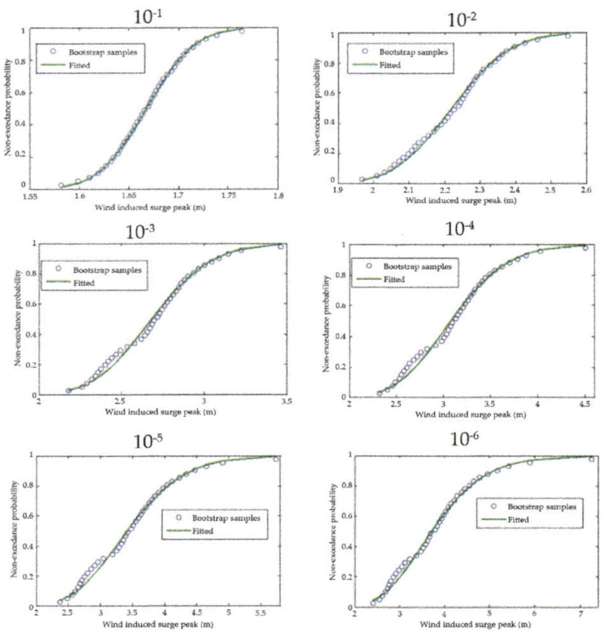

Figure 6: The fitted Log-normal distributions at the exceedance probability of 10^{-1}, 10^{-2}, 10^{-3}, 10^{-4}, 10^{-5}, 10^{-6} of wind induced surge h_{smax}.

The Log-normal parameters can be made continuous functions of the exceedance probability of the marginal variables. The estimated log-normal parameter σ for the exceedance probability is shown in Figure 7. The parameter σ can be modeled as a third-degree polynomial of the exceedance probability in Equation (5) in the region shown.

$$\sigma(P) = a_1\log_{10}(P)^3 + a_2\log_{10}(P)^2 + a_3\log_{10}(P) + a_4 \tag{5}$$

here P is the exceedance probability of the marginal variable, a_1, a_2, a_3, a_4 are the parameters estimated by fitting a polynomial with the method of least squares. The estimates of a_1, a_2, a_3 and a_4 for each marginal variable are shown in Table 1.

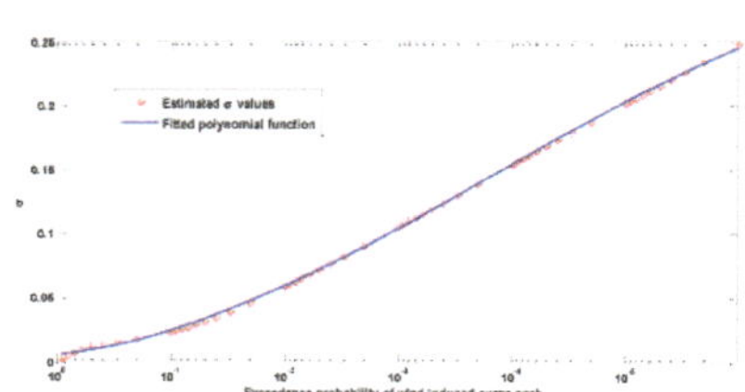

Figure 7: The Log-normal distribution's parameter σ is modeled as a function of the exceedance probability of the wind induced surge $h_{s\text{-}max}$.

Table 1. The estimates of a_1, a_2, a_3 and a_4 for the exceedance probabilities of each marginal variable.

σ	a_1	a_2	a_3	a_4
$\sigma(P(h_{smax}))$	0.0010	0.011	−0.0078	0.0065
$\sigma(P(T_s))$	−0.0001	−0.0021	−0.0141	0.0189
$\sigma(P(Q_r))$	0.0017	0.0247	0.0069	0.0189

The other Log-normal parameter u can be estimated based on Equation (3).

Uncertainty-Incorporated Marginal Distributions

Through application of Equation (4), the statistical uncertainties can be incorporated into the marginal distributions. The results are shown in Figure 8, Figure 9 and Figure 10.

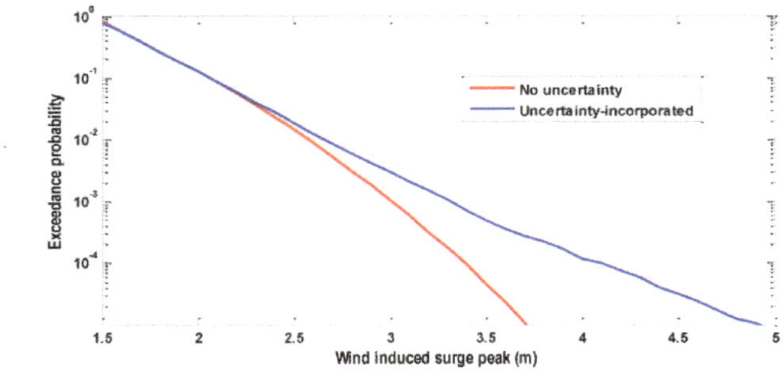

Figure 8: The uncertainty-incorporated marginal distribution of the wind induced surge h_{smax}.

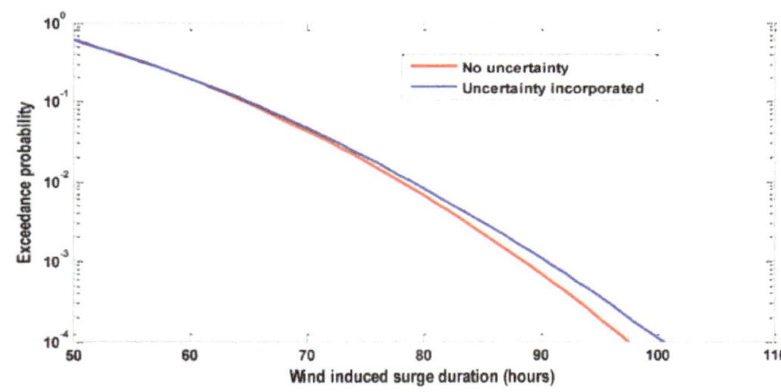

Figure 9: The uncertainty-incorporated marginal distribution of the wind induced storm surge duration *Ts*.

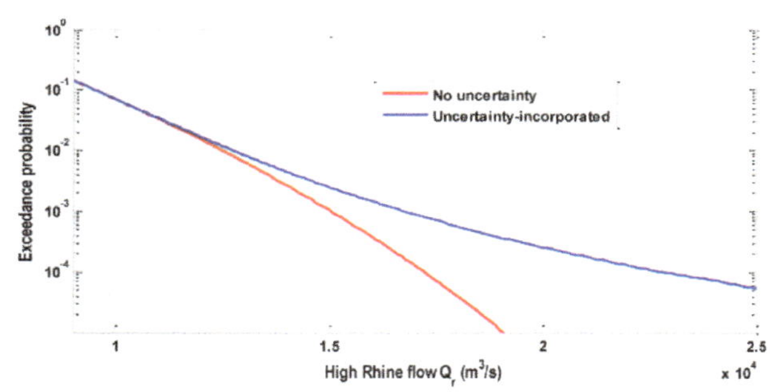

Figure 10: The uncertainty-incorporated marginal distribution of the high Rhine flow *Qr*.

For higher exceedance probabilities (lower return periods), the uncertainty has a negligible effect on the uncertainty-incorporated distribution, mainly due to the small variance of the uncertainty for higher exceedance probabilities (seeFigure 7). For lower exceedance probabilities (higher return periods), the variance increases, and the effect on the distributions becomes more substantial.

The results indicate that uncertainty incorporated marginal distributions result in more extreme values of the marginal variables at low exceedance probabilities.

However, considering the physical conditions the statistical uncertainty in the marginal distributions needs to be constrained at extreme return periods in order to avoid unreal situations.

Impact on the High Water Level Frequency

The impact of the statistical uncertainty on the high water level frequency can be addressed by the differences between the high water level frequency curves derived from two types of marginal distributions, as indicated in Section 2.3. The results are shown in Figure 11, Figure 12 and Figure 13 respectively.

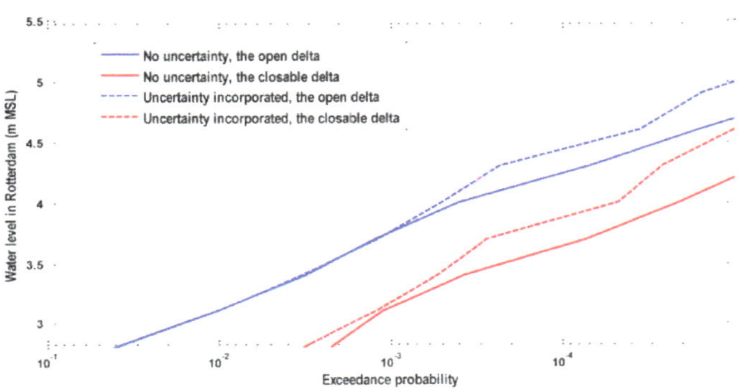

Figure 11: The high water level frequency curve considering the statistical uncertainty in the marginal distribution of the wind induced surge h_{smax}.

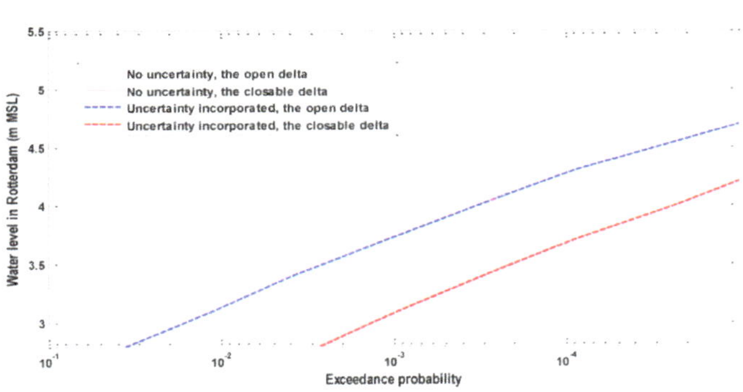

Figure 12: The high water level frequency curve considering the statistical uncertainty in the marginal distribution of the wind induced storm surge duration *Ts*.

In Figure 11, the high water level frequency curves in Rotterdam are estimated from two different marginal distributions of h_{smax} respectively. Incorporating the statistical uncertainty in the marginal distribution of h_{smax} significantly increases the high water level frequency in Rotterdam for the exceedance probabilities lower than 10^{-3}.

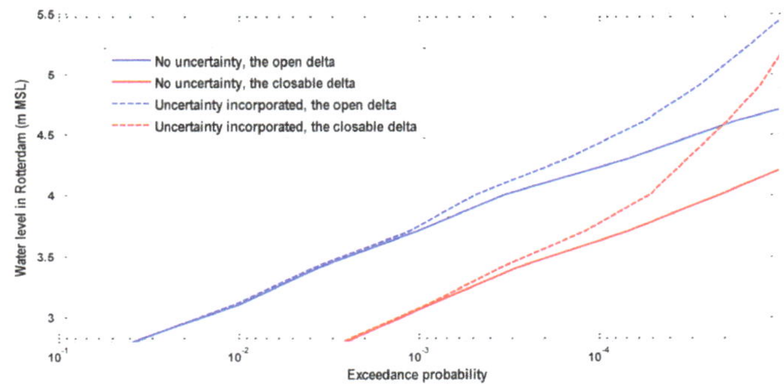

Figure 13: The high water level frequency considering the statistical uncertainty in the above two marginal distributions.

In Figure 12, incorporating the statistical uncertainty in the marginal distribution of *Ts* does not affect the high water level frequency in Rotterdam.

In conclusion, the statistical uncertainty in each marginal distribution differs and its impact on the high water level frequency also varies.

In Figure 13, incorporating the statistical uncertainty in two marginal distributions significantly increases the high water level frequency in Rotterdam. Generally for higher exceedance probabilities (lower return periods), the statistical uncertainty has a negligible effect on the high water level frequency, while for lower exceedance probabilities (higher return periods), the effect on the high water level frequency becomes more substantial.

The design water level in Rotterdam is regarded as the water level with an exceedance frequency of 1/10,000; and its present value is 3.60 m above mean sea level (in brief MSL) [25]. As can be seen in Figure 12, the design water level of Rotterdam corresponds to 3.60 m MSL without considering the statistical uncertainty in three marginal distributions, while it corresponds to 3.75 m with considering the statistical uncertainty in three marginal distributions.

CONCLUSIONS AND RECOMMENDATIONS

Quantifying high water level frequency is critical but complex in deltas or estuaries. In this article, the joint probability approach using a deterministic hydrodynamic model is applied to estimate the high water level frequency. This study aims at investigating statistical uncertainty of marginal distributions of the joint probability distribution and its impact on the high water level frequency. In the Rhine Delta, the results show that incorporating the statistical uncertainty in the marginal distributions will increase the high water level frequency because the probability of extreme hydraulic boundary conditions increases. The statistical uncertainty in each marginal distribution differs and its impact on the high water level frequency also varies.

Generally only a limited amount of data available when calculating the high water level frequency or flood frequency, however, this is what we have to work with. Hence, better assessment of the statistical uncertainty due to insufficient data would be helpful. In condition, considering real physical conditions, this statistical uncertainty should be constrained. For example, Rhine discharges in Lobith and sea levels in Hook of Holland have an physical upper limit. This constrain should be elaborated in future research.

ACKNOWLEDGMENTS

We appreciate financial supports from National Natural Science Foundation of China (51479222, 41471427, 41101510) , the 111 Project (B08048).

APPENDIX A

The Conceptual Model of the Rhine Delta

Considering the characteristics of the Rhine Delta, the conceptual model is illustrated. Two states of Equal Level Curves are introduced: one for the open delta where all dams and barriers are open, and the other for the closable delta where all dams and barriers along the coast are closed, as can be seen in Figure A1 and Figure A2. The closable delta can be open to the sea except that the delta can be closed with the help of hydraulic structures during the extreme weather conditions (storm surges). The conceptual model of the Lower Rhine Delta was introduced in Zhong *et al.* [16].

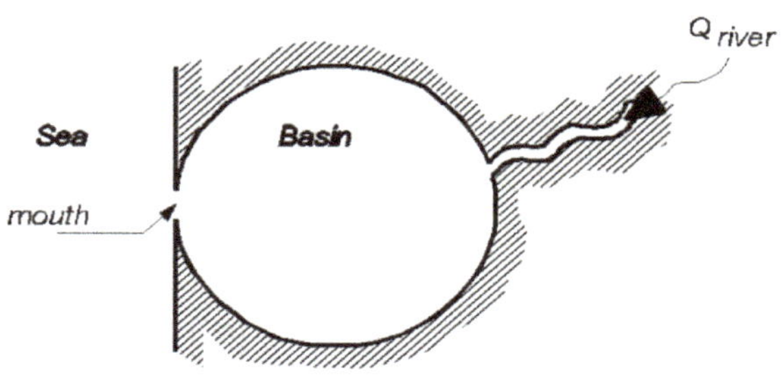

Figure A1: The conceptual model of the open delta.

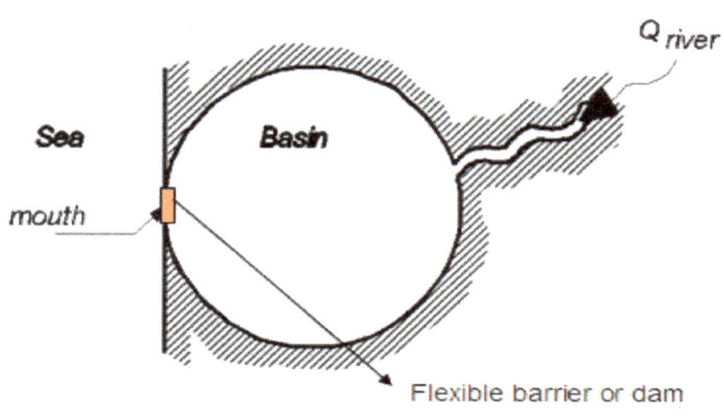

Figure A2: The conceptual model of the closable delta.

Equal Level Curves are a simple steady state function, which can simulate the highest water level at Rotterdam by boundary conditions during one tidal period. For the open delta, Rotterdam water levels can be modeled by the Equal Level Curves with the boundary conditions of the Rhine flow at Lobith and the sea level at Hook of Holland, see Equation (A1):

$$h_R = h_{hvh} + \left(\frac{Q_r + Q_m}{\mu A}\right)^2 \cdot \frac{1}{2 \cdot g} \quad \text{(A1)}$$

here h_R is the water level at Rotterdam; while h_{hvh} stands for the sea water level at Hook of Holland. Q_r is the Rhine flow at Lobith, and Q_r is the Meuse flow at Borgharen. μ is the discharge coefficient, A stands for the surface area of the cross section in Hook of Holland and g is the gravitational acceleration. The parameters of μ and A can be estimated by the linear regression method with the selected historical flood event, and μ·A is estimated to be 3620 m^2.

$$h_{hvh} = h_s + h_a + h_0 \quad \text{(A2)}$$

here h_0 is the mean sea level, h_a is the astronomical tide level, h_s is the wind induced surge level. From a statistical point of view, the occurrence of the astronomical tide component is independent of the occurrence of the wind induced storm surge component at the mouth of the Rhine Delta. However, these two components can interact with each other when they propagate into the delta. Their nonlinear interaction generally increases the surge height at a rising astronomical tide and decreases the surge height at a high astronomical tide. Quantifying the nonlinear effect is beyond the scope of this study. For the sake of convenience, it can be assumed that the wind induced storm surge is independent of the astronomical tide.

Equal Level Curves with the open delta are shown in Figure A3.

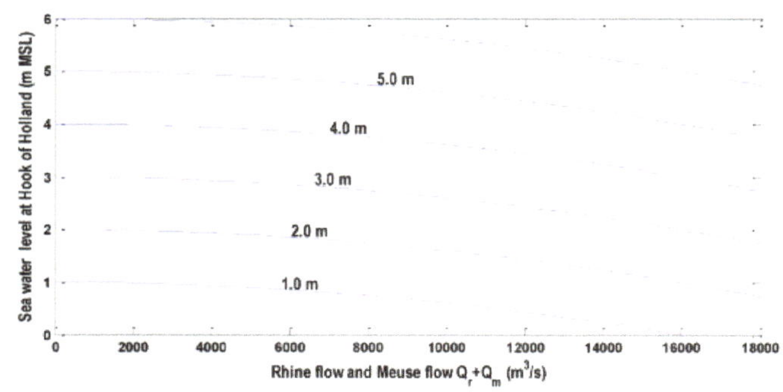

Figure A3: Equal Level Curves with the open delta at Rotterdam.

At present, the Rhine delta can be kept open always via the New Waterway and be closed by closing the Maeslant Barrier and the Haringvliet dams when facing storm surges.

Being a closable delta, there are two states for the Equal Level Curves: the state of open can be described by Equation (A1), the state of closure can be described by Equation (A3). After the closure of the delta, the water level behind the barrier rises because the Rhine flow cannot be discharged into the North Sea and will accumulate. In this process the heights of the surrounding dikes are assumed to be infinite high and no dike breaches occur.

$$h_R = h_{R,c} + \frac{(Q_r + Q_m) \cdot \Delta T}{B} \quad \text{(A3)}$$

where h_R is Rotterdam water level after the closure duration ΔT; $h_{R,c}$ is the average water level behind the Maeslant barrier at the closure time; Q_r is the Rhine flow at Lobith; Q_m is the Meuse flow at Borgharen; B is the surface area of the delta where water can be stored.

$h_{r,c}$ can be estimated from the average water level of four locations (Rotterdam, Goidschalxoord, Dordrecht, and Moerdijk) at the moment of the Maeslant barrier closing.

The water levels at these four locations at time t depend on the sea water level of a few hours ago at Hook of Holland and on the Rhine and Meuse discharges of 24 hours ago at Lobith and Borgharen. The functions are shown in Equations (A4)–(A7). The parameters of the below equations are estimated from system identification of historical measurements. The time unit of t is an hour.

$$h_R^t = 0.9735 \cdot h_{hvh}^{t-1} + 7.781 \cdot 10^{-9} \cdot (Q_r^{t-24} + Q_m^{t-24})^2 \tag{A4}$$

$$h_G^t = 0.7335 \cdot h_{hvh}^{t-1} + 1.013 \cdot 10^{-8} \cdot (Q_r^{t-24} + Q_m^{t-24})^2 \tag{A5}$$

$$h_D^t = 0.6065 \cdot h_{hvh}^{t-2} + 1.737 \cdot 10^{-8} \cdot (Q_r^{t-24} + Q_m^{t-24})^2 \tag{A6}$$

$$h_M^t = 0.3753 \cdot h_{hvh}^{t-3} + 1.458 \cdot 10^{-8} \cdot (Q_r^{t-24} + Q_m^{t-24})^2 \tag{A7}$$

The parameter B can be estimated by inverting Equation (A3) into:

$$B = \frac{(\frac{8}{9}Q_r + Q_m) \cdot \Delta T}{h_R - h_{R,c}} = \frac{(\frac{8}{9}Q_r + Q_m) \cdot \Delta T}{\Delta h} \tag{A8}$$

The factor 8/9 comes from the distribution of the Rhine River inflow of which 1/9 flows north towards the IJsselmeer. The water level can rise by Δh after the closure time ΔT. In the closure event of 2007, the Rhine discharge Q_r was 1171 m^3/s, the Meuse discharge Q_m was 148 m^3/s and the water level at Rotterdam rose from 0.70 to 1.12 m after 15 hours closure. In Figure A4, the derivative is constant and the value of B is estimated to be 152 km^2.

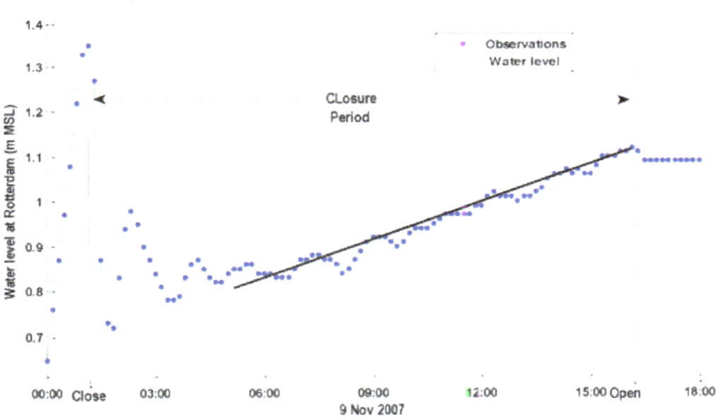

Figure A4: Rotterdam water level during the first closure event of 9 November 2007.

The closure duration ΔT depends on the operational control of closing and opening of the delta. Two hydraulic structures: the Maeslant storm surge barrier and the Haringvliet dams are mainly responsible for closing the Rhine delta. When a high Rhine flow coincides with a storm surge, the present operational control of the Haringvliet dams keep the Rhine delta partly open, depending on the quantity of the Rhine flow in Lobith. However, to keep the analysis simple, it is assumed that when the Maeslant Barrier closes, the Haringvliet dams fully close. This assumption will overestimate the water level when a storm surge coinciding with a high Rhine flow occurs. Therefore, the closure duration ΔT depends on the operational control of the Maeslant Barrier responding to the hydraulic boundary conditions.

The control system of the Maeslant Barrier (named BOS, in Dutch: *Beslissing & Ondersteunend Systeem*; in English, Decision and Support System) has the responsibility to close the barrier completely autonomously [18]. To keep the analysis simple, it is assumed that only one control parameter is considered: the closing decision level H_d, in the operational control of the barrier. When Rotterdam water level is predicted to exceed the closing decision level H_d (3.0 m MSL in Rotterdam), the barrier is assumed to close at two hour before. In reality, the barrier closes at the moment the current starts to change from seaward direction to landward direction, however, to calculate this moment needs extra information which is not the main concern in this article. The water level in Hook of Holland drops after the storm surge, and when the water level in Rotterdam is higher than the water level in Hook of Holland, the barrier re-opens. Considering the time the procedure of close and re-open takes, the minimum closure duration is 6 hours. The closure decision making of the Maeslant Barrier is illustrated in Figure A5.

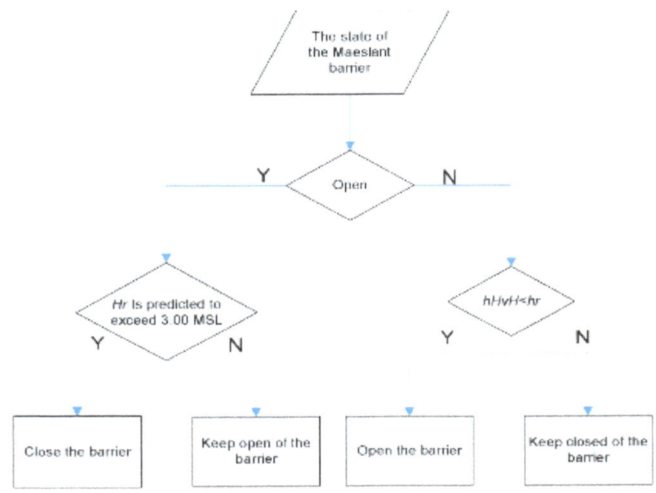

Figure A5: The closure decision making of the Maeslant Barrier.

Equal Level Curves for the closable delta are shown in Figure A6 in which Equal Level Curves in the shaded area are presented in Figure A7.

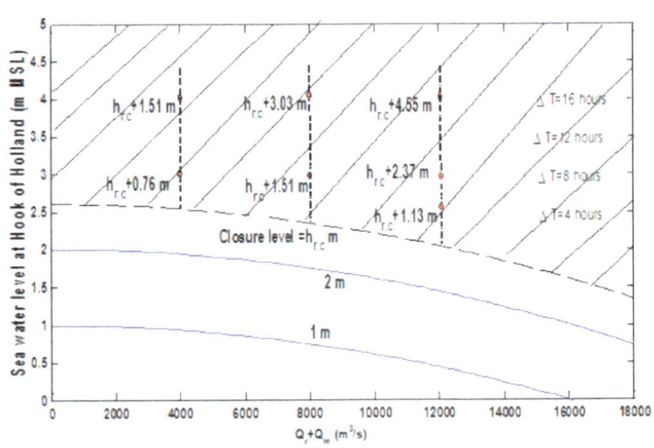

Figure A6: Equal Level Curves with the closable delta at Rotterdam.

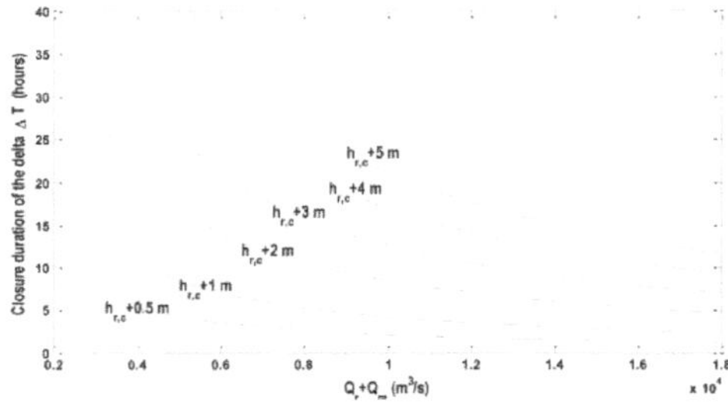

Figure A7: Equal Level Curves with the closable delta at Rotterdam after the delta closes at $h_{r,c}$.

APPENDIX B

Probability Distributions

The wind induced storm surge peak level h_{smax} fits the generalized Pareto distribution (GPD):

$$f(h_{\text{smax}}) = \frac{1}{\sigma}(1+\xi\frac{h_{\text{smax}}-\mu}{\sigma})^{-(\frac{1}{\xi}+1)} \tag{B1}$$

In this equation the shape parameter ξ is −0.0677; the scale parameter σ is 0.3140 m; the location parameter u is 1.0 m.

The wind induced storm surge duration T_s fits the Weibull distribution

$$f(T_s) = \frac{k}{\lambda}(\frac{T_s}{\lambda})^{k-1} e^{-(\frac{T_s}{\lambda})^k} \tag{B2}$$

In this equation, $T_s > 0$, k is the shape parameter, 2.5237; λ is the scale parameter, 38.0887 hours.

The high Rhine discharge Q_r fits the generalized Pareto distribution:

$$f(Q_r) = \frac{1}{\sigma}(1+\xi\frac{Q_r-\mu}{\sigma})^{(-\frac{1}{\xi})-1} \tag{B3}$$

In this equation, ξ is the shape parameter; σ is the scale parameter; u is the location parameter; and the parameters' values are -0.0667, 1629.7 m^3/s and 6000 m^3/s respectively.

The high Meuse discharge Q_m fits the Log-normal distribution:

$$f(Q_m) = \frac{1}{\sigma \cdot Q_m \cdot \sqrt{2\pi}} e^{-\frac{(\ln Q_m-\mu)}{2\cdot\sigma^2}} \tag{B4}$$

In this equation, u is the mean value, 6.8667; σ is the stand deviation value, 0.3752.

The joint cumulative probability distribution of high Rhine flow and high Meuse flow fits a Gumbel Copula function:

$$F_{Q_r,Q_m}(Q_r,Q_m) = C_\alpha(u,v) = \exp\{-[(-\ln u)^\alpha] + (-\ln v)^\alpha]^{\frac{1}{\alpha}}\} \tag{B5}$$

where α is estimated as 1.7158; F_r is the marginal distribution of high Rhine flow; F_m is the marginal distribution of the associated Meuse flow.

The joint probability distribution of storm surges and high Rhine flows

$$P(h_R^*) = \frac{9}{70} \cdot \iiint\iiint I(*)p(h_{\text{smax}})p(h_{HW})p(T_s)p(u)p(Q_r,Q_m)dh_{\text{smax}}dh_{HW}dT_s du dQ_r dQ_m$$
$$I = 1 : h_R^* <= h_R(h_{\text{smax}}, h_{HW}, T_s, u, Q_r, Q_m)$$
$$I = 0 : h_R^* > h_R(h_{\text{smax}}, h_{HW}, T_s, u, Q_r, Q_m) \tag{B6}$$

here I is an indicator function and h_R is the highest Rotterdam water level calculated from the specific input variables using the 1-D model; 9/70 is the occurrence probability per year for the combination event of storm surges and high Rhine flows; h_{smax} stands for the wind induced storm surge peak, and its unit is m; T_s is the wind induced storm surge duration, and its unit is hours; h_{HW} is the high tide level and h_{LW} is the low tide level, their unit is m MSL; u is the time shift between peaks of tide and surge residual; Q_r is Rhine discharge and Q_m is Meuse discharge, their unit is m^3/s; p(Qr,Qm) is the joint probability density distribution of Rhine discharge and Meuse discharge. More detailed information is in Zhong *et al.* [7].

REFERENCES

1. Mantz, P.A.; Wakeling, H.L. Forecasting flood levels for jointevents of rainfall and tidal surge flooding using extreme valuestatistics. *Proc. Institut. Civil Eng. Res. Theor.* **1979**, *67*, 31–50.
2. Acreman, M.C. Assessing the joint probability of fluvial and tidalfloods in the River Roding. *J. Inst. Water Environ. Manage.* **1994**, *8*, 490–496.
3. Gorji-Bandpy, M. The joint probability method of determining theflood return period of a tidally affected pond, Iran.*J. Sci. Technol.* **2001**, *25*, 599–610.
4. Samuels, P.G.; Burt, N. A new joint probability appraisal offlood risk. *Proc. Institut. Civil Eng. Water Maritime Eng.***2002**, *154*, 109–115.
5. Adib, A.; Vaghefi, M.; Labibzadeh, M.; Rezaeian, A.; Tagavifar, A.; Foladfar, H. Predicting extreme water surface elevation intidal river reaches by different joint probability methods. *J. FoodAgric. Environ.* **2010**, *8*, 988–991.
6. Lian, J.J.; Xu, K.; Ma, C. Joint impact of rainfall and tidal level on flood risk in a coastal city with a complex river network: A case study for Fuzhou city, China. *Hydrol. Earth Syst. Sci. Discuss.* **2012**, *9*, 7475–7505.
7. Zhong, H.; van Overloop, P.J.; van Gelder, P. A joint probability approach using a 1-d hydrodynamic model for estimating high water level frequencies in the lower Rhine delta. *Nat. Hazard. Earth Syst. Sci.* **2013**, *13*, 1841–1852.
8. Bruno, M.; Annegret, H.T. Separating natural and epistemic uncertainty in flood frequency analysis. *J. Hydrol.* **2005**,*309*, 114–132.
9. Efron, B. Bootstrap methods: Another look at the jackknife. *Ann. Statist.* **1979**, *7*, 1–26.
10. Efron, B.; Tibschirani, R.J. *An Introduction to the Bootstrap*; Chapman&Hall: New York, NY, USA, 1993.
11. Davison, A.C.; Hinkley, D.V. *Bootstrap Methods and Their Application*; Cambridge University Press: Cambridge, UK, 1997.
12. Dunn, P.K. Bootstrap confidence intervals for predicted rainfall quantiles. *Int. J. Climatol.* **2001**, *21*, 89–94.

13. Kysely, J. Coverage probability of bootstrap confidence intervals in heavy-tailed frequency models, with application to precipitation data. *Theor. Appl. Climatol.* **2010**, *101*, 345–361.

14. Roscoe, K.L.; Diermanse, F. Effect of surge uncertainty on probabilistically computed dune erosion. *Coast. Eng.* **2011**,*58*, 1023–1033.

15. Vrijling, J.K.; van Gelder, P.H.A.J.M. *Probabilistic Design in Hydraulic Engineering*; Delft University of Technology: Delft, The Netherlands, 1996.

16. Zhong, H.; van Overloop, P.J.; van Gelder, P.; Rijcken, T. Influence of a storm surge barrier's operation on the high water level frequency in the Rhine delta area. *Water* **2012**, *4*, 474–493.

17. van der Made, J. Design levels in the transition zone between the tidal reach and the river regime reach, hydrology of deltas. In Proceedings of the Bucharest Symosium, Bucharest, Rumania, 6–14 May 1969; pp. 246–257.

18. Bol, R. Operation of the "Maeslant Barrier": Storm surge barrier in the Rotterdam new waterway. In *Flooding and Environmental Challenges for Venice and Its Lagoon: State of Knowledge*; Cambridge University Press: Cambridge, UK, 2005; pp. 311–316.

19. van Overloop, P.J. *Operational Water Management of the Main Waters in the Netherlands*; Delft Unversity of Technology: Delft, The Netherlands, 2009.

20. van Overloop, P.J. *Optimization of "Everything", Prediction and Control of the entire Delta and River System of The Netherlands*; Technical Report for Water INNovation (WINN); Delft University of Technology: Delft, The Netherlands, 2011.

21. Dantzig, D.; Hemelrijk, J.; Kriens, J.; Lauwerier, H. *Rapport Deltacommissie*; Staatsdrukkerij En Uitgeverijbedrijf: 's-Gravenhage, The Netherlands, 1960. (In Dutch)

22. Jorigny, M.; Diermanse, F.; Hassan, R.; van Gelder, P. Correlation analysis of water levels along dike-ring areas.*Develop. Water Sci.* **2002**, *47*, 1677–1684.

23. Klerk, W.J.; Winsemius, H.C.; van Verseveld, W.J.; Bakker, A.M.R.; Diermanse, F.D. The co-incidence of storm surges and extreme discharges within the Rhine-Meuse Delta. *Environ. Res. Lett.* **2015**, *10*, 035005.

24. De Bruijn, K.M.; Klijn, F. Risky places in the netherlands: A first approximation for floods. *J. Flood Risk Manag.* **2009**, *2*, 58–67.

25. Ministerie van Verkeer en Waterstaat. *Hydraulische Randvoorwaarden Primaire Waterkeringen, Voor de Derde Toetsronde 2006–2011 (hr 2006)*; Ministerie van Verkeer en Waterstaat: Den Haag, The Netherlands, 2007. (In Dutch)

CHAPTER 11

POLLUTANT REMOVAL AND HYDRAULIC REDUCTION PERFORMANCE OF FIELD GRASSED SWALES DURING RUNOFF SIMULATION EXPERIMENTS

Terry Lucke, Mohamed Ansaf Kachchu Mohamed and Neil Tindale

Keywords: Swales, stormwater pollution, total suspended solids (TSS), particle size distribution (PSD), stormwater treatment train

INTRODUCTION

Grassed swales are increasingly being used in a variety of engineering applications to transport polluted stormwater runoff to downstream catchments in an efficient, economic and aesthetically pleasing way. In addition, swales reduce runoff pollutants, require little maintenance, and can be easily incorporated in projects that require a cost-effective stormwater conveyance system. This can often make swales a better choice than traditional curb-and-gutter systems [1]. It has been demonstrated that grassed swales minimise stormwater runoff pollution levels [2] by reducing stormwater flow velocities, which decreases peak outlet discharges and allows filtering and sedimentation processes to occur within the swale.

Research has shown swales can be used as stormwater runoff pre-treatment systems which reduce the need for downstream treatment facilities [3]. Water Sensitive Urban Design (WSUD) guidelines recommend cost-effective and sustainable non-point-source stormwater pollution treatment options. These can include incorporating swales into urban catchments for treating polluted stormwater runoff prior to discharge into receiving waters [4]. WSUD is about integrating water cycle management into urban planning and design. The principles of WSUD are similar to those of Sustainable Urban Drainage System (SUDS) design in Europe and Low Impact Development (LID) in America [5].

Despite significant literature sources reporting the benefits of grassed swales to treat urban stormwater runoff, a fully comprehensive understanding of the design and performance characteristics of swales is still not apparent [6]. This investigation builds on previous swale research with a particular focus on swale length, and how it influences the stormwater attenuation and pollution removal capabilities of grassed swales.

PREVIOUS RESEARCH

Numerous researchers have reported that swales substantially reduce runoff volumes. Ackerman and Stein [7] demonstrated that grassed swales reduce mean runoff volume by approximately 52.5%. Barrett [8] reported that swales may have the potential to infiltrate up to 50% of the runoff volume, provided the soil is permeable and the initial moisture content of that soil is low. Fassman and Liao [3] monitored field swales in New Zealand under natural storm conditions and concluded that, on average, 73.6% of the peak flow discharge was dampened by swales, while 63.7% of the total volume was captured. Bäckström *et al.* [9], and Fassman and

Citation: Terry Lucke, Mohamed Ansaf Kachchu Mohamed and Neil Tindale, Pollutant Removal and Hydraulic Reduction Performance of Field Grassed Swales during Runoff Simulation Experiments, Water 2014, 6, 1887-1904; doi:10.3390/w6071887.

Liao [3] noted complete capture of runoff by swales when rainfall events of less than 2 mm occurred. Yousef *et al.* [10] and Deletic [11] also reported significant runoff reduction by swales. Barrett [8] concluded that the reduction in runoff volume also meant that the total pollutant constituent load was reduced, including nutrient loads, which generally exhibit little change in concentration due to treatment by swales. Increased stormwater retention time, and reduced peak flow rates by swales has the potential to significantly improve the quality of stormwater runoff [7,8].

Previous research agrees that swales remove pollutants through the processes of sedimentation, filtration by grass blades, infiltration into the subsurface zone and bio-chemical processes [2,6,12]. Previous research also reports that swale length, slope, vegetation cover and soil type, all factor into pollution removal performance. Pollutant properties, such as the sediment particle size distributions (PSD) and concentrations, and the amount of particulate bound pollutants also directly affects the pollutant treatment efficiency of swales [1,13,14].

Due to their ability to trap sediments, and consequently pollutant constituents attached to particulate matter, many researchers have measured the pollution reducing efficiency of swales based on total suspended solids (TSS) removal. A summary of previous studies on the TSS removal performance of grassed swales is listed in Table 1.

Table 1: Previous studies on the Total Suspended Solids (TSS) removal performance of swales.

Literature Source	TSS Removal Performance of Grassed Swales (%)		Remarks
	Range	Mean (Median)	
Ackerman and Stein (2008) [7]	41–84	70.6 (72)	* Review of ten different swales studies
	80–99	89 (87)	* Review from five different peer reviewed swale study sources; ** TSS load reduction
Barrett *et al.* (1998) [2]	85–87	86	* Studied two field swales of 1055 m and 356 m long tested under real runoff events (n = 34); ** TSS concentration (EMC) reduction
Deletic and Fletcher (2006) [15] (review section)		72 (76)	* Review of 18 swale study sources
Deletic and Fletcher (2006) [15]	61–86		* A 6.2 m field grass channel studied with runoff simulation; ** TSS concentration (EMC) reduction
		69	* A 65 m long field swale with runoff simulation; ** TSS load reduction
Bäckström (2002) [13]	79–98		* Simulation study on nine different swales of 5–10 m long; ** TSS concentration (EMC) reduction
Yu *et al.* (2001) [12]	67.2–94		* From two field swale studies, one with a 30 m long swale using runoff simulation and other swale of 274.5 m with real time events (n = 4); ** mass sediment removal
Lloyd *et al.* (2001) [16]		74	* A 35 m long swale tested with runoff simulation; ** TSS load removal
Bäckström *et al.* (2006) [9]		15	* Field swale of 110 m long under real storm events (n = 7); ** TSS EMC removal; *** *few negative TSS removals were also observed in the study*
Kaighn and Yu (1996) [1]	29.7–49		* Results from two 30 m long field swales studied under real storm events (n = 8); ** TSS EMC removal
Scheuler (1994) [17]	65–98		* Results from three 61 m long field swales tested under real storm events; *** *one swale showed negative TSS removal due to erosion, which was not given in the range). This finding was verified by Winston et al. (2012) [18] who also found that erosion within a swale caused negative percent reductions for TSS.*
Stagge *et al.* (2012) [6]	44.1–82.7		* Two field swales of 198 m and 138 m long tested with different configurations under real events (n = 45); ** mass TSS removal
Mean	***61.3–86.4***	***67.9 (78.3)***	***Arithmetic mean of the listed literature performance data***

Notes: * type of swales used in the study and experimental method used and number of real storm events (n) sampled; ** TSS measurement method employed in the respective study; *** any specific observations noted; blank cells mean relevant data wasn't available; EMC—event mean concentration of pollutants.

Bäckström [13] found sedimentation of the coarse particles (>25 μm) within the first few metres of the swale length was the most significant factor in removing TSS from runoff, followed by filtration by grass blades predominantly in shallower flow regimes that often correspond to low to moderate intensity rainfall events. Bäckström [13] also reported that laboratory tests on swales generally performed better than field swale tests in sediment trapping. Five metre long field swales showed efficient removal of particles coarser than 25 μm. However, when the lengths of the swales were doubled, particles smaller than 25 μm were also trapped. Bäckström *et al.* [9] confirmed his earlier findings of sediment trapping using a 110 m long roadside grassed swale in Södra Hamnleden, Sweden, under different real rainfall and runoff events. This study revealed that particles larger than 25 μm were effectively trapped by the swale. However, this study found that sediments finer than 25 μm were not retained, and were transported out of the swale, which was in contrast to the earlier study results [13]. Bäckström *et al.* [9] attributed the export of finer sediment to higher flow rates that occurred under real runoff conditions. They concluded [9] that further studies are needed to improve the understanding of the capacity of swales to trap finer particles. Deletic's [14] experimental study on swales concluded that a substantial proportion of sediment particles larger than 57 μm in size were trapped by grassed swales. She also found that the removal efficiency of grassed swales was very low for particles smaller than 5.8 μm.

Previous studies have also looked into the nutrient removal performance by swales. Nutrients such as nitrogen and phosphorous were mostly considered in those studies due to their impact on urban waterways. Removal of total nitrogen (TN) in swales was found to be variable [15,16,17]. Other researchers reported that the removal of TSS particles finer than 150 μm would increase removal of total phosphorous (TP), because approximately 70% of the TP present in urban runoff is bound to particulates [6]. It has been suggested that relevant chemical or biological processes need to take place to significantly remove these nutrients, particularly the dissolved components [12]. However, it is unclear whether swale systems provide adequate Hydraulic Retention Time for these processes to occur [18]. Table 2 and Table 3 list previous research results on the TN and TP removal performance by swales respectively.

It appears from the literature reviewed above that there are significant knowledge gaps relating to the ability of swales to remove pollutants from stormwater runoff. This study investigated the pollutant removal performances of field swales under simulated runoff conditions. As swales convey runoff to downstream water bodies, the main focus of the study was to investigate the level of pollution removal performance that can be expected from grass swales used to treat stormwater runoff before it reaches receiving waters. The study focussed on the three most common pollutants of concern to WSUD practitioners, namely: TSS, TN and TP. The particle size distributions (PSD) of the sediment trapped by the swales, runoff volume reduction, and peak discharge attenuation were also investigated in the study.

Table 2: Previous studies on the Total Nitrogen (TN) removal performance of swales.

Literature Source	TN removal performance of grassed swales (%)		Remarks
	Range	Mean (Median)	
Deletic and Fletcher's [15] review (2006)		45 (50)	* Review of 13 swale study sources
Deletic and Fletcher (2006) [15]		56	* A 65 m long field swale with runoff simulation; ** TN load reduction
Yu *et al.* (2001) [12]	13.8–23.1		* From two field swale studies, one with a 30 m long swale with runoff simulation and other swale of 274.5 m with real time events (n = 4); ** mass TN removal
Lloyd *et al.* (2001) [16]		Nil	* A 35 m long swale tested with runoff simulation; ** TN load removal
Scheuler (1994) [17]	(−X)–46.5		*Results from three 61 m long field swales tested under real storm events; (*** *one swale showed TN export of a certain negative percentage*)
Stagge *et al.* (2012) [6]	(−25.6)–85.6		* Two field swales of 198 m and 138 m long tested with different configurations under real events (n = 45); ** mass TN removal
Yousef *et al.* (1987) [10]	(−7)–11		* From two field swales of 53 m and 170 m long under simulated runoff events; ** EMC reduction
Mean	***−6.3–41.2***	***33.7 (50)***	***Arithmetic mean of the listed literature performance data***

Notes: * type of swales used in the study and experimental method used and number of real storm events (n) sampled; ** TN measurement method employed in the respective study; *** any specific observations noted; −X is an unknown negative value; blank cells mean relevant data wasn't available; EM—event mean concentration of pollutants.

Table 3: Previous studies on the Total Phosphorous (TP) removal performance of swales.

Literature Source	TP removal performance of grassed swales (%) Range	Mean (Median)	Remarks
Barrett *et al.* (1998) [2]	34–44	39	* In two field swales of 1,055 m and 356 m long + tested under real runoff events (n = 34); ** TP concentration (EMC) reduction
Deletic and Fletcher's [15] review (2006)		52 (55)	* Review of 20 swale study sources
Deletic and Fletcher (2006) [15]		46	* A 65 m long field swale with runoff simulation; ** TP load reduction
Yu *et al.* (2001) [12]	28.8–98.6		* From two field swale studies, one with a 30 m long swale with runoff simulation and other swale of 274.5 m with real time events (n = 4); ** mass TP removal
Lloyd *et al.* (2001) [16]		55	* A 35 m long swale tested with runoff simulation; ** TP load removal
Kaighn and Yu (1996) [1]	(−0.4)–33		* Results from two 30 m long field swales tested under real storm events (n = 8); ** EMC removal
Scheuler (1994) [17]	18–41		* Results from three 61 m long field swales tested under real storm events
Stagge *et al.* (2012) [6]	(−49.6)–68.7		* Two field swales of 198 m and 138 m long tested with different configurations under real events (n = 45); ** mass TP removal
Yousef *et al.* (1987) [10]	3–25		* From two field swales of 53 m and 170 m long under simulated runoff events; ** EMC reduction
Mean	***5.6–51.7***	***48 (55)***	***Arithmetic mean of the listed literature performance data***

Notes: * type of swales used in the study and experimental method used and number of real storm events (n) sampled; ** TP measurement method employed in the respective study; *** any specific observations noted; blank cells mean relevant data wasn't available; EMC—event mean concentration of pollutants.

STUDY OBJECTIVES

The main goal of this study was to evaluate the overall performance of grass swales in improving urban stormwater runoff quality and mitigating runoff quantity. Four different grassed swales on the Sunshine Coast in Australia were studied using controlled stormwater runoff simulation experiments to evaluate their pollution removal performance. The specific objectives of this research project were to:

- Correlate the overall TSS removal efficiency of the swales to their length;
- Determine the relationship between the trapping efficiency of various sediment size fractions and swale length;
- Evaluate the nutrient removal performance of swales relative to their length;
- Understand the effects of varying influent pollutant concentrations on the swale pollution removal performance; and
- Evaluate the hydrological control characteristics of swales.

EXPERIMENTAL METHODOLOGY

The stormwater pollutant removal performance of four different field swale installations was monitored during 24 controlled field runoff simulation experiments. Controlled field runoff simulations were selected for the study because of their reliability and the difficulties in sampling real time precipitation runoff events. The experiments were designed to compare selected water quality parameters in the influent and effluent runoff. Three different pollutants were tested, namely: TSS, TN and TP. TSS was sampled every 5 m along the swale length and the nutrients TN and TP were tested every 10 m. Four different pollutant concentrations were used in the experiments as shown in Table 4. The reduction in flow rates due to infiltration along the swales was also measured. It must be noted that the pollution loads for the C and D tests are much higher than typical nutrient and sediment concentrations in stormwater runoff in Australia and these were included to ensure that differences in results could be measured.

Table 4: Synthetic runoff pollutant constituents and test types used in simulation experiments.

Pollutant constituents	Test types and design pollutant mix concentrations (mg/L) Test A (TA)	Test B (TB)	Test C (TC)	Test D (TD)	Concentrations observed at swale inlets (mg/L) Test A (TA)	Test B (TB)	Test C (TC)	Test D (TD)
Total suspended solid (TSS)–Silica	0	150	750	1500	0–19	67–96	283–451	511–1211
Total nitrogen (TN)–KNO_3	0	1.000	5.000	10.000	0.115–0.209	1.120–1.270	4.926–5.384	9.495–10.520
Total phosphorous (TP)–KH_2PO_4	0	1.000	5.000	10.000	0.088–0.261	0.947–1.245	3.868–5.145	8.570–11.650

Three of the swales tested were located on the campus of the University of the Sunshine Coast (identified as USC, IC, and CPB in Table 5). The fourth swale was located in Caloundra, Sunshine Coast (identified as SC in Table 5). The swale size, shape, length and slope are also given in Table 5. The four swales were between 30 and 35 m in length. Figure 1 shows the CPB swale that was used in simulation experiments. All four study swales had similar characteristics with the grass type of kikuyu (*Pennisetum clandestinum*). Experiments were performed in swales within seven days of mowing, and the grass heights were varied between 10 and 60 mm.

Table 5: Study swale characteristics.

Swale Name	Length (m)	Shape	Dimensions (m)	Slope (%)	Grass type and grass height (mm)
		Swale characteristics			
USC Engineering (USC)	35	Triangular	b = 4.0, h = 0.16	<1	Kikuyu, 10–60
Sports Complex (SC)	35	Triangular	b = 6.1, h = 0.44	<1	Kikuyu, 10–60
Innovation Centre (IC)	35	Triangular	b = 3.0, h = 0.35	1	Kikuyu, 10–60
Car Park–B (CPB)	30	Triangular	b = 4.3, h = 0.49	1	Kikuyu, 10–60

Notes: * b—top width of swales in metres; h—mid height of swales in metres; all swales were tested under recently mowed conditions (within seven days of mowing) under which grass heights were varying between 10 and 60 mm.

Figure 1: Car Park-B (CPB) swale used in simulation experiments.

Experiments were conducted in 2012 and 2013, identified as R1 (Round 1) and R2 (Round 2) in Table 6. The experiments were conducted at least one day apart in order to allow the soil moisture to stabilise between tests. A runoff simulation approach similar to that used by Deletic and Fletcher [15] was employed in this field study. Each round (R1 and R2) had 12 individual experimental runs. To simulate the rainfall events, a 2000 L tank filled with clean water was used. The first set of experiments in 2012 (R1) were conducted using an average inflow rate of approximately 1.6 L/s for 21 min. The selected flow rate and the duration were limited by the capacity of the tank. However, this flow rate was considered to be appropriate to simulate a one year, 21 min, average recurrence interval (ARI), naturally occurring storm event (rainfall intensity = 29.3 mm/h) typically experienced on the Sunshine Coast.

Table 6: Experimental arrangements and tested parameters.

Test Name	Swale Name	Experiment	Inflow	IVMC (%)	Outflow (%)	WQ Tests
Round—R1 (Experiments performed in 2012)	USC Engineering (USC)	R1-USC-TA	2000 L of runoff delivered into the swales at an approximate average flow rate of 1.6 L/s (simulating 21 min runoff events)	NM	NM	TSS, TN & TP (Samples collected at every 5 m along swales and analysed for these WQ parameters)
		R1-USC-TB				
		R1-USC-TC				
		R1-USC-TD				
	Sports Complex (SC)	R1-SC-TA				
		R1-SC-TB				
		R1-SC-TC				
		R1-SC-TD				
	Innovation Centre (IC)	R1-IC-TA				
		R1-IC-TB				
		R1-IC-TC				
		R1-IC-TD				
Round—R2 (Experiments performed in 2013)	USC Engineering (USC)	R2-USC-TA	2000 L of runoff delivered into the swales under varying flow rates of 0.5–2.0 L/s (simulating 30 min runoff events)	39.5	NM	TSS, TN, TP & PSD (TSS and PSD analysis performed on samples collected at every 5 m along swales and nutrient tests were performed with every 10 m samples)
		R2-USC-TB		45.6	NM	
		R2-USC-TC		10.2	46.5	
		R2-USC-TD		34.3	53.5	
	Sports Complex (SC)	R2-SC-TA		47.2	NM	
		R2-SC-TB		27.6	NM	
		R2-SC-TC		11.3	0	
		R2-SC-TD		19.5	0	
	Car Park–B (CPB)	R2-CPB-TA		35.0	42.1	
		R2-CPB-TB		52.0	68.1	
		R2-CPB-TC		48.5	75.0	
		R2-CPB-TD		52.3	88.3	

Notes: * IVMC—initial volumetric moisture content of swales, Outflow—outflow measured as a percentage of inflow, NM—not measured, WQ tests—water quality tests performed in respective experiments, experiment names should be read as Round#-Swale name-Test type as shown in Table 5 (e.g., R1-USC-TA).

In 2013 (R2), the Innovation Centre (IC) swale was replaced by Car Park-B (CPB) swale for field simulation experiments due to non-accessibility to the IC swale. For the second set of experiments (R2) conducted in 2013, the swale inflow rates were varied from between 0.5 and 2.0 L/s based on the hydrograph from a one year, 30 min ARI rainfall event. Inflow rates were regulated using an electromagnetic flow meter to measure and a valve at the 2000 L tank outlet. The PSD of the water sample sediment was also analysed. In R2, initial swale moisture contents of swales were measured, and swale outflow measurements were also performed (Table 6). The moisture content of the swale soil profile was measured at different locations using a moisture probe that records volumetric moisture content of the soil matrix. The average measured moisture value can be seen in Table 6 in the IVMC column. The flow rate at the outlet was measured continuously throughout the event using a sharp edged V-notch weir during R2 experiments.

Synthetic pollutant constituents were used in the experiments to simulate runoff pollutant levels, which allowed finer control of their concentration levels. Both R1 and R2 experiments were designed with different pollutant concentrations in order to help understand the performance of swales under varying pollutant discharges. This can be related to conditions occurring during the "first flush" phenomenon events, as well as typical pollutant loads. Each swale was tested under four different inlet pollutant loading conditions referred to as TA, TB, TC and TD as in Table 4.

Test-A (TA) was a control experiment, with no added pollutants, to determine the background concentration of the pollutant constituents in each swale tested. All other test runs were carried out with the addition of pollutants as shown inTable 4. Test-B (TB) was designed to simulate urban runoff pollutant concentrations typically found in Australian urban catchments [15,19]. Commercially available silica powder, which closely resembled the PSD of sediment found in urban runoff, was used as the synthetic TSS constituent in the simulated stormwater inside the tank. Chemical reagents KNO_3 and KH_2PO_4 were used to simulate the TN and TP loads respectively. Test-C (TC) and Test-D (TD) were comprised of pollutant concentrations five times and ten times higher than typical Australian urban stormwater pollutant concentrations.

To ensure a relatively homogeneous water column inside the tank, and to maintain constant concentrations of influent pollutant concentrations at the swale inlet, a stirring system using a submersible pump was used inside the 2000 L tank. Synthetic runoff water inside the tank was mixed for 30 min before each experiment, and during each runoff simulation.

Marginal variations in the swale inlet TN and TP concentrations could be attributed to the compound effect of nutrients attached to settled sediments within the tank, and from residual nutrients inside the tank or water (Table 4). Swale inlet pollutant concentrations found during Test-A experiments represent the background pollutants present in the clean water. Samples from the tank outlet every five minutes revealed that outflow concentrations of TSS could vary by up to ±10% during simulations, possibly due to settling of larger particles in the tank.

The release of runoff into the swale was adjusted to different inflow hydrographs and took place for 21 min in R1 experiments and for 30 min in R2 experiments.

Manual grab samples were collected at selected sampling points located along the length of the swales. Figure 2 shows the conceptual swale testing setup and typical swale testing locations. Water samples were collected at the inlet, the outlet and at every either 5 m (for TSS, TN & TP in R1 and for TSS in R2) or 10 m (for TN & TP in R2) along the length of the swale (Table 5, Figure 2). Three individual samples of 300 mL were collected at each of the sampling locations at between 10 and 15 min intervals during the experiments. The three samples were later mixed together to form composite samples for each sampling point. Sampling was undertaken carefully to avoid disturbing the swale bed.

Collected samples were taken to the USC analytical lab within three hours of the field collection. Each water sample was preserved in accordance with the Standard Methods for the Examination of Water and Wastewater and then analysed for TSS, TN and TP according to APHA/AWWA/WEF [20]. Each sample was analysed for PSD of the suspended solid contents using a laser particle sizer—Malvern Mastersizer 3000 [21].

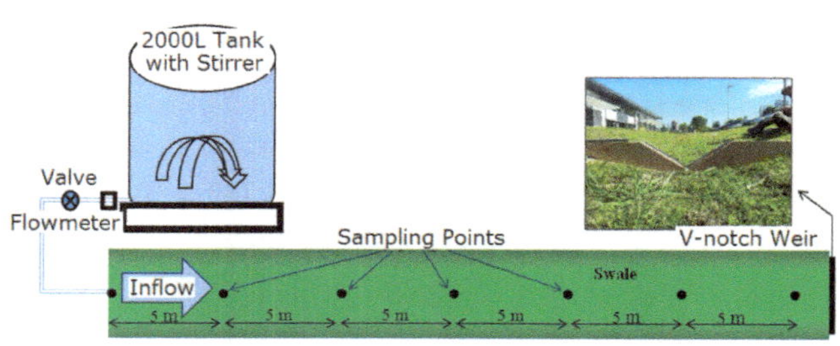

Figure 2: Conceptual swale testing setup.

RESULTS AND DISCUSSION

Hydraulic Performance of Swales

Figure 3 shows the swale inlet hydrograph, and the outflow hydrographs measured at the swale outlet during the different R2 experiments. Flow rates at the outlet reached a peak after approximately 20 min and then decreased exponentially to nearly zero flow after approximately 80 min. Figure 3 shows that a lag time of approximately 13 min was recorded after the start of the tests before any flow was measured at the outlet. While the flow into the swales ceased after 30 min, trickle flows were recorded at the outlet for up to 120 min after the start of the tests. Flow measurements demonstrated a mean total flow reduction of 52% in the 30 m long swales studied, with a peak flow reduction of 61% occurring in one of the study swales. As expected, more infiltration (and hence greater flow reduction) was observed in swales with low initial soil moisture contents.

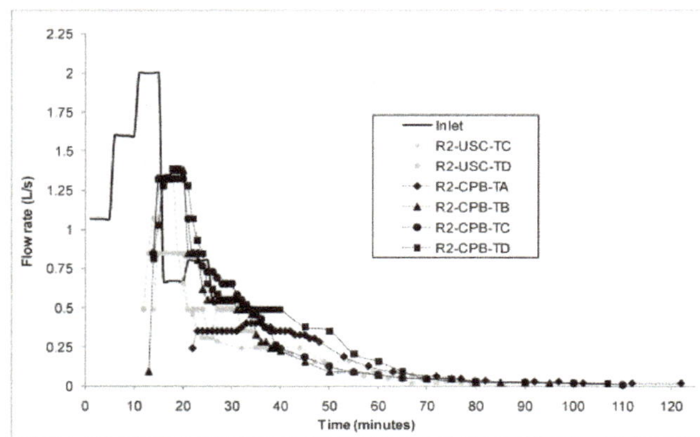

Figure 3: Inlet and outlet hydrographs of flow measured experiments in R2.

The results in Figure 3 demonstrate that swales can be used successfully to attenuate peak stormwater flow rates and to significantly reduce runoff volumes to downstream water courses. The increased runoff retention and peak flow reduction shown by the swales in this study have also suggested that they have the potential to significantly improve the quality of stormwater.

TSS Removal Performance of Swales

The average TSS concentrations measured at 5 m intervals for 30 m along the four study swales for tests TA, TB, TC & TD are shown in Figure 4. The figure clearly shows an exponential decay of TSS concentration along the swale, particularly at the higher pollutant loading tests, TC and TD. This trend agrees with previous research by Deletic [14] who reported an exponential decline of TSS concentration corresponding to swale length. Test-B data points also show an exponential decay of TSS concentration along the swale length, although this was less pronounced.

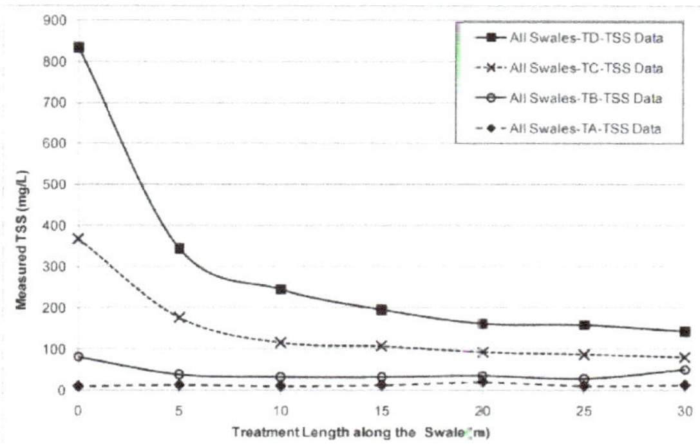

Figure 4: Measured TSS concentrations along the swale length.

TSS concentrations measured during the TA tests showed that the swales had background TSS concentration values of between 0 and 40 mg/L. This agrees with pervious research finds [15,16,18]. It was hypothesised that these background TSS concentrations may have been due to disturbance of the swale bed during sampling or potential scouring of sediments by the runoff along the swale. Measured TSS concentrations below 40 mg/L for all other tests therefore led to variability in the results with some values showing slight increases along the swale length. The study results demonstrate that a swale's TSS removal performance is highly dependent on the inlet concentrations as was shown in previous research [15,16,18]. Results of TA and TB have demonstrated the difficulty in quantifying the efficiency of stormwater treatment devices with very low inlet pollutant concentrations (<40 mg/L).

Figure 4 demonstrates that the swales tested in the study were effective in reducing the higher TSS concentrations in the TC and TD tests. The results of the TC and TD tests also show that swales can treat higher pollution loads typically associated with the "first flush" phenomenon. Results showed that between 50% and 80% of the TSS was generally removed within the first 10 m of the swales. A further 10% to 20% reduction in TSS concentrations can be expected in swales up to 30 m long. Figure 4 also shows that there was a substantial decline in the TSS removal rate after the initial between 10 and 15 m length of the swales and the removal rate becomes very low from that point on. The results of TSS removal by swales in this study generally agreed with previous research results (Table 1).

Sediment Particle Size Removal Efficiency of Swales

PSDs of the swale synthetic sediment (silica) inlet samples were relatively uniform for all the experiments. Figure 5 shows the variation in PSDs of the samples collected at 5 m intervals along the length of the swale for the R2-USC-TD experiment. There is a substantial difference between the PSD samples collected at the swale inlet and the samples collected within the first five to ten metres suggesting that the larger sediment particles were removed in this initial length. However the PSD of samples collected after 15 m show little difference in size. PSD curves followed a relatively similar pattern for the other R2 experimental runs performed under TB, TC and TD test conditions.

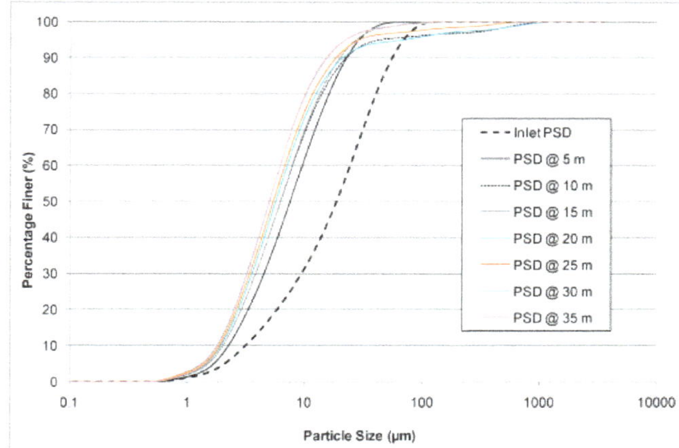

Figure 5: PSD0 of the samples collected during R2-USC-TD experiment.

The variation in the D_{10}, D_{50} and D_{90} sediment sizes along the swales under TC conditions is shown in Figure 6. D_{50} and D_{90} values decreased rapidly within the first 10 m of the swale length and then continued to slowly decrease. D_{90} values declined steadily from approximately 58 µm at the inlet, to approximately 18 µm at the 25 m swale length point. It then increased slightly to approximately 27 µm at the 30 m point. The reason for the slight rise was not confirmed. However, it was hypothesised that it may have been due to the soil erosion occurring in the swale, or due to some minor disturbances potentially caused to the swale bed during sampling. No noticeable change was recorded in the D_{10} values along the length of the swale for the TC tests.

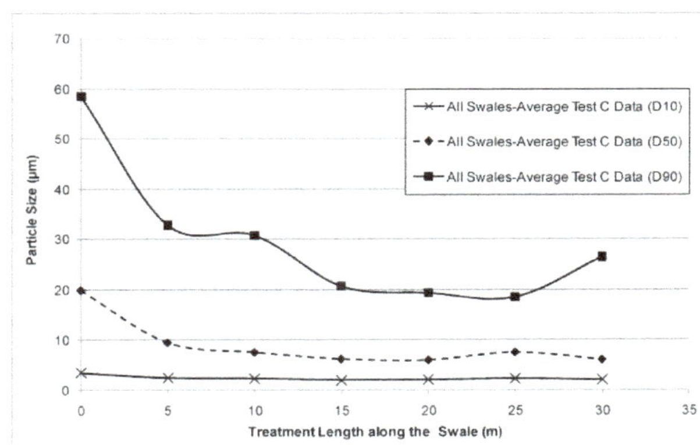

Figure 6: Average D_{10}, D_{50} and D_{90} values along the swale length during Test-C experiments.

The variation of the particle size parameters D_{10}, D_{50} and D_{90} along the swales under TD conditions is shown in Figure 7. D_{50} and D_{90} values decreased rapidly within the first 10 m of the swale length and then continued to slowly decrease. D_{90}values declined steadily from approximately 65 µm at the inlet, to approximately 24 µm at the 10 m swale length point. It then continued to decrease slightly along the rest of the swale length to a value of 15 µm at the 30 m point. A slight decrease in the D_{10} values from approximately 3 µm to 2 µm was measured after a length of 5 m, after which time the value remained relatively constant.

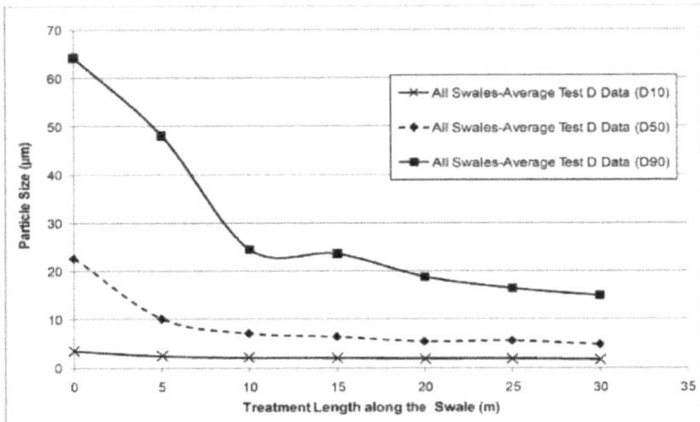

Figure 7: Average D_{10}, D_{50} and D_{90} values along the swale length during Test-D experiments.

The trends and variations of PSD along the swale suggested a clear relationship with the TSS removal (Figure 4). Sedimentation processes and removal of larger sediment particles may explain the higher TSS removal rates within the first 10–15 m of the swales (Figure 4). Similarly, the minimal changes in D_{50} and D_{90} after the first 15 m mirrored the TSS reduction occurring in the initial part of the swale. The results shown in Figure 7 also showed that swales evaluated in this study were not effective in capturing particles finer than 20 µm. However, potential scouring and minor disturbances to the soil during sampling may have led to the varying sediment performance shown in the TA and TB experiments.

The study has found that swales can be used effectively as a primary treatment measure to remove larger sediment from stormwater runoff. The results showed that the first 15 m of the swale length is the most effective in treating the bulk of the TSS. This suggests that the installation of unnecessarily long swales to treat TSS pollutants may not be the optimal solution.

The selection of swales as a primary stormwater treatment measure could significantly affect the design requirements of downstream (or secondary) treatment systems. As the swales were generally shown to be successful in removing particles larger than 20 µm, this suggests that swales could be used in a stormwater treatment train as a pre-treatment to prevent clogging in downstream treatment systems. Results of this study also confirm that a comprehensive understanding of TSS removal and PSD reduction along the swale length is important in the design and sizing of swales, particularly when planning the construction of an urban stormwater runoff treatment train system.

Nutrient Removal Performance of Swales

The average TN concentrations measured at 10 m intervals along the four swales during the four tests (TA-TD) are shown in Figure 8. The results demonstrate that there was no reduction in TN levels measured along the length of the swales for any of the four tests. Indeed, for the TB and TA experiments, the TN concentrations appeared to increase. However, the measured TN levels were low in comparison to runoff from other sites such as highways and carparks and it was hypothesised that leaching of nitrogen components from the swales may have caused the TN increases measured during the TA and TB experiments.

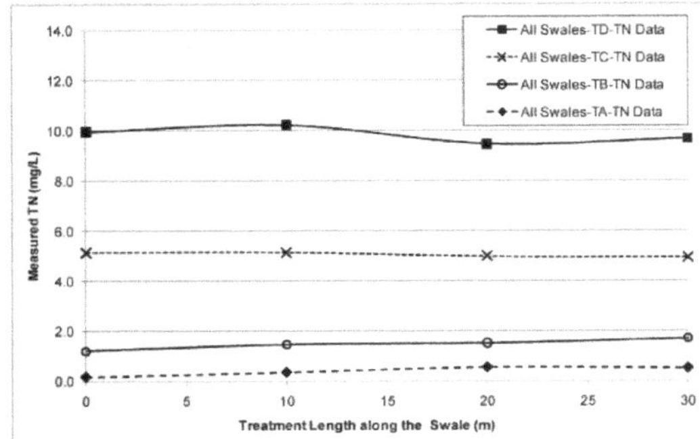

Figure 8: Average TN concentrations for all study swales measured at 10 m intervals.

The experimental results demonstrated that the swales were not effective in removing TN from the synthetic stormwater runoff used in the study. This result is in contrast to a number of previous study results (Table 3). However, the sediment reduction achieved in the swales may also result in a reduction in the overall amount of TN leaving the swales as nutrients are known to attach to sediment particles.

The average TP concentrations measured at 10 m intervals along the four swales during the four tests (TA-TD) are shown in Figure 9. The results show that there was between 20% and 23% reduction in measured TP levels between the inlet and the outlet for the TB, TC and TD tests. The uptake of TP along the swale during the simulation experiments may be attributed to several phosphorous trapping mechanisms that can occur when high TP concentrations are present. Other than direct removal of TP onto the surface of grass and soil within the swale, phosphorous from the simulated runoff may have been adsorbed by finer sediments that settled while flowing in the swales. In addition, the high TSS removal rates shown by swales (Figure 4) may have also assisted in the TP removal performance [6]. However, the results in Figure 9 show there was a substantial increase (61%) in the TP levels between the inlet and the outlet for the TA tests. This was presumably due to leaching of phosphorous components along the swales. The residual of the fertilizers that was used in the tested swales to maintain grass growth may have been contributed to this phosphorous leaching.

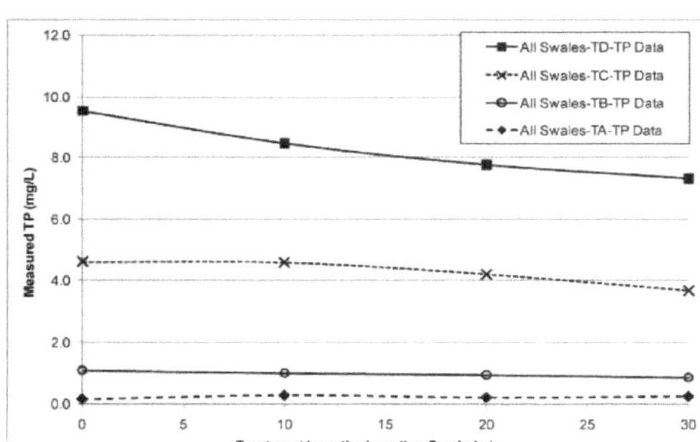

Figure 9: Average TP concentrations for all study swales measured at 10 m intervals.

Differences in the nutrient removal performance of the swales used in this study, compared to previous study results may be attributed to a number of causes, including the testing conditions under which the experiments were performed. For example, the synthetic nutrients (*i.e.*, chemical reagents) used in this study to replicate runoff nutrients were fully dissolved in the simulated stormwater. Real stormwater runoff also contains nutrients in particulate form and the methodology used in this study did not account for these pollutant types.

CONCLUSIONS

Four different field swales were tested during 24 standardised synthetic runoff simulation experiments under varying pollutant loading conditions to evaluate their performance in removing TSS, TN and TP from stormwater runoff. Hydraulic reduction capability of the swales was also assessed by flow measurements carried out at the outlet of the swale during some of the experiments.

Flow measurements demonstrated a mean total flow reduction of 52% in the 30 m long swales studied, with a peak flow reduction of 61%. The initial soil moisture content of a swale was shown to affect infiltration rates, total flow volumes and peak discharges. The study results have demonstrated that swales can be used successfully to attenuate peak stormwater flow rates and to substantially reduce runoff volumes to downstream water courses which can significantly improve the quality of stormwater runoff.

The study has shown that swales were effective in reducing the higher TSS concentrations used in the tests. However, the results demonstrate that a swale's TSS removal performance is highly dependent on the inlet concentrations. Results showed that between 50% and 80% of the TSS was generally removed within the first 10 m of the swales. A further 10% to 20% reduction in TSS concentrations can be expected in swales up to 30 m long. The study also demonstrated that swales can be used to treat higher pollution loads typically associated with the "first flush" phenomenon.

The study has found that swales can be used effectively as a primary treatment measure to remove larger sediment from stormwater runoff. The results showed that the first 15 m of the swale length is the most effective in treating the bulk of the TSS. This suggests that the installation of unnecessarily long swales to treat TSS pollutants may not be the optimal solution. The results suggest that swales could be used in a stormwater treatment train as a pre-treatment to prevent clogging in downstream treatment systems.

The study found no reduction in TN levels in any of the four tests that could be attributed to treatment by the swales. This was in contrast to previous study results. However, the study demonstrated a reduction in measured TP levels of between 20% and 23% between the inlet and the outlet for the TB, TC and TD tests. This reduction is within the range of TP removal reported in previous studies. Differences in nutrient removal performance by swales from this study and other studies may be attributed to the differences in testing conditions and pollutant constituents.

The overall study findings suggest that swales can be used effectively to reduce stormwater runoff pollution, particularly runoff with high concentrations of TSS and TP. Selection of swales as a primary stormwater treatment measure could significantly affect the design requirements of downstream treatment systems. The results from this study will assist designers to estimate the appropriate length of swale required to achieve specific TSS and TP pollution reductions in urban stormwater runoff.

ACKNOWLEDGMENTS

The authors would like to thank the USC laboratory technicians, Hugh Allen and Bernhard Black, and Liam Owen for all their help with this project. This project was jointly funded by the Sunshine Coast Council, the University of the Sunshine Coast, and the Australian Research Council (LP 120200678).

REFERENCES

1. Kaighn, R.; Yu, S. Testing of roadside vegetation for highway runoff pollutant removal. *Transp. Res. Rec.* **1996**, *1523*, 116–123.
2. Barrett, M.E.; Walsh, P.M.; Malina, J.F., Jr.; Charbeneau, R.J. Performance of vegetative controls for treating highway runoff. *J. Environ. Eng.* **1998**, *124*, 1121–1128.
3. Fassman, E.A.; Liao, M. Monitoring of a Series of Swales within a Stormwater Treatment Train. In Proceedings of the 32nd Hydrology and Water Resources Symposium, Newcastle, Australia, 30 November–3 December 2009; Engineers Australia: Barton, Australia, 2009; pp. 368–378.
4. Melbourne Water. WSUD Approach. Available online: http://www.melbournewater.com.au/Planning-and-building/Stormwater-management/Water-Sensitive-Urban-Design/Pages/The-WSUD-approach.aspx (accessed on 26 March 2014).
5. Chahar, B.; Graillot, D.; Gaur, S. Storm-water management through infiltration trenches. *J. Irrig. Drain. Eng.* **2012**, *138*, 274–281.
6. Stagge, J.H.; Davis, A.P.; Jamil, E.; Kim, H. Performance of grass swales for improving water quality from highway runoff. *Water Res.* **2012**, *46*, 6731–6742.
7. Ackerman, D.; Stein, E. Evaluating the effectiveness of best management practices using dynamic modeling. *J. Environ. Eng.* **2007**, *134*, 628–639.
8. Barrett, M.E. Comparison of BMP performance using the international BMP database. *J. Irrig. Drain. Eng.* **2008**, *134*, 556–561.
9. Bäckström, M.; Viklander, M.; Malmqvist, A. Transport of stormwater pollutants through a roadside grassed swale. *Urban Water J.* **2006**, *3*, 55–67.
10. Yousef, Y.A.; Hvitved-Jacobsen, T.; Wanielista, M.P.; Harper, H.H. Removal of contaminants in highway runoff flowing through swales. *Sci. Total Environ.* **1987**, *59*, 391–399.
11. Deletic, A. Modelling of water and sediment transport over grassed areas. *J. Hydrol.* **2001**, *248*, 168–182.
12. Yu, S.; Kuo, J.; Fassman, E.; Pan, H. Field test of grassed-swale performance in removing runoff pollution. *J. Water Resour. Plan. Manag.* **2001**, *127*, 168–171.
13. Bäckström, M. Sediment transport in grassed swales during simulated runoff events. *Water Sci. Technol.* **2002**, *45*, 41–49.
14. Deletic, A. Sediment transport in urban runoff over grassed areas. *J. Hydrol.* **2005**, *301*, 108–122.
15. Deletic, A.; Fletcher, T.D. Performance of grass filters

used for stormwater treatment—A field and modelling study. *J. Hydrol.* **2006**, *317*, 261–275.

16. Lloyd, S.D.; Fletcher, T.D.; Wong, T.H.F.; Wootton, R. Assessment of pollutant removal in a newly constructed bio-retention system. In Proceedings of the 2nd South Pacific Stormwater Conference, Auckland, New Zealand, 27–29 June 2001; pp. 20–30.
17. Schueler, T.R. Performance of grassed swales along east coast highways. *Watershed Prot. Technol.* **1994**, *1*, 122–123.
18. Winston, R.; Hunt, W.; Kennedy, S.; Wright, J.; Lauffer, M. Field Evaluation of Storm-Water Control Measures for Highway Runoff Treatment. *J. Environ. Eng.* **2012**, *138*, 101–111.
19. Duncan, H.P. *Urban Stormwater Quality: A Statistical Overview*; Report No. 99/3. Cooperative Research Centre for Catchment Hydrology: Melbourne, Australia, 1999; p. 80.
20. American Public Health Association, American Water Works Association and Water Environment Federation (APHA/AWWA/WEF). *Standard Methods for the Examination of Water and Wastewater*, 21st ed.; American Public Health Association, American Water Works Association and Water Environment Federation: Washington, DC, USA, 2005.
21. Malvern. Mastersizer 3000 Basic Guide. Malvern Instruments Limited: Malvern, Worcestershire, UK. Available online:http://www.malvern.com/en/products/product-range/mastersizer-range/mastersizer-3000/default.aspx (accessed on 26 March 2014).

CHAPTER 12 AN APPROACH USING A 1D HYDRAULIC MODEL, LANDSAT IMAGING AND GENERALIZED LIKELIHOOD UNCERTAINTY ESTIMATION FOR AN APPROXIMATION OF FLOOD DISCHARGE

Younghun Jung, Venkatesh Merwade, Kyudong Yeo, Yongchul Shin and Seung Oh Lee

Keywords: Discharge approximation, GLUE, Landsat, likelihood measure, data-poor environment

INTRODUCTION

It is recognized that floods are one of the most severe and frequent globally-occurring natural hazards which can limit human activities, and, in this respect, continuous efforts and investments are leading to a better understanding of the nature of floods. Among flood properties, the measurement of river discharge provides fundamental information for flood risk management, and gauging stations play a critical role in collecting data of discharge, water level, and water quality, which are essential information to address real time flood forecasting and reservoir operation. Good gauging station networks allow flow estimations for various return periods, using extreme value statistics based on long gauge records [1,2]. However, the use of subjectivity in relation to the decision of theoretical probability distribution functions for data and the inference methods for parameters, can critically affect the estimation of flow for a rare extreme event [3,4,5,6]. Moreover, lack of the observed data gives a critical difficulty in planning flood risk through computer modeling, which requires hydrologic data, including discharge data that cannot be easily obtained in ungauged basins.

Flood discharge estimation gives significant contribution in identifying the design flood as fundamental information for flood risk management at a planning level. Therefore, flood estimation has been conducted in ungauged basins using hydrologic approaches such as regionalization techniques and conceptual models. Regionalization techniques are used to apply hydrologic information retrieved from neighboring basins, where the required data are collected, to a targeted ungauged basin [7,8,9,10]. The techniques are based on the hydrologic homogeneity between neighboring gauged and ungauged regions. Sometimes, the rational method is also used. The rational method is a simple transformation process of rainfall into runoff using the relationship between the runoff coefficient, rainfall intensity, and basin size [11]. However, estimating the runoff coefficient is difficult because it can vary depending on the rainfall intensity or the geometric properties of a basin. For these reasons, regionalization techniques and conceptual models can only estimate the design flow, but have difficulties in considering distinct attributes for individual flood events.

The availability of spatio-temporal data using remote sensing devices can provide an information-rich environment for analyzing flood properties [12,13,14]. In particular, satellite remote sensing imagery with specific cycles (e.g., daily or every 16 days), can be used to detect the spatiotemporal changes of floods [15,16]. Geographic information systems

Citation: Younghun Jung, Venkatesh Merwade, Kyudong Yeo, Yongchul Shin, and Seung Oh Lee, An Approach Using a 1D Hydraulic Model, Landsat Imaging and Generalized Likelihood Uncertainty Estimation for an Approximation of Flood Discharge, Water 2013, 5, 1598-1621; doi:10.3390/w5041598.

(GIS), also help to build a database, covering spatiotemporal flood evidence obtained from satellite imagery, and allows easy access for the analysis of flood properties with other geographic data (e.g., digital elevation models and land use maps). As a result, the extent of a flood or the water surface elevation obtained from satellite imagery, have played a role in the calibration of data in hydrologic and hydraulic modeling [17,18,19,20]. In addition of obtaining data, there have been great efforts to improve the quality of GIS and RS data in order to extend their applicability [21,22,23,24]. Due to these advantages, the application of satellite imaging and GIS has been considered as one of the most practical and cost-effective approaches for hydraulic and hydrologic analysis in data-poor, or even in ungauged basins, without limitations of spatial boundaries.

The generalized likelihood uncertainty estimation (GLUE) method, proposed by Beven and Binley [25], is one of the first methods to represent prediction uncertainty in hydrologic and hydraulic modeling. The GLUE method uses the Monte Carlo simulations in conjunction with Bayesian theory to produce parameter distributions conditioned on available data and associated uncertainty bounds. The parameter distributions are generated based on multi-parameter sets that can produce acceptable model outputs in comparison with observed data. The criterion for an acceptable model is based on the definition of a user-specified likelihood function. The GLUE method has found widespread implementation in various studies related to uncertainty analysis in environmental, hydraulic, and hydrologic modeling, including flood mapping [26,27,28]. However, the use of formal and informal likelihood measures for model calibration and uncertainty estimations have been debated in certain literatures [29,30,31]. Here, formal and informal mean statistical likelihood measure and non-statistical, respectively. In this study, we acknowledge limitations related to the subjectivity of GLUE's informal likelihood measure, but assume that the GLUE methodology has the practical potential for incorporating the uncertainty in flood modeling in a data-poor environment, where a formal likelihood is not available. However, the use of the GLUE methodology in flood discharge estimation can make it difficult to forecast flood discharge in real time application.

Informal likelihood measures to be used in the GLUE methodology can be estimated by using flood information derived from remote sensing and water level information. In this context, the objectives of this study are to: (1) identify the flooded area and the water surface elevation from satellite imagery (Landsat TM); (2) propose a methodology for flood discharge estimate by using satellite imagery and GLUE; and (3) verify the approximate flood discharge for different informal likelihood measures using the observed discharge from gauge stations. These three objectives are accomplished with the use of satellite image processing, 1D hydraulic modeling, and the informal likelihood measurement estimation for two study reaches in the United States, including a reach along the Missouri River near Nebraska City, Nebraska, and the Wabash River near Montezuma, Indiana. The selected case studies are characterized by the presence of gauged sections where discharge data are available and are used as a benchmark for verifying the discharge values provided by the proposed methodology.

STUDY AREA AND DATA SET

Two river reaches are selected as case studies: one along Missouri River near Nebraska City, Nebraska, USA, henceforth referred to as "Nebraska reach", and another along the Wabash River near Montezuma, Indiana, USA, named as "Montezuma reach", as shown in Figure 1. The two reaches are selected because of the difference in their physical conditions, *i.e.*, in their geometry, size, and topography (Table 1). Satellite imagery taken by the Landsat 5 Thematic Mapper (TM) imaging sensors are available from the USGS Landsat Missions [32] for flood events that occurred after 1982, including the ones corresponding to the catastrophic floods that occurred on 11 June 2008 in Indiana, and on 9 July 2011, in Nebraska.

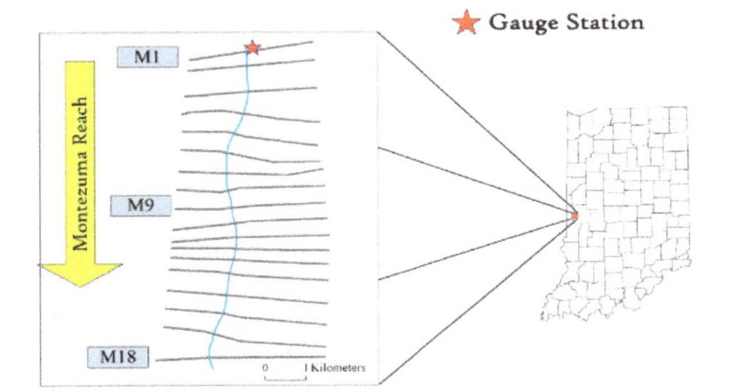

(a)

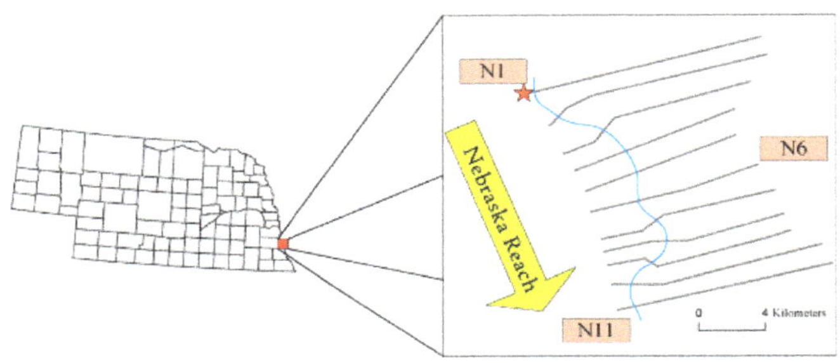

(b)

Figure 1. Study reaches. (**a**) The Wabash River at Montezuma, IN, USA (The Montezuma reach); (**b**) The Missouri River at Nebraska City, NE, USA (The Nebraska reach).

Table 1: Geo-morphological properties of the two selected river reaches for hydraulic modeling.

Study Reach	River length (km)	Mean bed slope (m/km)	Number of cross-section	Mean width of cross-section (km)	Mean spacing cross-section (km)
Montezuma	9	0.25	18	3.61	0.53
Nebraska	19	0.21	11	12.84	1.90

In this study, the use of Hydrologic Engineering Center-River Analysis System (HEC-RAS), a 1D hydraulic model, is justified because the flood extent on the selected Landsat imagery distributes in both of the main channel and floodplain. The geometric data for HEC-RAS consist of eighteen cross sections for the Montezuma reach, and eleven cross sections for the Nebraska reach is commonly extracted from a 10 m × 10 m resolution USGS digital elevation model (DEM), by using HEC-geoRAS. The Manning's n values for the HEC-RAS model are extracted from the National Land Cover Database 2001 [33]. As shown in Table 2, Moore [34] assigned Manning's n roughness coefficients for each land use category in NLCD 2001, based on established literature [35,36].

Table 2: Manning's *n* values for 2001 National Land Cover Database (NLCD) classification.

2001 NLCD Classification	Manning's *n*			Source
	Minimum	Normal	Maximum	
Open Water	0.025	0.030	0.033	[35]
Developed, Open Space	0.010	0.013	0.160	[36]
Developed, Low Intensity	0.038	0.050	0.063	[36]
Developed, Medium Intensity	0.056	0.075	0.094	[36]
Developed, High Intensity	0.075	0.100	0.125	[36]
Barren Land	0.025	0.030	0.035	[35]
Deciduous Forest	0.100	0.120	0.160	[35]
Evergreen Forest	0.100	0.120	0.160	[35]
Mixed Forest	0.100	0.120	0.160	[35]
Scrub/Shrub	0.035	0.050	0.070	[35]
Grassland/Herbaceous	0.025	0.030	0.035	[35]
Pasture/Hay	0.030	0.040	0.050	[35]
Cultivated Crops	0.025	0.035	0.045	[35]
Woody Wetlands	0.080	0.100	0.120	[35]
Emergent Herbaceous Wetland	0.075	0.100	0.150	[35]

Each reach has distinct topographic and geomorphic settings which provide good test beds for comparison: The Wabash River, one of the major rivers in Indiana, drains to the Mississippi River, and another reach, the White River, drains to Wabash River. The selected Montezuma reach along the Wabash River is relatively straight and has a length of about nine kilometers. The floodplain of the Montezuma reach is rough and bumpy and is in a relatively U-shaped valley (Figure 2a). The Missouri River is the longest river (3767 km) in North America, and enters the Mississippi River. The selected Nebraska reach along the Missouri River meanders, and is about 19 km long. It has a flat floodplain and lies in a relatively rectangular valley (Figure 2b).

Discharge data are available at gauged sections, which are only used for verifying the methodology developed in this study. The discharge data available are from gauge stations USGS 03340500 [37] for the Montezuma reach, and USGS 06807000 [38] for the Nebraska reach (Table 3). In addition, both gauge stations provides discharge and gauge height for peak flow as well as daily, monthly, and annual stream flow.

Table 3: The observed discharge at USGS gauge stations for flood discharge on Landsat and peak flow in a flood event.

Study Reach	USGS gauge station	For flood discharge on Landsat imagery		For a peak flow of the flood event	
		Date of Landsat image	Discharge at gauge station (m^3/s)	Date of peak flow	Discharge at gauge station (m^3/s)
Montezuma	03340500	11 June 2008	1450	8 June 2008	2197
Nebraska	06807000	9 July 2011	6031	7 July 2011	6258

Note: the peak flow belongs to a flood event including the flood discharge on Landsat image.

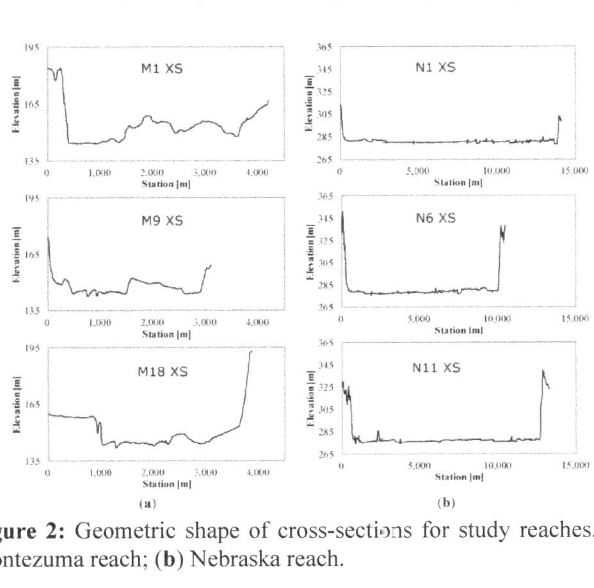

Figure 2: Geometric shape of cross-sections for study reaches. (**a**) Montezuma reach; (**b**) Nebraska reach.

METHODOLOGY

The proposed methodology involves the following steps: (1) identifying the flooded area from Landsat 5 TM satellite imagery using Iterative Self-Organizing Data Analysis (ISODATA); (2) running Monte Carlo simulations with the HEC-RAS model; and (3) approximating the flood discharge using different likelihood measures in the GLUE framework. A brief description of each process used in our methodology follows.

Extraction of the Observed Data from Landsat 5 TM Satellite Imagery

Image classification categorizes values, which indicate specific properties of pixels in a digital image. There are two main methods of classification used for image processing: supervised classification and unsupervised classification. In this study, the process of extracting the water-body from Landsat imagery is conducted using the image processing algorithm ISODATA (Iterative Self-Organizing Data Analysis), which is one of the unsupervised classification methods that uses cluster algorithms to categorize unknown pixels in the image data, without any foreknowledge of the classes [39,40,41,42,43]. In the processes of ISODATA, a number of points indicating cluster centers are randomly placed on the imagery. The pixels are then located within the given Euclidean distance from the placed points, and a cluster is developed. If the standard deviation for the pixel values in a cluster is greater than a certain threshold, the clusters are split, and if the distance between two clusters is less than the threshold, the clusters are merged. The iterative new cluster centers work as a standard to further divide or merge the cluster in the next iteration. Iterations are performed until: (1) the mean distance between adjacent clusters falls to a threshold in a successive iteration; or (2) the maximum number of iterations is reached. However, it should be noted that there are a few subjective decisions, which need to be made, such as the number of classifications and the number of iterations when using ISODATA. In this study, ISODATA classification is performed for 20 classes and 3 iterations [44].

Landsat 5 TM image has seven spectral bands with different wavelengths, and the resolution of all bands is 30 m, except band 6 with a resolution of 120 m. Bands in Landsat imagery can also be differently combined to classify a specific category, such as a water-body, urban land and farmland. In this study, the ISODATA technique was applied to a combination of the spectral bands 1, 4, and 7 of the Landsat 5 TM image, which was categorized into the 20 groups of land use. This specific combination of the spectral bands is considered for the water-body extraction due to the following reasons: (1) blue light (band 1) with the shortest wavelength penetrates clear water; (2) near infrared (band 4) is strongly absorbed by water, and reflected for soil and vegetation; and (3) mid infrared (band 7) has the distinct nature of absorption of water and reflectance for soil and rock [45]. The classified groups are then compared with the satellite imagery and visible spectral bands (bands 1–3), and the colors placed in water-bodies such as lakes, rivers, and reservoirs are then selected as the water-bodies to be extracted from the imagery. Among the selected clusters, ones, which are falsely classified as water-bodies were manually removed, compared with orthography data, and the rest are merged using the GIS tool. In this study, the waterbodies extracted are considered as a flood inundation map for a flood event, and are used to estimate the water surface elevation at each cross-section with DEM. The flood inundation map and the water surface elevation are then used as observations to estimate the likelihood measures in the approximation of flood discharge using the GLUE methodology.

Approximation of Flood Discharge Using HEC-RAS and the GLUE Methodology

Monte Carlo Simulation Using HEC-RAS

The Hydrologic Engineering Center-River Analysis System (HEC-RAS) is a one-dimensional (1D) model developed by the Hydrologic Engineering Center (HEC), of the United States Army Corps of Engineers (USACE) [46]. The HEC-RAS has the ability to simulate water surface elevation for both steady and unsteady flow conditions in a river channel. Water surface elevations using HEC–RAS are calculated using the energy equation [Equation (1)] and the Manning's equation [Equation (2)]:

$$Y_2 + Z_2 + \frac{\alpha_2 V_2^2}{2g} = Y_1 + Z_1 + \frac{\alpha_1 V_1^2}{2g} + h_e \tag{1}$$

where: Y_1, Y_2 are the depth of water; V_1, V_2 are the average velocities; Z_1, Z_2 are the elevation of the main channel; $_1$, $_2$ are the velocity weighting coefficients; g is the gravitational acceleration; and h_e is the energy head loss. In addition, 1 and 2 refer to the upstream and downstream cross section, respectively.

$$Q = \frac{1}{n} A R^{\frac{2}{3}} S_f^{\frac{1}{2}} \tag{2}$$

where, n is manning's roughness coefficient; A is the flow area; S_f is the friction slope; and R is the hydraulic radius.

In this study, the hydraulic model HEC-RAS is used assuming that the main river and the floodplains can be simulated as a single channel and that the floodplain only provides a path of flow parallel to the river centerline not acting as a storage area. As with all the hydraulic models, HEC-RAS needs boundary conditions and input data, including cross-section geometry, roughness coefficients, and river discharge. The HEC-RAS needs boundary conditions and input data, including cross-section geometry, roughness coefficients and flow. HEC-GeoRAS, a GIS tool, helps with preprocessing to run the HEC-RAS model. The HEC-GeoRAS can extract the geometry of digitized cross sections with DEM, and export this geometric information to HEC-RAS.

In Monte Carlo simulations using HEC-RAS, A total of 10,000 HEC–RAS simulations are conducted by assuming a steady-state flow condition for each study reach. The model set-up is performed as follows: (1) Discharge is considered as a single input to the model, and, thus, only one random discharge value is generated to conduct one simulation; (2) The prior probability distribution function (PDF) of unknown discharge are assumed as a uniform because there is no available information on discharge in places where no gauge station exists [47,48,49]; (3) The random discharge for the Monte Carlo simulation is uniformly generated in ranges from 100 m^3/s to 5000 m^3/s for the Montezuma reach, and from 1000 m^3/s to 10,000 m^3/s for the Nebraska reach. Here, the ranges of discharge for both study reaches are determined to sufficiently cover the discharge on the target flood events, by considering the flood extend width and the water depth obtained from satellite imagery and DEM; (4) Manning's n value is generally considered as a primary parameter in calibrating HEC-RAS [50,51]. However, in this study, the roughness coefficients extracted from NLCD 2001 (Table 2) are used for calculation of the water surface elevation without calibration of the model, due to the assumption that there is no gauged information regarding the discharge relationship. Downstream boundary condition is represented by normal depth.

Approximation of Flood Discharge Using the GLUE Methodology

The concept of the GLUE methodology is the basis of the Hornberger-Spear-Young (HSY) global sensitivity analysis [52,53], which involves Monte Carlo simulations using different datasets randomly selected from a feasible range. The GLUE methodology has the objective of classifying behavioral and non-behavioral models with feasible parameter datasets [25]. The results from Monte Carlo simulations are used to calculate likelihood measures, which describe the degree of fitness between the simulations and the observations. Generally, likelihood measures in the GLUE methodology are based on the Bayes equation [Equation (3)], and are estimated using several likelihood functions, including the inverse of the sum of the squared error, the inverse of the sum of absolute error, and the Nash-Sutcliffe efficiency:

$$P_L(X < x) = \sum (L[M(\Theta, I)] \mid X < x)$$

$$L[M(\Theta, I)] = \begin{cases} 0, & P_i < r \\ \frac{1}{P_i}, & P_i \geq r \end{cases} \tag{3}$$

where: P_L is posterior likelihood values; and X is the value of x simulated by model; $L[M(\Theta,I)]$ is a likelihood measure by model prediction (M) for given parameter (Θ) and set of input data (I); P_i is a penalty function; i is iteration; and r is a cut-off threshold.

As shown in Equation (3), a high likelihood measure means a good fit between the simulation and the observations, and*vice versa*. The classification of either a behavioral model or a non-behavioral model among likelihood measures is determined by the user-specific cutoff threshold. However, there is no obvious definition or criteria to select the cut-off threshold for classifying simulations into behavioral and non-behavioral datasets [54]. The cut-off threshold can be determined in terms of an absolute value (e.g., Nash Sutcliffe Efficiency >0.5), or as a percentage of total simulation (e.g., a top 30% of the 10,000 likelihood measures sorted from the highest to the lowest value). The selected likelihood measures as behavioral models are then rescaled from 0 to 1 by Equation (4):

$$RSL_i = \frac{L_i - L_{MIN}}{L_{MAX} - L_{MIN}} \tag{4}$$

where, the ith rescaled likelihood measure (RSL_i) is calculated by the ith likelihood measure (L_i), the minimum likelihood measure (L_{MIN}), and the maximum likelihood measure (L_{MAX}). The rescaled likelihood measures are used to obtain the probability density function (PDF) and cumulative density function (CDF) of the output prediction. Generally, the median of the CDF based on the rescaled likelihood measures, represents the deterministic model prediction, and the uncertainty bound can be represented with the 90% confidence interval bounded by 5% and 95% confidence levels [55,56,57]. In this study, the behavioral model is based on a cut-off threshold, by taking the top 30% of likelihood measures [58], and is used to estimate 90% of uncertainty bounds for approximation of a flood discharge.

In this study, the informal likelihood measure in the GLUE methodology is based on water surface elevations and continuous flood extents obtained from Landsat imagery and GIS. Equation (5) means the sum of the squared error (SSE) of the water surface elevations used in the GLUE methodology to calculate likelihood measures (E likelihood measure), using the difference between the observations and the simulations in flood modeling [59,60,61,62].

$$E_i = P_i = \sum_{j=1}^{N} \left(E_{m,i} - E_o\right)^2 \quad (5)$$

where, $E_{m,i}$ and E_o represent the ith iteration of the modeled water surface elevation and the observed water surface elevation, respectively, for the jth among a total N of cross-sections.

Despite the limitation that water surface level at gauge stations as an observation cannot capture the spatial distribution of flooding, the water surface level still plays an important role, within a spatial data-poor area, in estimating the likelihood measure for the observed flood inundation maps, (due to the lack of observed flood inundation maps). In recent decades, the extent of flood inundation extracted from satellite imagery has been used as observations for capturing the spatial uncertainties in flooding [44,63,64,65]. F-statistic [Equation (6)] is a typical likelihood function using the spatial distribution of flooding, which means the goodness of fit between the observed flood area and the simulated flood area, based on their overlapped area [66,67,68,69,70].

$$F_i = \frac{1}{P_i} = \left(\frac{A_{op,i}}{A_o + A_{p,i} - A_{op,i}} \right) \times 100 \quad (6)$$

where A_o indicates the observed inundation area; A_p refers to the predicted flood inundation area; and A_{op} represents an overlap of both the observed and the predicted inundation areas.

In this study, the simulated flood area (A_p) of Equation (6) is based on results obtained from HEC-RAS, which provides water surface elevations only along cross sections. To produce the flood inundation area from these discrete model outputs, the DEM is subtracted from the water surface area, which is derived by the inverse distance weight (IDW) interpolation of the water surface elevations at each cross section. Equation (7) (EF likelihood measure) is a likelihood measure proposed in this study, which considers both the vertical and spatial differences due to the availability of the water surface elevation and the flood extents as observations.

$$EF_i = a \times RSLE_i + (1-a) \times RSLF_i \quad (7)$$

where, EF indicates the coupled likelihood measure based on the rescaled E likelihood measure (*RSLE*) and the rescaled F-statistic (*RSLF*); i is iteration; and   is the weight coefficient ranging from 0 to 1.

In this study, the value of   in Equation (7) is selected with 0.5, under the condition that the vertical and spatial effects of likelihood measure are the same on the extracted flood inundation map. As mentioned in the Introduction, the use of informal likelihood measure in the GLUE methodology has been criticized due to its subjectivity, but different informal likelihood functions, based on the water surface elevation and the extent of flood inundation, can be used to approximate flood discharge bounds, due to the unavailability of formal likelihood measure in this study.

RESULTS AND DISCUSSION

Extraction of the Observed Data from Landsat 5 TM Satellite Imagery

The water-bodies for both study reaches are extracted from spectral imageries consisting of bands 1, 4, and 7 of the Landsat 5 TM data. Figure 3 shows the water bodies extracted using the ISODATA technique. Considering the distribution of the flood extent for both study reaches, the result shows that the flow direction does not follow the original river center for extreme flood events because the flood streamflow flowed over the main channel and the floodplain, while the original flow direction in only main channel located inside floodplain is relatively meandering (Figure 4). This restrains the application for high flow conditions using the model calibrated for low flow conditions, because the downstream slope of flow can be changed by water level corresponding to the magnitude of streamflow. The total area of the extracted water-body is 13.2 km^2 for the Montezuma reach and 105.0 km^2 for the Nebraska reach, and is used as the observed flood inundation maps to estimate the F likelihood measures in the GLUE methodology.

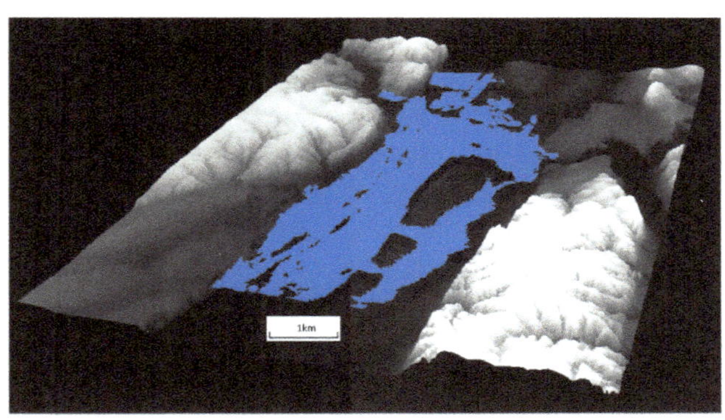

Figure 3: Three-dimensional representations of water-body overlaid on DEM extraction using ISODATA (Montezuma reach, 11 June 2008). Water-body (blue solid) extracted from Landsat image on DEM.

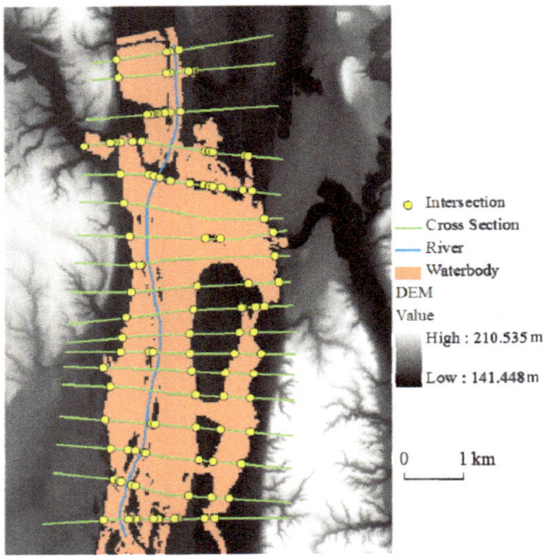

Figure 4: Intersection points for reading water surface elevation from DEM (Montezuma reach, 11 June 2008).

Another observation data obtained from Landsat image is the water surface elevation at each cross section used in the 1D model. To derive the water surface elevation from DEM, the intersection points between boundaries of the flood inundation extents and the cross-sections are required, because the water surface elevation simulated by HEC-RAS is calculated for each cross-section. The use of GIS enhances the accuracy of point locations and reduces the time in finding the intersection points and in reading the elevations of those points from the DEM. Figure 4 shows the cross-sections, extracted flood extent boundary, and intersections between the two overlaid features on DEM of the Montezuma reach. The results for the observed water surface elevations are shown in Figure 5. As a result, there is fluctuation in elevations of intersection points belonging to a cross-section. Thus, an observed water surface elevation at each cross-section is determined by taking an average of elevations of intersection points at the cross-section. Points at which the water-body is too small, or which cloud covers completely, are excluded in the calculation of the average elevation. For 18 cross-sections of the Montezuma reach, the average elevations range from 144.47 m to 148.00 m, and the standard deviations range from 0.03 m to 2.62 m. In the case of the Nebraska reach, the average elevations are in the range of 277.45 m to 281.27 m, and the standard deviations are between a minimum of 0.82 m and a maximum of 1.42 m, respectively. As a result obtained from Landsat and DEM, the Montezuma reach, with its relatively narrow and rough floodplain has a greatly fluctuating water level from upstream to downstream, compared with the Nebraska reach which has a wide and flat floodplain. From these results, the coarse resolution of Landsat imagery produces relatively greater uncertainty in reading the elevation from DEM in a small reach than in a large reach. In particular, compared with the Nebraska reach, rough floodplain of the Montezuma reach (Figure 2) would lead to high storage effect or water level deviation from the extracted flood inundation and DEM because the coarse resolution of Landsat data and the horizontal and vertical error of DEMs can be more propagated to the rough floodplain than to flat floodplain. With coarse resolution of Landsat data and horizontal and vertical errors in DEM, various uncertainty sources, such as the image processing technique, and the geometric characteristics in the processes of flood information acquisition should be considered to improve the accuracy of the flood information obtained from GIS and RS data.

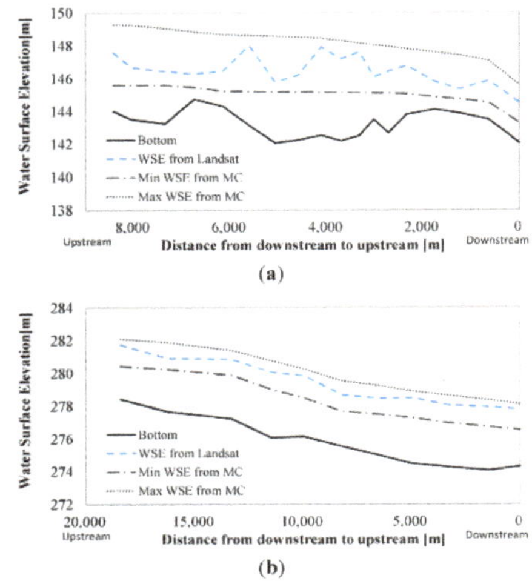

Figure 5: The water surface elevation (WSE) obtained from Landsat image and water surface elevation from Monte Carlo (MC) simulations. (**a**) For the Montezuma reach; (**b**) For the Nebraska reach.

Approximation of Flood Discharge Using HEC-RAS and the GLUE Methodology

Results from Monte Carlo simulations for river discharge are presented in Figure 5. Results show that the water surface elevations obtained using satellite Landsat image and DEM for each study reach, are contained within boundaries of minimum and maximum outputs, based on a total of 10,000 simulations carried out by randomly sampling the steady discharge from the identified variation range and assuming a uniform probability distribution. The simulated water surface elevations over all cross-sections range from 143.26 m to 149.28 m for the Montezuma reach, and from 276.52 m to 281.86 m for the Nebraska reach, respectively. The simulation for the Nebraska reach is conducted with a wider range of discharge than that of the Montezuma reach, but results in a smaller deviation of the simulated water surface elevations. These surface elevations obtained from each simulation are used to calculate E likelihood measures [Equation (5)]. Water surface elevations produced from HEC-RAS are used to produce flood inundation areas using IDW interpolation. The minimum and maximum areas are 7.2 km^2 and 16.8 km^2 for the Montezuma reach, and 63.9 km^2 and 144.1 km^2 for the Nebraska reach, respectively. The Nebraska reach has a twice-longer river than the Montezuma reach. However, although minimum random discharge (1000 m^3/s) in the Nebraska reach is much less than maximum (5000 m^3/s) in the Montezuma reach, the minimum inundated area (63.9 km^2) of the Nebraska reach is about four times larger than the maximum (16.8 km^2) of the Montezuma reach. From these results, it can be expected that the flood inundation depth in the Nebraska reach is shallower than that in the Montezuma reach.

This study used NED DEMs with 10 m resolution for estimation of flood discharge, but available DEMs in most ungauged basins all over the world have low resolutions. Therefore, application using satellite-based DEMs such as SRTM, ASTER can enhance the value of the suggested methodology in ungauged basins for flood discharge estimation. In this context, Sanders [71] evaluated the effect of on-line DEMs such as NED (3 m, 10 m, and 30 m), SRTM (1 s and 3 s), IfSAR (3 m), and LiDAR (3 m) for flood inundation modeling, and obtained the result that the 3 s SRTM (90 m) produced a flood zone only 12% larger than NED DEMs (3 m, 10 m, and 30 m) predictions in the unsteady flow, and a 25% larger flood zone in the steady flow. In addition, the effect of DEM on flood inundation maps could further decrease when the discharge is high or inundated area is large. Thus, the topographic details extracted from high resolution DEM can over-estimate or under-estimate floodwater [72]. These results support the applicability of other satellite-based DEMs in the suggested methodology for flood inundation. However, it should be noted that this methodology has limitations corresponding to the uncertainty sources such as assumption of flow condition and the use of normal roughness values. If errors in the roughness value and flow condition are added, the range of the estimated discharge will be much wider due to uncertainty propagated from them. The assumption of flow condition that the main river and the floodplains can be simulated as a single channel leads to no storage effects in flood inundation modeling. Although the assumption is applied to the flood inundation modeling using hydraulic models [73,74,75], it still has a weakness that the simulated water levels using 1D hydraulic model can be overestimated or underestimated because the stored water in floodplain is not considered. In addition, the roughness is a key uncertainty source in flood inundation modeling. In particular, the HEC-RAS model is based on the Manning's equation [Equation (2)], in which roughness is one of variables. In this study, normal roughness values [34] are used because calibration of the roughness is difficult without the observed discharge. However, we can expect that lower roughness will typically produce lower water level by Manning's equation. In addition, there are additional uncertainty sources such as DEM, rating curve, modeling type, model set-up and assumption, model parameter, and lack of model calibration data in flood inundation modeling [76,77].

The results from Figure 6 show the PDF based on the rescaled likelihood measures of discharge based on three different likelihood functions (F, E, and EF) through the MC simulations. The rescaled value indicates the probability of discharge weighted by each likelihood measure.

For both reaches, the distribution of F likelihood measure, considering only spatial distribution of flood extents, commonly show that they slope very steeply upwards in low discharge and very mildly downward (or almost constant) in high discharge (Figure 6a1,b1). This result is related to the shape of the valley in both study reaches. The U-shaped or rectangular valley has geometric characteristics where a variation of discharge leads to an abrupt change in the inundation area in low flow conditions and a slight change in high flow conditions. For E likelihood measure, considering only elevational differences, likelihood measures of discharge turn very steeply upwards in low discharge, and after the peak they turn very steeply downward in high discharge (Figure 6a2,b2). In this study, the peak in distribution of the E likelihood measure is obviously detected. The distribution of the likelihood measures is almost symmetrical for the Nebraska reach and positively skewed for the Montezuma reach. The distribution of the likelihood measures are likely to be skewed due to the side slope of the valley and the observed surface elevations. The EF likelihood measure of discharge are calculated by taking of 0.5 in Equation (7), which is the average of F and E likelihood measures rescaled from 0 to 1. Figure 6a3,b3 shows that the PDFs of discharge based on EF likelihood measures follow the shape of the ones based on the E likelihood. In actual, in EF likelihood measures can be another uncertainty source in the GLUE methodology. However, research about will be performed in future because is a factor to explain how physical characteristics of flood inundation affect to informal likelihood measure. For example, in U-shaped valley will be greater than 0.5 for more accurate flood discharge approximation because the variation in discharge around fully filled water in the U-shaped valley brings very small change in flood inundation area.

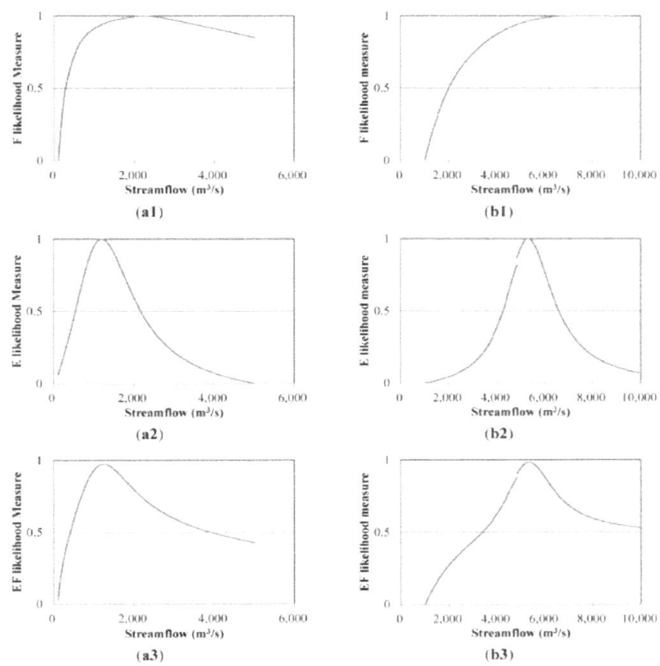

Figure 6: PDFs of discharge based on the rescaled likelihood measure. F likelihood measure: (**a1**) Montezuma Reach; (**b1**) Nebraska Reach; E likelihood measure: (**a2**) Montezuma Reach; (**b2**) Nebraska Reach; The combination of E and F likelihood measure (EF likelihood measure): (**a3**) Montezuma Reach; (**b3**) Nebraska Reach.

Figure 7 and Table 4 show the results from GLUE that the boundaries of the approximated flood discharge is compared with the observations including water surface elevation and flood extents obtained from Landsat 5 TM imagery. The flood discharge is approximated on the CDFs of discharge, which are based on the behavioral models selected by taking the top 30% of each likelihood measure [52]. Figure 7 and Table 4 show the 5%, 50%, and 95% boundaries of the approximated flood discharge using different likelihood measures. The approximation boundary for the Nebraska reach is in the range of 6938 to 8862 m³/s for the F likelihood measure, 4471 to 6344 m³/s for the E likelihood, and 4576 to 6412 m³/s for the EF likelihood measure. The 50% of the CDF giving the deterministic model output, ranges from 5346 to 7911 m³/s. Considering the gauged discharge of 6030 m³/s, the relative errors of approximated discharge are in the range of 10% (617 m³/s) to 31% (−2830 m³/s). In the case of the Montezuma reach, the flood discharge over all likelihood measures is approximated in the range of 687 m^3/s to 2769 m^3/s. Considering the gauged discharge of 1450 m^3/s, the relative errors range from 14% (205 m^3/s) to 55% (−797 m^3/s). From these results, E likelihood based on the elevational difference, commonly produced a much better approximation of discharge than the F likelihood measure, based on spatial differences for both study reaches. The use of the EF likelihood measure improves the approximation of discharge, by reducing the relative error of 1% for the Nebraska reach and 4% for the Montezuma reach. A relative error of 10% in the Nebraska reach indicates a better approximation of discharge than the 14% in the Montezuma reach. The gauged discharge used in this study is estimated from a stage-discharge rating curve involving uncertainty. For comparing results from the suggested method with uncertainty in rating curves, the regression equation for the stage-discharge rating curve based on peak flows provided by USGS is developed assuming a Student's t-distribution (Figure 8). Using a 95% confidence interval, uncertainty for discharge of 1450 m^3/s at USGS gauge station in Montezuma ranges from 938 to 2241 m^3/s and uncertainty for discharge of 6030 m^3/s at USGS gauge station in Nebraska city rages from 2263 to 16,048 m^3/s. These results show that almost all discharge estimated by the suggested method are in the bound of uncertainty in discharge estimated from rating equations (Table 4). Figure 9 shows the simulated flood inundation maps for the approximated discharge, which are deterministic representation values (50% of CDF) based on the EF likelihood measure. Among the three different likelihood measures used in this study, EF likelihood measures commonly produced the best approximation of discharge for both study reaches.

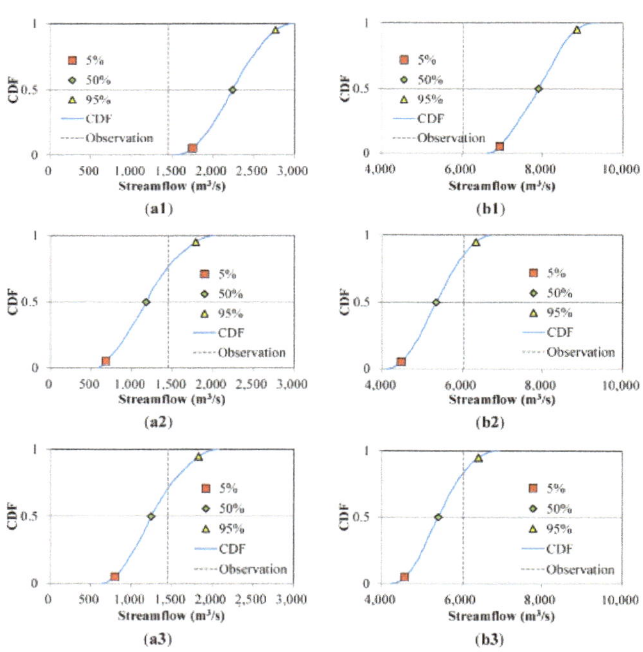

Figure 7: Cumulative density function (CDF) obtained by taking top 30% of likelihood measure and the 5%, 50%, and 95% boundaries of the approximated flood discharge. Based on F likelihood measure: (**a1**) Montezuma Reach; (**b1**) Nebraska Reach; Based on E likelihood measure: (**a2**) Montezuma Reach; (**b2**) Nebraska Reach; Based on the combination of E and F likelihood measure (EF likelihood measure): (**a3**) Montezuma Reach: (**b3**) Nebraska Reach.

Table 4: Boundaries of the approximated discharge for each likelihood measures.

Likelihood Measure	CDF	Discharge (m^3/s)	
		Montezuma	Nebraska
F	0.05	1752	6938
	0.50	***2247***	***7911***
	0.95	2769	8862
E	0.05	687	4471
	0.50	***1179***	***5346***
	0.95	1788	6344
EF	0.05	801	4576
	0.50	***1245***	***5413***
	0.95	1827	6412
Observation		**1450**	**6030**

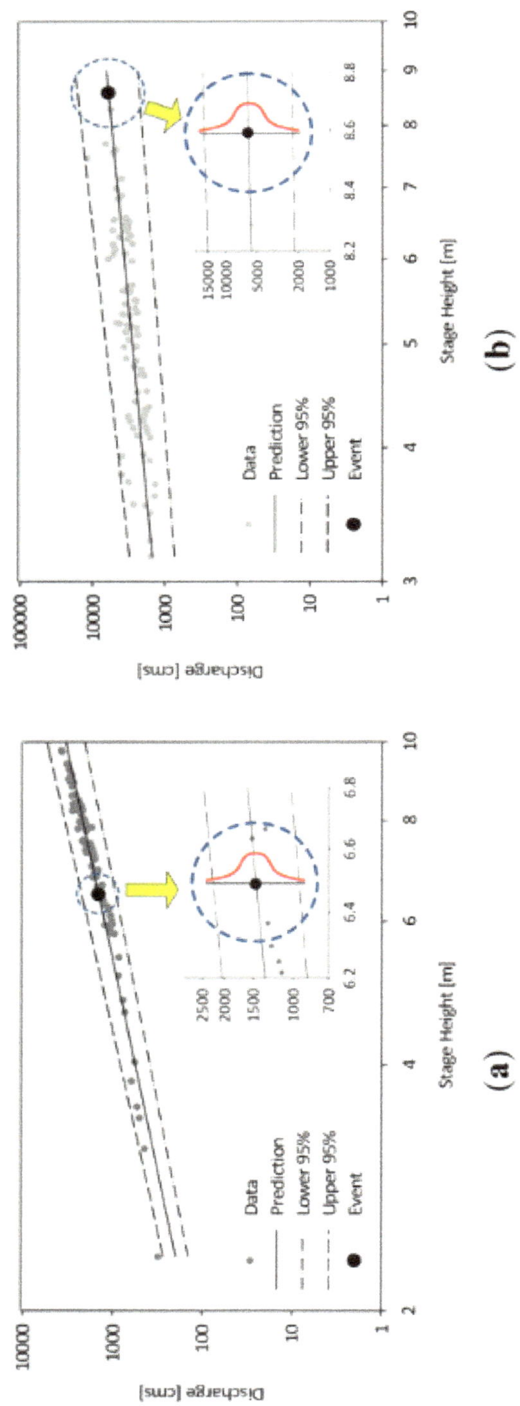

Figure 8: Stage-discharge rating curve based on peak flows provided by USGS. (**a**) Montezuma reach; (**b**) Nebraska reach.

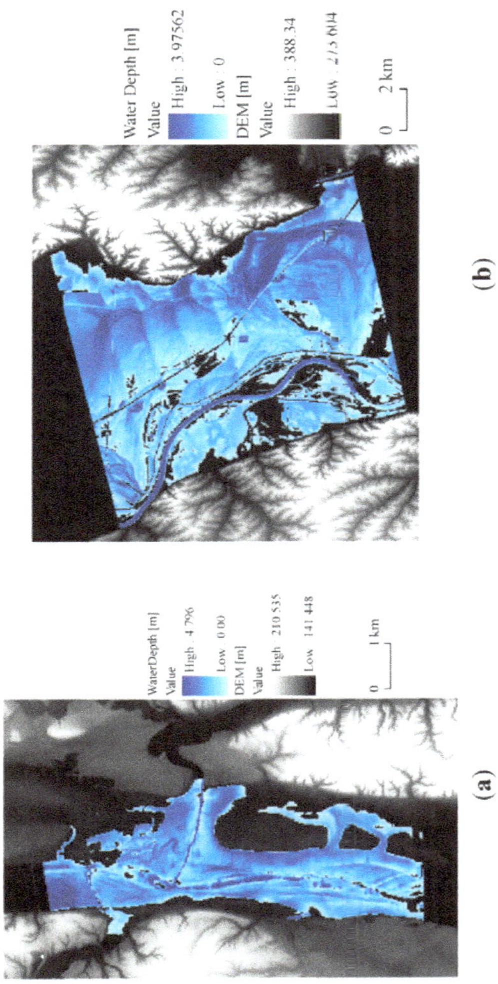

Figure 9: The simulated flood inundation maps for the approximated discharge (50% of CDF based on EF likelihood measure). (**a**) Montezuma (Q = 1245 m³/s); (**b**) Nebraska (Q = 5413 m³/s).

SUMMARY AND CONCLUSIONS

Obtaining flood information in data-poor regions is a difficult and uncertain task, but is very valuable for flood risk management in order to protect human civilization and the environment. The development of GIS, launching devices, and optical sensors have helped to investigate and analyze spatiotemporal flood attributes for past flood events, and the use of remote sensing data can therefore supplement flood information in poorly gauged area. In this paper, we have demonstrated the potential of Landsat image and the GLUE methodology in obtaining hydro-information in an ungauged basin, where formal likelihood measures are unavailable. To demonstrate the approach suggested in this study, Landsat images and the GLUE methodology are used to approximate the flood discharge without information from the gauge stations, which existed in the area.

The following conclusions are drawn from this study:

- This study demonstrates that Landsat imagery can be used as secondary source for discharge estimation in a data-poor environment. The water-body extracted from the Landsat imagery can be used in conjunction with a hydraulic model to estimate flood discharge. However, the use of Landsat imagery in a small-scale study can produce relatively more uncertainty in reading water surface elevation from a DEM (10 m × 10 m) than in a large-scale study, due to the coarse resolution (30 m × 30 m) of a Landsat image. Therefore, flood information obtained from Landsat imagery in planning flood risk management in a data-poor environment is more appropriate for larger rivers. The approximated flood discharge estimated for the Nebraska reach is 5413 m³/s, and 1245 m³/s for the Montezuma reach. The relative errors between the gauged data and the approximations are 10% for the Nebraska reach and 14% for the Montezuma reach, respectively.
- In the GLUE methodology, the different results between E Likelihood measure and F likelihood measure showed subjectivity on the selection of criteria meeting informal likelihood measure. However, when considering the physical conditions of the study reach, such as the shape of the valley, size of the reach, and the flood intensity, the informal likelihood measure in the GLUE methodology can enhance the ability of finding improved flood information in data-poor environment. In addition, each likelihood measure is differently responded corresponding to the random discharge, but produces common results in flood discharge estimation for both study reaches. For example, the approximated discharge for both reaches is overestimated for F likelihood measure on the spatial flood extent and underestimated for E likelihood measure on the observed water surface elevation. In addition, the combination (EF likelihood) of two likelihood measures estimates discharge closest to the observed discharge at gauge station.

This study shows that the use of satellite imagery provides an economical way of obtaining flood information at a planning level in ungauged basins. The approach suggested in this study uses relatively high quality DEM in flood discharge approximation at only two study reaches. Therefore, application of this methodology using available DEMs, such as STRM and ASTER, in most ungauged areas all over the world would give a significant contribution to the applicability of the suggested method. In addition, understanding uncertainty source such as roughness and channel geometry will enhance the accuracy of flood damage estimation in ungauged basin. Moreover, Sufficient RS data with high quality for flood events will provide useful information such as the relationship between flood inundation area and flood level to flood risk management in ungauged basin [16,78]. In this regard, it needs the improvement in sensors and easy access in stationary satellites. Although the use of GLUE involves uncertainty in the subjective decisions on the selection of informal likelihood, the effect of the uncertainty can be reduced by combination of informal likelihood measures considering the surrounding conditions (e.g., EF likelihood measure). The approach suggested in this study is based on flood discharge approximation at only two study reaches. Therefore, more case studies with various physical conditions, higher dimensional hydraulic models, and different remote sensing data, would be helpful in generalizing the role of remote sensing data and GLUE in estimating flood discharge in data-poor environments.

ACKNOWLEDGMENTS

This study was supported financially by the Construction Technology Innovation Program (08-Tech-Inovation-F01), through the Research Center of Flood Defense Technology for the Next Generation, at the Korea Institute of Construction & Transportation Technology Evaluation and Planning (KICTEP), in the Ministry of Land, Transport and Maritime Affairs (MLTM).

REFERENCES

1. Stedinger, J.R.; Vogel, R.M. Foufoula-Georgiou E. Frequency Analysis of Extreme Events. In *Handbook of Hydrology*; Maidment, D.R., Ed.; McGraw-Hill: New York, NY, USA, 1993.
2. Ramachandra, R.A.; Hamed, K.H. *Flood Frequency Analysis*; CRC Press: Boca Raton, FL, USA, 2000; p. 350.
3. Vogel, R.M.; McMahon, T.A.; Chiew, F. Floodflow frequency model selection in Australia. *J. Hydrol.* **1993**, *146*, 421–449.
4. Klemeš, V. Tall tales about tails of hydrological distributions. *J. Hydrol. Engrg. ASCE* **2003**, *5*, 232–239.
5. Mitosek, H.T.; Strupczewski, W.G.; Singh, V.P. Three procedures for selection of annual flood peak distribution. *J. Hydrol.* **2006**, *323*, 57–73.
6. Laio, F.; di Baldassarre, G.; Montanari, A. Model selection techniques for the frequency analysis of hydrological extremes. *Water Resour. Res.* **2009**, *45*, W07416.
7. Dalrymple, T. *Flood Frequency Analyses*; USGS Water Supply Paper 1543-A; USGS: Reston, VA, USA, 1960.
8. Viviroli, D.; Zappa, M.; Schwanbeck, J.; Gurtz, J.; Weingartner, R. Continuous simulation for flood estimation in ungauged mesoscale catchments of Switzerland—Part I: Modelling framework and calibration results. *J. Hydrol.* **2009**,*377*, 191–207.
9. Viviroli, D.; Mittelbach, H.; Gurtz, J.; Weingartner, R. Continuous simulation for flood estimation in ungauged mesoscale catchments of Switzerland—Part II: Parameter regionalisation and flood estimation results. *J. Hydrol.* **2009**,*377*, 208–225.
10. Grimaldi, S.; Petroselli, A.; Serinaldi, F. Design hydrograph estimation in small and ungauged watersheds: Continuous simulation method versus event-based approach. *Hydrol. Process.* **2012**, *26*, 3124–3134.
11. Chow, V.T.; Maidment, D.R.; Mays, L.W. *Applied Hydrology*; McGraw-Hill: New York, NY, USA, 1988.
12. Oberstadler, R.; Honsch, H.; Huth, D. Assessment of the mapping capabilities of ERS-1 SAR data for flood mapping: A case study of Germany. *Hydrol. Process.* **1997**, *10*, 1415–1425.
13. Sivapalan, M.; Takeuchi, K.; Franks, S.W.; Gupta, V.K.; Karambiri, H.; Lakshim, V.; Liang, X.; McDonnell, J.J.; Mendiondo, E.M.; Connell, O.; *et al.* IAHS decade on predictions in ungauged basins (PUB), 2003–2012: Shaping an exciting future for the hydrological sciences. *Hydrol. Sci. J.* **2003**, *48*, 857–880.
14. Montanari, M.; Hostache, R.; Matgen, P.; Schumann, G.; Pfister, L.; Hoffmann, L. Calibration and sequential updating of a coupled hydrologic–hydraulic model using remote sensing-derived water stages. *Hydrol. Earth Syst. Sci.* **2009**, *13*, 367–380.
15. Smith, L.C. Satellite remote sensing of river inundation area, stage and discharge: A review. *Hydrol. Process.* **1997**, *11*, 1427–1439.
16. Pan, F.; Nichols, J. Remote sensing of river stage using the cross sectional inundation area—River stage relationship (IARSR) constructed from digital elevation model data. *Hydrol. Process.* **2012**.
17. Bates, P.D.; Horritt, M.S.; Smith, C.N.; Mason, D. Integrating remote sensing observations of flood hydrology and hydraulic modelling. *Hydrol. Process.* **1997**, *11*, 1777–1796.
18. Sun, W.C.; Ishidaira, W.H.; Bastola, S. Prospects for calibrating rainfall-runoff models using satellite observations of river hydraulic variables as surrogates for *in situ* river discharge measurements. *Hydrol. Process.* **2012**, *26*, 872–882.
19. Bjerklie, D.M.; Moller, D.; Smith, L.C.; Dingman, S.L. Estimating discharge in rivers using remotely sensed hydraulic information. *J. Hydrol.* **2005**, *309*, 191–209.
20. Getirana, A.C.V.; Peters-Lidard, C. Estimating water discharge from large radar altimetry datasets. *Hydrol. Earth Sys. Sci.* **2013**, *17*, 923–933.
21. Durand, M.; Andreadis, K.M.; Alsdorf, D.E.; Lettenmaier, D.P.; Moller, D.; Wilson, M. Estimation of bathymetric depth and slope from data assimilation of swath altimetry into a hydrodynamic model. *Geophys. Res. Letters* **2008**, *35*, L20401.
22. Neal, J.; Schumann, G.; Bates, P.; Buytaert, W.; Matgen, P.; Pappenberger, F. A data assimilation approach to discharge estimation from space. *Hydrol. Process.* **2009**, *23*, 3641–3649.
23. Hostache, R.; Lai, X.; Monnier, J.; Puech, C. Assimilation of spatially distributed water levels into a shallow-water flood model. Part II: Use of a remote sensing image of Mosel river. *J. Hydrol.* **2010**, *390*, 257–268.
24. Matgen, P.; Montanari, M.; Hostache, R.; Pfister, L.; Hoffmann, L.; Plaza, D.; Pauwels, V.R.N.; de Lannoy, G.J.M.; de Keyser, R.; Savenije, H.H.G. Towards the sequential assimilation of SAR-derived water stages into hydraulic models using the particle filter: Proof of concept. *Hydrol. Earth Syst. Sci.* **2010**, *14*, 1773–1785.
25. Beven, K.J.; Binley, A.M. The future of distributed models: Model calibration and uncertainty prediction. *Hydrol. Process.* **1992**, *6*, 279–298.
26. Romanowicz, R.; Beven, K.J. Dynamic real-time prediction of flood inundation probabilities. *Hydrol. Sci. J.* **1998**, *43*, 181–196.
27. Pappenberger, F.; Beven, K.J.; Hunter, N.M.; Bates, P.D.; Gouweleeuw, B.T.; Thielen, J.; de Roo, A.P.J. Cascading model uncertainty from medium range weather forecasts (10 days) through a rainfall-runoff model to flood inundation predictions within the European Flood Forecasting System (EFFS). *Hydrol. Earth Syst.*

Sci. **2005**, *9*, 381–393.

28. Pappenberger, F.; Matgen, P.; Beven, K.; Henry, J.; Pfister, L.; de Fraipont, P. Influence of uncertain boundary conditions and model structure on flood inundation predictions. *Adv. Water Resour.* **2006**, *29*, 1430–1449.
29. Mantovan, P.; Todini, E. Hydrological forecasting uncertainty assessment: Incoherence of the GLUE methodology. *J. Hydrol.* **2006**, *330*, 368–381.
30. Beven, K.J.; Smith, P.J.; Freer, J.E. Comment on "Hydrological forecasting uncertainty assessment: Incoherence of the GLUE methodology" by Pietro Mantovan and Ezio Todini. *J. Hydrol.* **2007**, *338*, 315–318.
31. Beven, K.J.; Smith, P.J.; Freer, J.E. So just why would a modeller choose to be incoherent? *J. Hydrol.* **2008**, *354*, 15–32.
32. USGS Landsat Missions Home Page. Available online: http://landsat.usgs.gov/ (accessed on 26 September 2013).
33. The National Map Viewer and Download Platform. Available online: http://seamless.usgs.gov/ (accessed on 26 September 2013).
34. Moore, M.R. Development of a High-Resolution 1D/2D Coupled Flood Simulation of Charles City, Iowa. Master's thesis, University of Iowa, Iowa city, IA, USA, May 2011.
35. Chow, V.T. *Open-Channel Hydraulics*; McGraw-Hill: New York, NY, USA, 1959.
36. Calenda, G.C.; Mancini, C.P.; Volpi, E. Distribution of extreme peak floods of the Tiber River from the XV century.*Adv. Water Resour.* **2005**, *28*, 615–625.
37. National Water Information System: Web Interface. USGS Water Resources. USGS 03340500. Available online: http://waterdata.usgs.gov/usa/nwis/uv?03340500 (accessed on 26 September 2013).
38. National Water Information System: Web Interface. USGS Water Resources. USGS 06807000. Available online: http://waterdata.usgs.gov/usa/nwis/uv?site_no=06807000 (accessed on 26 September 2013).
39. Tou, J.T.; Gonzales, R.C. *Pattern Recognition Principles. Isodata Algorithm, Pattern Classification by Distance Functions*; Addison-Wesley: Reading, MA, USA, 1974; pp. 97–104.
40. Richards, J.A. *Remote Sensing Digital Image Analysis: An Introduction*, 2nd ed.; Springer-Verlag: Berlin, Germany, 1993.
41. Castañeda, C.; Herrero, J.; Casterad, M.A. Facies identification within the playalakes of the Monegros Desert, Spain, with field and satellite data. *Catena* **2005**, *63*, 39–63.
42. Lang, M.W.; McCarty, G.W. LiDAR intensity for improved detection of inundation below the forest canopy. *Wetlands***2009**, *29*, 1166–1178.
43. Thomas, N.E.; Huang, C.; Goward, S.N.; Powell, S.; Rishmawi, K.; Schleeweis, K.; Hinds, A. Validation of north American forest disturbance dynamics derived from Landsat time series stacks. *Remote Sens. Environ.* **2011**, *115*, 19–32.
44. Khan, S.I.; Hong, Y.; Wang, J.; Yilmaz, K.K.; Gourley, J.J.; Adler, R.F.; Brakenridge, G.R.; Policelli, F.; Habib, S.; Irwin, D. Satellite remote sensing and hydrologic modeling for flood inundation mapping in Lake Victoria basin: Implications for hydrologic prediction in ungauged basins. *IEEE Trans. Geosci. Remote Sens.* **2011**, *49*, 85–95.
45. Song, C.; Woodcock, C.E.; Seto, K.C.; Pax-Lenney, M.; Macomber, S.A. Classification and change detection using Landsat TM data: When and how to correct atmospheric effects. *Remote Sens. Environ.* **2001**, *75*, 230–244.
46. U.S. Army Corps of Engineers (USACE). *HEC–RAS Hydraulic Reference Manual*; Hydrologic Engineering Center: Davis, CA, USA, 2006.
47. Pappenberger, F.; Beven, K.; Horritt, M.; Blazkova, S. Uncertainty in the calibration of effective roughness parameters in HEC-RAS using inundation and downstream level observations. *J. Hydrol.* **2005**, *302*, 46–69.
48. Horritt, M.S. A methodology for the validation of uncertain flood inundation models. *J. Hydrol.* **2006**, *326*, 153–165.
49. Fewtrell, T.J.; Bates, P.D.; Horritt, M.; Hunter, N.M. Evaluating the effect of scale in flood inundation modelling in urban environments. *Hydrol. Process.* **2008**, *22*, 5107–5118.
50. Hicks, F.E.; Peacock, T. Suitability of HEC-RAS for flood forecasting. *Can. Water Resour. J.* **2005**, *30*, 159–174.
51. Meselhe, E.A.; Pereira, J.F.; Georgiou, I.Y.; Allison, M.A.; McCorquodale, J.A.; Davis, M.A. Numerical Modeling of Mobile-Bed Hydrodynamics in the Lower Mississippi River. American Society of Civil Engineers. In Proceedings of the World Environmental and Water Resources Congress (EWRI), Providence, RI, USA, 16–20 May 2010; pp. 1433–1442.
52. Hornberger, G.M.; Spear, R.C. An approach to the preliminary analysis of environmental systems. *J. Environ. Manag.***1981**, *12*, 7–18.
53. Young, P.C. *The Validity and Credibility of Models for Badlydefined Systems. Uncertainty and Forecasting of Water Quality*; van Straten, B., Ed.; Springer: Berlin, Germany, 1983; pp. 69–98.
54. Beven, K.J. *Environmental Modelling: An Uncertain Future?* Routledge: London, UK, 2009.
55. Wang, X.; Frankenberger, J.R.; Kladivko, E.J. Uncertainties in DRAINMOD predictions of subsurface drain flow for an Indiana silt loam using the GLUE methodology. *Hydrol. Process.* **2006**, *20*, 3069–3084.
56. Blasone, R.S.; Madsen, H.; Rosbjerg, D. Uncertainty assessment of integrated distributed hydrological models using GLUE with Markov chain Monte Carlo sampling. *J. Hydrol.* **2008**, *353*, 18–32.
57. Blasone, R.S.; Vrugt, J.A.; Madsen, H.; Rosbjerg, D.; Robinson, B.A.; Zyvoloski, G.A. Generalized likelihood uncertainty estimation (GLUE) using adaptive Markov chain Monte Carlo sampling. *Adv. Water*

Resour. **2008**, *31*, 630–648.

58. Hornberger, G.M.; Cosby, B.J. Selection of parameter values in environmental models using sparse data: A case study.*Appl. Math. Comp.* **1985**, *17*, 335–355.
59. Kiczko, A.; Pappenberger, F.; Romanowicz, R.J. Flood Risk Analysis of the Warsaw Reach of the Vistula River. In *ERB 2006: Uncertainties in the 'Monitoring-Conceptualisation-Modelling' Sequence of Catchment Research*; European Network of Experimental and Representative Basins (ERB): Luxembourg, 2007.
60. Aronica, G.; Hankin, B.G.; Beven, K.J. Uncertainty and equifinality in calibrating distributed roughness coefficients in a flood propagation model with limited data. *Adv. Water Resour.* **1998**, *22*, 349–365.
61. Hunter, N.M.; Bates, P.D.; Horritt, M.S.; de Roo, A.P.T.; Werner, M.G.F. Utility of different data types for calibrating flood inundation models within a GLUE framework. *Hydrol. Earth Syst. Sci.* **2005**, *9*, 412–430.
62. Schumann, G.; Cutler, M.; Black, A.; Matgen, P.; Pfister, L.; Hoffmann, L.; Pappenberger, F. Evaluating uncertain flood inundation predictions with uncertain remotely sensed water stages. *Int. J. River Basin Manag.* **2008**, *6*, 187–199.
63. Horritt, M.S.; Bates, P.D. Evaluation of 1D and 2D numerical models for predicting river flood inundation. *J. Hydrol.***2002**, *268*, 87–89.
64. Schumann, G.; Bates, P.D.; Horritt, M.S.; Matgen, P.; Pappenberger, F. Progress in integration of remote sensing derived flood extent and stage data and hydraulic models. *Rev. Geophys.* **2009**, *47*.
65. Di Baldassarre, G.; Schumann, G.; Bates, P.D. A technique for the calibration of hydraulic models using uncertain satellite observations of flood extent. *J. Hydrol.* **2009**, *367*, 276–282.
66. Horritt, M.S.; Bates, P.D. Predicting floodplain inundation: Raster-based modelling versus the finite element approach. *Hydrol. Process.* **2001**, *15*, 825–842.
67. Aronica, G.; Bates, P.D.; Horritt, M.S. Assessing the uncertainty in distributed model predictions using observed binary pattern information within GLUE. *Hydrol. Process.* **2002**, *16*, 2001–2016.
68. Bates, P.D.; Horritt, M.S.; Aronica, G.; Beven, K.J. Bayesian updating of flood inundation likelihoods conditioned on flood extent data. *Hydrol. Process.* **2004**, *18*, 3347–3370.
69. Pappenberger, F.; Frodsham, K.; Beven, K.J.; Romanowicz, R.; Matgen, P. Fuzzy set approach to calibrating distributed flood inundation models using remote sensing observation. *Hydrol. Earth Syst. Sci.* **2007**, *11*, 739–752.
70. Cook, A.; Merwade, V. Effect of topographic data, geometric configuration and modeling approach on flood inundation mapping. *J. Hydrol.* **2009**, *377*, 131–142.
71. Sanders, B.F. Evaluation of on-line DEMs for flood inundation modeling. *Adv. Water Resour.* **2007**, *30*, 1831–1843.
72. Zheng, T.; Wang, Y. Mapping flood extent using a simple DEM-inundation model. *North Carol. Geographer.* **2007**, *15*, 1–19.
73. Bates, P.D.; Horritt, M.S.; Hunter, N.M.; Mason, D.C.; Cobby, D.M. Numerical Modelling of Floodplain Flow. In*Computational Fluid Dynamics:Applications in Environmental Hydraulics*; Bates, P., Ferguson, L., Eds.; John Wiley and Sons Ltd.: Chichester, UK, 2005.
74. Fewtrell, T.J.; Neal, J.C.; Bates, P.D.; Harison, P.J. Geometric and structural river channel complexity and the prediction of urban inundation. *Hydrol. Process.* **2011**, *25*, 3173–3186.
75. Jung, Y.; Merwade, V. Uncertainty quantification in flood inundation mapping using generalized likelihood uncertainty estimate and sensitivity analysis. *J. Hydrol. Eng.* **2012**, *17*, 507–520.
76. Bales, J.D.; Wagner, C.R. Sources of uncertainty in flood inundation maps. *J Flood Risk Manag.* **2009**, *2*, 139–147.
77. Merwade, V.M.; Olivera, F.; Arabi, M.; Edleman, S. Uncertainty in flood inundation mapping—Current issues and future directions. *ASCE J. Hydrol. Eng.* **2008**, *13*, 608–620.
78. Pan, F.; Liao, J.; Li, X.; Guo, H. Application of the inundation area-lake level rating curves constructed from the SRTM DEM to retrieving lake levels from satellite measured inundation areas. *Comput. Geosci.* **2013**, *52*, 168–176.

CHAPTER 13

RAINFALL VARIABILITY IN THE HUANGFUCHUANG WATERSHED AND ITS RELATIONSHIP WITH ENSO

Thelma Dede Baddoo , Yiqing Guan , Danrong Zhang and Samuel A. Andam-Akorful

Keywords: ENSO, Huangfuchuan, China, wavelet, precipitation, Yellow River

INTRODUCTION

The Yellow River, the second largest river in China and the sixth longest in the world is the most important freshwater resource for northern and north-western China, supporting about 107 million people [1,2,3]. The Yellow River has long been regarded as the "Mother River of China" because human inhabitants have existed in this region since prehistoric times, and, therefore, water shortage issues have experienced particular attention throughout the world [4,5]. Many studies have indicated that the Yellow River basin is dominated by decreasing precipitation, which may further deteriorate the stress conditions on water resources [2,6,7]. The source areas/regions of a river form its major tributaries, contributing to its flow. The Yellow River is divided into three major reaches; the upper, middle and lower reaches [8]. The middle reaches of the Yellow River basin contribute significantly to the total streamflow and sediment discharge of the Yellow River [9]. Zhang *et al.* [6] indicated that there is evidence of precipitation decline in the middle portion of the Yellow River basin, particularly in the Loess Plateau. The Huangfuchuan watershed is one of the most important source areas in the middle reach of the Yellow River [8,10]. Studies by Sui *et al.* [11] observed that long-term precipitation depth in the Huangfuchuan watershed had a decreasing trend. Additionally, Zhou *et al.* [12] demonstrated that the climate aridity in the Huangfuchuan watershed has increased in the recent three decades.

Precipitation is considered as the primary source of water in China and, therefore, most water resource projects are designed and implemented based on the historical pattern of water availability, quality, and demand, assuming a constant climate [7,13,14,15]. In addition, precipitation is the exclusive source of the Yellow River as the river flows through arid, semi-arid and semi-humid regions; its distribution and trend affect the water resources in this basin to a large extent [7]. Similarly, precipitation is a direct natural factor influencing runoff in the Huangfuchuan watershed due to its semi-arid climatic conditions. However, the distribution of precipitation is uneven, and, therefore, a better understanding of precipitation variability on basin and regional scales is of great significance to management of water resources by assisting in the implementation of proper water management policies for sustainable ecological conservation and environmental protection [14,16].

The El Niño-Southern Oscillation (ENSO) is in recent times recognized as the most prominent mode of climate variability that operates on seasonal-to-interannual time scales influencing hydrology in many parts of the world [17,18]. Through its effect on the global atmospheric circulation, ENSO influences patterns of temperature and precipitation, including extreme events such as, drought, floods, and tropical cyclones, in many regions of the world [18]. ENSO accounts for the largest contribution to interannual climate variability globally with a 2–7 year period [19,20,21,22]. The warm phase of ENSO, El Niño is associated with rising sea surface temperature (SST), while the atmospheric convection zones of the tropical Pacific expand and merge. On the other hand, La Niña, the cold

Citation: Thelma Dede Baddoo, Yiqing Guan, Danrong Zhang, and Samuel A. Andam-Akorful, Rainfall Variability in the Huangfuchuang Watershed and Its Relationship with ENSO, Water 2015, 7, 3243-3262; doi:10.3390/w7073243.

phase of ENSO is associated with low SST near the equator, with atmospheric convection zones that are isolated from each other [23]. The influence of ENSO on rainfall variability in China has been the subject of many studies (e.g., [3,24,25,26]). Previous studies by Ropelewski and Halpert [27] showed that there is no steady relationship between El Niño and rainfall over China as a whole. Hui *et al.* [28] also pointed out that the connection between mean summer precipitation and the preceding winter SST anomalies (ENSO) in China have weakened since the 1980s, making regional prediction much more difficult. However, Lau and Weng [25] stated that during the summer of 1997, southern China faced severe flooding, while northern China experienced one of the driest seasons on record, which can be linked to the 1997/98 strong El Niño event. Furthermore, Tong *et al.* [29] showed that El Niño activity is closely related to flood episodes in the lower and middle Yangtze catchments, whereas dry spells are associated with it in the upper catchment of the river. Major hydrological events on the other hand, tend to exhibit a spurious relationship with La Niña in the basin [29]. In addition, Ronghui and Yifang [30], Ding [31], Zhang *et al.* [32], in investigating the impacts of ENSO events on summer rainfall variability in northern China, reported a shortfall in summer rainfall during the El Niño development and maturity stages, whereas abundant rainfall was observed during the weakening phase of an El Niño event.

Rainfall deficits in the Yellow river have been linked to El Niño events [3,33]. Fu *et al.* [4] reported that, the average annual precipitation in the Yellow River basin in an El Niño year is less than that in a La Niña year. Additionally, Wang and Li [26] found that, there is a significant correlation between monthly precipitation of semiarid northern China and ENSO in the frequency bands of 2–3 years; in that, during ENSO years, rainfall was mostly below normal with some occurrences of drought in this region. Furthermore, Wang *et al.* [34] confirmed that ENSO events often resulted in decreased precipitation in the source regions of the Yellow River.

To the best of our knowledge, studies on ENSO influences on northern China have usually focused on the Yellow River basin as a whole, and not much on the sub-basins which form its major tributaries. In this study, we investigate the relationship between rainfall and ENSO in the Huangfuchuan watershed, an important sub-catchment in the middle region of the Yellow River. The Yellow River also spans from a semi-humid zone into an arid zone and, though we know the impacts of ENSO on the basin as a whole, this study goes ahead to study the impacts of ENSO on a sub-basin in the semi-arid zone of the Yellow River basin to determine if there is any difference and at what timescale. The question here is what is the relationship between ENSO and rainfall variability in the middle source regions of the Yellow River basin, and to what extent is this relationship; in terms of different timescales? The main objectives of this study, therefore, are: (a) to correlate the findings of other studies to the declining rainfall trend; (b) to examine the relationship and impact of ENSO events on rainfall variability in terms of timescales, on the Huangfuchuan, which is one of the major middle source sub-basins of the Yellow River in northern China over the period 1954 to 2010. To this end, the current study adopts exploratory data analysis (EDA) and wavelet transforms to investigate trends and variability in rainfall patterns over the Huangfuchuan basin. Wavelet analysis has found many uses in hydrology, because it provides an effective approach for hydrologic time series analysis by revealing characteristics under multi-temporal scales. This means that wavelet analysis demonstrates the scale contents of a signal and its variability in time [35,36,37,38]. Here, wavelet analysis was employed to investigate the variations in the dominant rainfall modes/signals with time in the Huangfuchuan watershed. Additionally, we study the co-varying relationships between rainfall over the basin and ENSO using cross wavelet and wavelet coherence analysis [39,40]. Cross wavelet and wavelet coherence analysis have been used in many studies to identify covariance and coherency in geophysical time series (see, e.g., [7,37,41,42,43]).

Following a brief introduction in Section 1, the arrangement of the rest of the paper is as follows. Section 2 gives an overview of the study area. Additionally, Section 3 describes the datasets and methodology used in the study and the results obtained are discussed in Section 4. Finally, the conclusions from the results are inferred in Section 5.

STUDY AREA

The Huangfuchuan watershed is the drainage area covered by River Huangfu in the north of the Hekou-Longmen region located between longitudes 110°18′ E and 111°12′ E and latitudes 39°12′ N and 39°54′ N spanning a surface area of 3246 km^2[16]. The river is 137 km long, a tributary and one of the most important source areas in the middle reach of the Yellow River originating from the transect of the Erdos Plateau and Loess Plateau in Northern China flowing from northwest to southeast into the Yellow River [8,11,44]. The watershed lies in the semi-arid climatic region in the northern Shaanxi province with the majority in the inner Mongolia region [11,45]. With a typical monsoon climate, the watershed has an average annual precipitation of 350 mm to 450 mm with more than 80% coming from June to September and a mean annual air temperature ranging from 6.9 and 9.7 °C [16,45,46]. The Huangfu River is representative of the "hill-gully" landscape of the northern Loess Plateau [45]. The Loess Plateau has a greatly varied and rolling surface with a large topographic relief, a deep layer of loess, a loose layer of soil, and a fragmented landform and sparse vegetation [11]. Due to this feature of the Loess Plateau, the soils are greatly susceptible to water and wind erosion making the Huangfuchuan watershed one of the main sources of sediments into the Yellow River [8,45]. The Huangfuchuan watershed is shown in Figure 1.

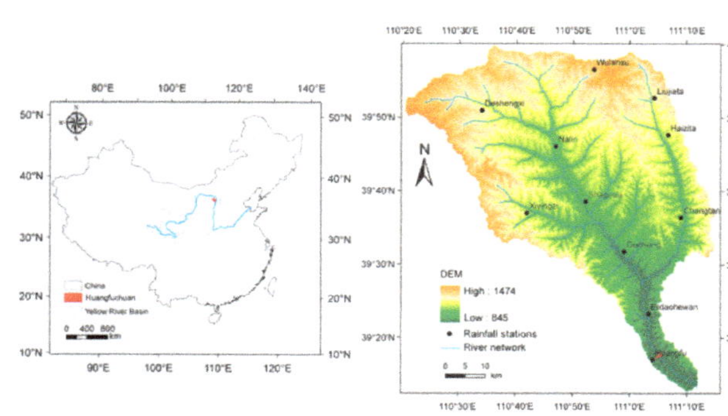

Figure 1: Location of Huangfuchuan catchment in China (**a**) and the watershed map showing the different rainfall stations (**b**); the red arrow on the watershed map showing the hydrological station of the

watershed. Retrieved from [47].

MATERIALS AND METHODS

Materials

Rainfall data

Daily areal mean rainfall data from 1954 to 2010 was obtained from the "Experimental research on vegetation water source partitioning of riparian zone in arid and semi-arid areas based on multi-tracer technique and its uncertainty" (NSFC51209064) project taking place in the watershed. The daily rates were accumulated to monthly and annual rates for the purposes of this study.

Multivariate ENSO Index (MEI)

The Multivariate ENSO Index (MEI) time series was acquired from the Physical Sciences Division of the Earth System Research Laboratory of the National Oceanic and Atmospheric Administration (NOAA) at [48]. MEI is obtained as the first unrotated principal component (PC) of six variables over the tropical Pacific, which includes: sea-level pressure (P), zonal (U) and meridional (V) components of the surface wind, sea surface temperature (S), surface air temperature (A), and total cloudiness fraction of the sky (C). Given its multivariate nature, the index is claimed to provide a more vivid representation of characterizing the ENSO phenomenon relative to other indices, which only considers either the oceanic or atmospheric components [49]. It defines ENSO events as five consecutive overlapping three-month periods at or above the +0.5° anomaly for El Niño events and at or below the −0.5° anomaly for La Niña events. The strength of an event is further described as weak (with a 0.5 to 0.9 SST anomaly), moderate (1.0 to 1.4) and strong (≥1.5); an event is said to have occurred if the threshold is equaled or exceeded for at least three consecutive overlapping three-month periods. This study focuses only on strong and moderate El Niño/La Niña events. Figure 2 presents the MEI over the study period, showing the respective El Niño/La Niña events according to their strengths.

Methods

Annual precipitation values were calculated from the daily rainfall data of the Huangfuchuan watershed acquired. Exploratory data analysis was performed on the raw data to understand the data and have a summary of the visible patterns in rainfall in the watershed before applying statistical tests. Homogeneity tests were performed on the annual rainfall series to check the quality of the data. Four methods: standard normal homogeneity test (SNHT) [50], Buishand range test [51], Pettitt test [52], and von Neumann ratio test [53] were chosen to detect the homogeneity of the collected Huangfuchuan rainfall data. The test results are classified as: Useful (the series that rejects one or none null hypothesis under the four tests at 5% significance level), Doubtful (the series that rejects two null hypotheses of the four tests at 5% significance level) and Suspect (when there are three or all tests reject the null hypothesis at 5% significance level) as in [54,55]. The null hypothesis (H_0) in these tests assume that the data is homogeneous as opposed to the alternative hypothesis, H_a, that the series consists of a break in the mean. Distribution-free CUSUM (cumulative sum) test was employed to determine abrupt changes in rainfall patterns from 1954 to 2010 in the Huangfuchuan watershed. Similarly, the trend in the time series was established using the Mann-Kendall test statistic. The statistic software utilized was TREND trend/change detection software, a product from the Cooperative Research Centre (CRC) for Catchment Hydrology's (CRCCH) Climate Variability Program [56]. Significance levels α = 0.01, 0.05 and 0.1 were used throughout.

Furthermore, wavelet power spectrum was employed to investigate the dominant cycles in rainfall pattern over the basin. Then, wavelet coherency (WTC) was used to identify where the two variables (rainfall and MEI) were correlated at a local scale. Furthermore, cross wavelet analysis (XWT) was adopted to investigate regions of high common power and further reveal information about the phase relationship of the two variables. The MatLab software package was used to perform all wavelet analysis. A MatLab software package for performing the XWT and WTC can be found at [57]. For more details on the wavelet analysis see [39].

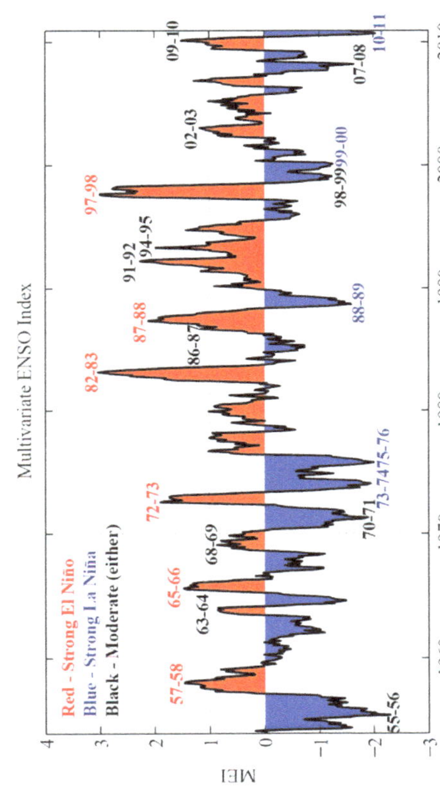

Figure 2: Multivariate ENSO indices from 1954 to 2010. El Niño (positive phase) in red; La Niña (negative phase) in blue. Numbers on the plot show El Niño/La Niña event years. Red and blue digits denote strong El Niño and La Niña events respectively. Black digits show the moderate years for either event.

Mann-Kendall Trend Analysis

This is a ranked nonparametric test developed by Mann [58] and Kendall [59]. The algorithm tests for monotonic trends and turning points in a series without a prior assumption of its distribution [60]. In this test, the null hypothesis H_0, assumes that the data is independent and arbitrarily ordered (that is, there is no trend and the observations are randomly ordered) and this is tested against the alternative hypothesis H_A, which assumes that there is a trend [61]. For details of the Mann-Kendall non-parametric test see [60].

Distribution Free Cumulative Sum (CUSUM) Test

This is a distribution free non-parametric test which determines whether the means in two parts of a record are different (for an unknown time of change) [56]. This rank-based test compares successive observations with the median of the series [62,63,64]. The test statistic V_k of a time series data is the maximum cumulative sum (CUSUM) of the signs of the difference from the median (a series of −1 or +1) starting from the beginning of the series [63,65]. More explanation on the distribution free CUSUM test can be found in [56].

Wavelet Analysis

The wavelet transform decomposes a series into time-frequency space, enabling the identification of both the dominant modes of variability and how those modes vary with time [66]. The wavelet transform as given by Labat [67] is:

$$W(\omega, \tau, x(t)) = \omega^{1/2} \int x(t)\Psi^*(\omega(t-\tau))dt \quad (1)$$

where ω is the scale factor (frequency); τ is time shift; t is time; x(t) is the time series; and Ψ^* is the complex conjugate of Ψ, the wavelet function. The continuous wavelet transform is used in this study to compute the wavelet power spectrum. The Morlet wavelet function [68] given as

$$\Psi(t) = e^{-i\omega_o t} e^{(-t^2/2)} \quad (2)$$

where Ψ(t) is the wavelet function; i is the imaginary symbol of complex number; ω_0 is the non-dimensional frequency (ω_0= 6 to satisfy admissibility condition [69]); and t is time, was chosen as the mother-wavelet in this study because it describes the shape of hydrological signals well, and provides a good balance between time and frequency localization [67,70,71]. The standardized anomaly R'_{ij}, which removes seasonal fluctuations and short-term biases, was employed to standardize the rainfall data in the continuous wavelet transform and spectrum analysis according to

$$R'_{ij} = \frac{(R_{ij} - \bar{R}_j)}{\sigma_j} \quad (3)$$

where R_{ij} is the monthly averaged observed data in the year i and the month j, and $\bar{R}_j$ and σ_j represent the longterm mean and standard deviation of each successive month j, respectively [72]. For more details of wavelet transform analysis, see [67,73]. Figure 4 shows the wavelet analysis of total monthly Huangfuchuan precipitation intensity.

Cross Wavelet Spectrum

Cross-wavelet power reveals areas with a high common power [74]. According to Torrence and Compo [73], given two time series *X* and *Y*, with wavelet transforms $W_n^X(s)$ and $W_n^Y(s)$, where *n* is the time index and *s* is the scale, one can define the cross-wavelet spectrum as:

$$W_n^{XY}(s) = W_n^X(s) W_n^{Y*}(s) \quad (4)$$

where $W_n^{Y*}(s)$ is the complex conjugate of $W_n^Y(s)$. The cross-wavelet spectrum is complex, and hence one can define the cross-wavelet power as $|W_n^{XY}(s)|$. The phase angle of W_{XY} describes the phase relationship between X and Y in time-frequency space [39,40,75].

Wavelet Coherence

Wavelet coherence discussed refers to the Fourier squared coherency used to identify frequency bands within which two time series are co-varying [76]. The wavelet coherence (wavelet squared coherency) given by Torrence and Webster [76] is defined as the absolute value squared of the smoothed cross-wavelet spectrum, normalized by the smoothed wavelet power spectra,

$$R_n^2(s) = \frac{|S(s^{-1}W_n^{XY}(s))|^2}{S(s^{-1}|W_n^X(s)|^2).S(s^{-1}|W_n^Y(s)|^2)} \quad (5)$$

where *S* is a smoothing operator. Grinsted *et al.* [39] gives the smoothing operator as

$$S(W) = S_{scale}\left(S_{time}(W_n(s))\right) \quad (6)$$

where S_{scale} denotes smoothing along the wavelet scale axis and S_{time} smoothing in time. For the Morlet wavelet, a suitable smoothing operator is given by Torrence and Webster [77] as

$$S_{time}(W)|_s = \left(W_n(s) * c_1^{\frac{-t^2}{2s^2}}\right)\Big|_s \quad (7)$$

$$S_{scale}(W)|_n = (W_n(s) * c_2 \Pi(0.6s))|_n \quad (8)$$

where c_1 and c_2 are normalization constants and is the rectangle function. The factor of 0.6 is the empirically determined scale decorrelation length for the Morlet wavelet [73]. The statistical significance level of the wavelet coherence is estimated using Monte Carlo methods [39]. Here, the wavelet coherence is used to identify both frequency bands and time intervals within which rainfall in the study area and ENSO are covarying.

RESULTS AND DISCUSSION

Exploratory Data Analysis (EDA)

A quick look at the annual precipitation data values of the Huangfuchuan watershed (Figure 3) indicates that, the years before the 1970s recorded much more precipitation than the post 1970s. The annual rainfall amount in the watershed has decreased since 1968 until 2003. It can be seen that between 1968 and 2002, the basin experienced decreasing precipitation until a sharp rise in 2003. From 2003, annual precipitation decreased steeply till 2006 and rose sharply again in 2007 with another steep decline in 2009. The highest precipitation in the watershed occurred in 1959 and the lowest was 1999. The trend line indicates that rainfall intensity in the Huangfuchuan watershed decreased during the period of study. Furthermore, the equation of the linear trend line of Figure 3 demonstrates that rainfall in the Huangfuchuan watershed has been decreasing by 1.056 mm each year from 1954 to 2010.

Homogeneity test results on the annual rainfall data are presented in Table 1.

From Table 1, it is evident that the collected rainfall data is homogeneous and useful, and can therefore be used for further analysis.

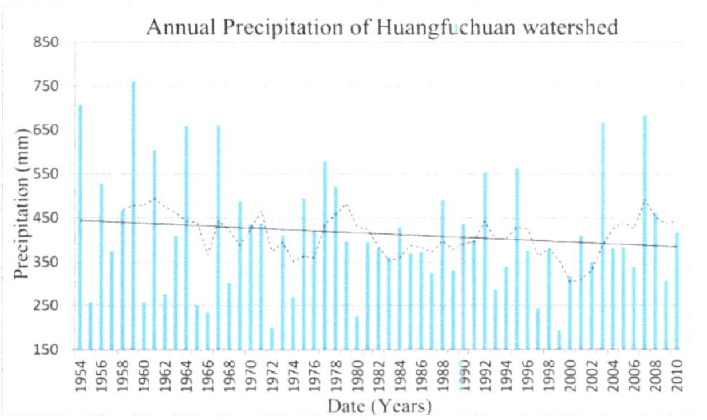

Figure 3: Annual precipitation of Huangfuchuan watershed. The dashed black line shows the five year moving average, while the straight black line shows the trend line.

Table 1; Homogeneity test results for annual rainfall data of the Huangfuchuan watershed.

Test Variables	SNHT	BR Test	Pettitt Test	VNR
Value	4.843	5.619	158.000	2.447
t (time of change)	1954	1964	1979	Does not determine time of change
p-value (two-tailed)	0.314	0.533	0.715	0.959
alpha (significance level)	0.05	0.05	0.05	0.05
Result	Accept the null hypothesis that the data is homogeneous	Accept the null hypothesis that the data is homogeneous	Accept the null hypothesis that the data is homogeneous	Accept the null hypothesis that the data is homogeneous

Statistical Analysis

The results of the Mann-Kendall statistical test performed on the annual rainfall time series of the Huangfuchuan watershed showed no statistically significant trend at $\alpha = 0.01$, 0.05 and 0.10. Although the trend is not significant, the negative value of the Z statistic shows a decreasing trend. This means that rainfall in the Huangfuchuan watershed is decreasing.

In addition, step change analysis from the distribution free CUSUM test on annual precipitation data showed no significant abrupt changes at significance levels of $\alpha = 0.01$, 0.05 and 0.1. Nevertheless, the results revealed a change point (year) in 1979 with a test statistic, V_k, of 8. The mean, median annual rainfall before and after 1979 were 439.6, 428.1 and 393.8, 379.8 respectively. Table 2 and Table 3 summarize the statistical test results performed on the annual precipitation data of the Huangfuchuan watershed.

Table 2: Mann-Kendall test results of Huangfuchuan watershed.

S Statistic	Z	Critical Values			Interpretation
		$\alpha = 0.1$	$\alpha = 0.05$	$\alpha = 0.01$	
−130	−0.888	1.645	1.96	2.576	NS

Notes: α = Significance level; NS = Not significant

Table 3: Distribution free CUSUM test results.

Test Statistic (V_k)	Change Point (year)	Critical Values			Interpretation
		$\alpha = 0.1$	$\alpha = 0.05$	$\alpha = 0.01$	
8	1979	9.211	10.268	12.306	NS

Wavelet Analysis

Figure 4 shows the results of wavelet analysis of monthly precipitation of Huangfuchuan watershed from 1954 to 2010.

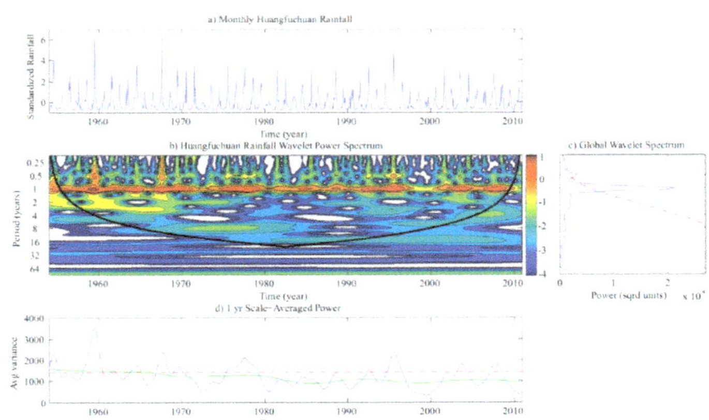

Figure 4: Wavelet analysis of monthly precipitation in the Huangfuchuan watershed; (**a**) is the standardized time series of precipitation; (**b**) is the wavelet power spectrum (WPS), the thick black arc delimits the cone of influence (COI) of the signal. High intensity is coded by the color red, whereas blue denotes the weaker intensity; (**c**) is the global wavelet spectrum for the WPS; and (**d**) is the averaged variance at the annual scale (blue). The dashed red line is the significant level. The green line shows the trend of the annual rainfall variance. Sqrd units indicates squared units.

From the global wavelet spectrum (Figure 4c), it is apparent that the dominant modes (highest intensity) of the precipitation signal is concentrated around the annual and semi-annual cycles, although the annual signal intensity is highest. The annual and semi-annual signals of 1959 have the highest power which corresponds to the high precipitation recorded in the region in that period. However, this above normal rainfall record corresponds to a four year lag after La Niña occurrence and a 17 month lag after an El Niño event in 1958, which suggests that the ENSO phenomenon may not have had a significant impact on rainfall during that period (see Figure 5a and Figure 6). The low precipitation in 1999 also shows relatively high signal intensity. This rainfall decline was preceded by the strong 1997/98 El Niño event and an intermediate La Niña event in 1998, also in Figure 6. Annual signals for years 1960–1965, 1966–1971, 1974–1979, 1984–1985, 1988–1989, 1991–1996, 2001–2004, and 2006–2009 are analogous and relatively strong, whilst those between 1965–1966, 1971–1974, 1979–1984, 1985–1989, 1989–1991, 1996–2001 and 2005–2006 indicate lower amplitudes (see Figure 4). The relatively strong similar signals correspond to rainfall increment in average annual rainfall variance in Figure 4d. The contrast is true for lower amplitude signals. The averaged-annual scale power (Figure 4d) presents a persistent reduction in rainfall amplitudes between 1980 and 1994. There is a brief peak in 1996, after which the lowest amplitude is observed around 1997–1999. There seems to be a recovery in the amplitudes after this time, however, they are consistently below amplitudes observed before 1980. These results correspond to the following: between the periods of 1980 to 1999, there was a higher occurrence of El Niño events of higher intensities than the preceding years (see Figure 2). The recovery in rainfall amplitudes after 1999 occurred due to a reduction in the El Niño strengths in that period. However, the rainfall amplitudes in the recovery years were consistently below amplitudes before 1980 because, albeit the decline in El Niño strengths, its occurrence was greater than before the 1980s.

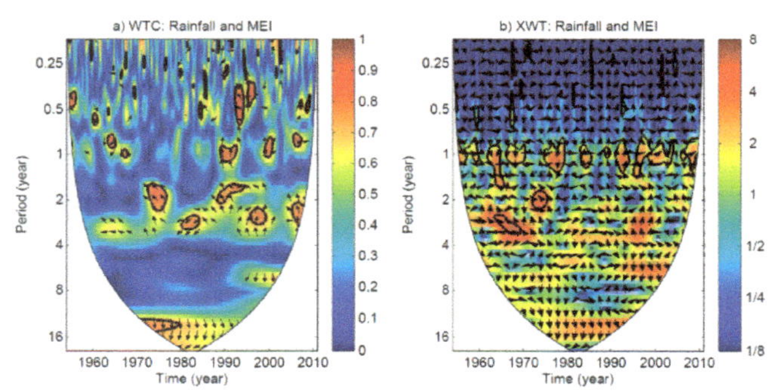

Figure 5: Wavelet coherency (**a**) and cross-wavelet spectrum (**b**) analysis showing the relations between Huangfuchuan rainfall and MEI. The thick black contour indicates the 95% confidence level against red noise and the cone of influence is shown as the lighter shade. The arrows (vectors) designate the phase difference between Huangfuchuan rainfall and MEI (with in-phase pointing right, anti-phase pointing left).

Results of the co-varying relationships between rainfall and the ENSO phenomenon over the Huangfuchuan basin, as depicted by the wavelet coherence and cross wavelet are presented in Figure 5.

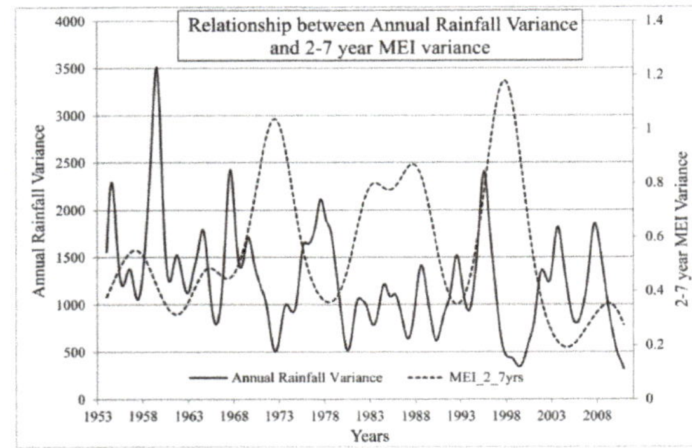

Figure 6: Relationship between Huangfuchuan rainfall and MEI showing the average annual rainfall variance and the 2–7 year variance of MEI. The black continuous curve shows rainfall and black dashed curve shows MEI.

The wavelet coherence in Figure 5a shows significant covariance in the 0.5–1 year bands from 1954–1968. However; the significant sections within the semi-annual and annual (0.5–1 year) bands from 1954 to 1968 provide inadequate results concerning the phase difference of Huangfuchuan rainfall and MEI. Significant cross wavelet power of approximately 0.85 in the semi-annual (0.5) band from 1992 to 1994 shows an anti-phase difference of roughly 330° corresponding to rainfall deficit lagging an El Niño event. In addition, similar significant coherence in the same band from 1995 to 1997 corresponds to an approximate annual lag in rainfall with El Niño incidences inducing dry conditions. Additionally; a significant coherency of 0.8 in the annual band from 1988 to 1993 demonstrates an anti-phase difference of roughly 360°. This implies an annual lag of ENSO impacts on rainfall variability in the study watershed. Similarly significant covariance in the annual band from 1998 to 2001 indicates anti-phase difference of a year and more, corresponding to a rainfall deficit lagging the strong 1997/98 El Niño event by a year with persisting impacts. Furthermore, similarly significant coherence in the annual band during 2005 to 2008 does not provide adequate results on the phase difference. Looking at the ENSO frequency band (2–7 year band), significant coherence is also seen only around the 2–4 year band. The significant covariance of around 0.85 in the two year band from 1972–1977 indicates a 300° anti-phase difference correlating to induced dry periods lagging the 1972/73 El Niño event. This period also relates to a rainfall increase in 1977 preceded by successive strong La Niña events in 1973/74 and 1975/76. Significant coherence of around 0.75 within the two year band from 1988 to 1995 shows anti-phase difference of approximately 330° to 360° which can be linked to the 0.8 significant covariance in the annual band during approximately the same period (1988–1993). Additionally, significant covariance in the three year band from 1980 to 1984 indicates a lag in rainfall deficit to El Niño events by approximately 240°. Moreover; similar coherency significance in the same band co-varies with about a 360° phase difference from 1996 to 2000 corresponding to an approximate annual lag in rainfall with El Niño occurrences inducing dry conditions. Likewise, the 0.75 significant covariance in the 2–4 year band from 2005 to 2008 correspond to an anti-phase difference of 11 to 12 months meaning rainfall lags ENSO events by that period. Finally, co-variance between Huangfuchuan rainfall and MEI within the ~14–16 year band from 1970 to 1980 shows anti-phase difference of 180° corresponding to a semi-annual lag in rainfall with ENSO events. This means that the seasonal impacts of ENSO on rainfall variability of the Huangfuchuan watershed occur on an annual scale, whereas, on a longer timescale; the impacts follow after six months and persist for a year or more. This result is representative of the fact that ENSO events usually occur in the boreal winter and the Huangfuchuan watershed experiences the majority of its rainfall in the summer (June; July and August) [16,78]. The persistence of the covariance from six months to a year can be attributed to the fact that ENSO signals can persist from one year to the next [79].

The cross-wavelet spectrum (XWT) of Huangfuchuan rainfall and MEI is shown in Figure 5b. It is noted that there is significant common power in the ~1–1.5 year band from 1955–1959, 1961–1970, 1974–1979, 1983–1986, 1988–1990, 1991–1994, 1995–2004 and 2005–2009. There is also highly significant common power in the 3–4 year band from 1963–1971 and in the 2 year band from 1972–1976. The arrows in the ~1–1.5 year band are randomly distributed and therefore, do not provide adequate results to conclude the phase difference of the two time periods. However, the arrows of the highly significant 2–4 year bands from 1963–1975 are relatively out of phase. That is the phase arrows in the two year band in the years 1972–1976 and the 3–4 year bands between 1963 and 1971 are ~300°–330° anti-phase meaning that the common power between the two time periods vary by 10 to 11 months. Figure 5b indicates that the two periods mainly have common power in the one-year band. The common power years in the one year band from 1955–1959 and 1974–1979 correspond to rainfall peaks in the study watershed. The rainfall peak from 1974–1979 correspond to strong La Niña events, whereas the ENSO-rainfall relationship from 1955–1959 are inconclusive. However, the common power years from 1961–1970, 1983–1986, 1988–1990, 1991–1994 and 1995–2004 mainly indicate troughs in the rainfall series corresponding to strong and intermediate El Niño events. The common power in the annual band from 2005–2009 also represents the recovery of rainfall in the study watershed correlating to intermediate El Niño events. This linkage is presented more clearly in Figure 6, which demonstrates the correlation between the ENSO events and rainfall in the Huangfuchuan catchment by extracting the 2–7 year and annual averaged variance respectively from the dominant modes of their wavelet analysis: 2–7 years for ENSO and annual (1 year) for rainfall.

Figure 6 illustrates the extracted dominant signals of rainfall in the Huangfuchuan watershed which is located at the annual scale and that of the Multivariate ENSO Index (MEI) which has been shown by [18,19,20,21] to be in the 2–7 year bands. Figure 6 is presented to further establish the relationship between annual Huangfuchuan rainfall and ENSO, or the modulating effects ENSO has on annual rainfall in the study watershed.

Figure 6 demonstrates that the warm tongue of the ENSO phenomenon (*i.e.*, El Niño) corresponds to rainfall decline, whereas the cold (La Niña) enhances rainfall. For examples, the strong 72/73, 87/88 and 97/98 El Niño events coincided with decreased rainfall amounts in the succeeding years; the 1997/98 event causing the driest period in the watershed [25]. The results of El Niño impacts on annual rainfall in the Huangfuchuan basin are similar to that of La Niña, although, El Niño impacts are more pronounced. For the study period, no strong La Niña events occurred before the 1970s, though the moderate occurrence in 1955/56 caused a decline in rainfall in 1957. Subsequent La Niña occurrences result in increasing rainfall in the watershed from the 1970s. For example, the strong La Niña events of 73/74, 75/76, 88/89 and 99/00 resulted in increasing rainfall in the years 1975, 1977, 1990 and 2001, respectively. From the above observations, it can be presented that El Niño events have a greater influence on the Huangfuchuan watershed, relative to La Niña.

Furthermore, we performed correlation tests between rainfall and MEI, as well as the extracted annual rainfall and 2–7 year MEI variances to further investigate the relationship(s) between the two. To this end, both the Pearson linear and Spearman rank correlations were employed at a 0.05 significance level as presented in Table 4. The coefficients for the original series (*i.e.*, rainfall and MEI) exhibit weak relationships with varying signs for correlation methods (−0.050 and 0.028 respectively for Pearson and Spearman), which is expected since both fields are non-linear and non-stationary. However, for the extracted series, both methods yielded a stronger correlation with the same coefficient ($R = -0.363$), demonstrating the efficacy of wavelets to accurately represent the dominant patterns in non-linear, non-stationary series. The negative correlation indicates an anti-phase relationship between rainfall and ENSO phenomenon in the Huanfuchuan watershed,*i.e.*, El Nino results in dryer than normal conditions, whereas the converse is true for La Nina events. Moreover, we employed the proportion of variance explained (PVE) to quantify the magnitude of rainfall explained by the ENSO phenomenon in the Huanfuchuan watershed. The result suggests that ENSO events accounts for ~13% of Huangfuchuan rainfall, which implies that, other factors also impact rainfall variability in the watershed.

Table 4: Comparison of correlation results of unfiltered and filtered rainfall and multivariate El Niño-Southern oscillation index (MEI) data.

Data	Pearson Correlation	Spearman Rank Correlation
Monthly rainfall and MEI data (original)	−0.05	0.028
Extracted annual rainfall and 2–7 year MEI	−0.363	−0.363

Additionally, to further investigate the behavior of the rainfall-ENSO relationship during earlier years and more recent times, a Spearman rank correlation was performed for the extracted signals before and from 1968. The result is shown inTable 5. The correlation results of Table 5 shows that ENSO impacts on Huangfuchuan rainfall were more pronounced from 1968 than in earlier years. This supports the observation in Figure 6 showing the weak relationship between Huangfuchuan rainfall and ENSO events before 1968.

Table 5: Correlation results of rainfall-El Niño-Southern oscillation index (ENSO) relationship before and from 1968.

Extracted Annual Rainfall and 2–7 Year MEI	Spearman Rank Correlation
Before 1968	−0.113
From 1968	−0.351

CONCLUSIONS

In this study, wavelet analysis was employed to investigate the impact of ENSO events on the variability of precipitation in the Huangfuchuan watershed. The following points were concluded from the study:

(1)Results from the wavelet power spectrum presented an annual trend in Huangfuchuan rainfall. The Mann-Kendall trend test and the average annual variance of the wavelet analysis showed a downward trend in the annual rainfall amounts, although this trend was found to be statistically insignificant.

(2)Cross-wavelet transform analysis indicated that, Huangfuchuan rainfall is dominated by the 1–1.5 year band which has common powers corresponding to peaks and troughs in the rainfall time series. The phase difference arrows of the watershed rainfall and MEI in the 1–1.5 year bands for the cross-wavelet transform are randomly distributed and, therefore, do not provide adequate results to conclude the phase difference of the two time periods. However, significant common power in the 2–4 year band demonstrates that the common power of the two time periods varies by 10 to 11 months.

(3)Wavelet coherency results establish a six-month to one year phase difference between MEI and rainfall in the Huangfuchuan watershed. This means that ENSO events lead rainfall in the watershed by that period.

(4)The wavelet spectral analysis revealed that the dominant patterns in rainfall series are located at the annual scale, while that of the Multivariate ENSO Index (MEI) is found at the 2–7 year frequency band. The extracted signals at the dominant scales indicated that no significant relationship was identified between the two series (*i.e.*, rainfall and MEI) before 1968 while subsequent years demonstrated that ENSO events have a modulating effect on rainfall, especially during the El Niño activity, which induces dry conditions. The wavelet coherence analysis showed that rainfall lags ENSO by six to twelve months. A Pearson correlation and Spearman rank correlation analysis of the extracted signals of rainfall in the Huangfuchuan watershed and MEI further demonstrated that El Niño events correspond to recurring drought periods and La Niña events are succeeded by increases in rainfall.

Overall, ENSO events impact Huangfuchuan rainfall with El Niño corresponding to rainfall decline and La Niña to rainfall increase with a semiannual to annual lag. However, the impacts of El Niño persist in the watershed, resulting in increasing aridity in the study area. Other precipitation influences such as the intra-seasonal changes, Indian summer monsoons, and many others, coupled with anthropogenic factors, may be at play, when MEI events do not correlate with rainfall intensities.

ACKNOWLEDGMENTS

This research paper is funded and supported by the National Natural Science Foundation of China (NSFC51209064).

REFERENCES

1. McVicar, T.R.; Van Niel, T.G.; Li, L.; Hutchinson, M.F.; Mu, X.; Liu, Z. Spatially distributing monthly reference evapotranspiration and pan evaporation considering topographic influences. *J. Hydrol.* **2007**, *338*, 196–220.
2. She, D.; Xia, J. The spatial and temporal analysis of dry spells in the Yellow River basin, China. *Stoch. Environ. Res. Risk Assess.* **2012**, *27*, 29–42.
3. Wang, H.; Yang, Z.; Saito, Y.; Liu, J.P.; Sun, X. Interannual and seasonal variation of the Huanghe (Yellow River) water discharge over the past 50 years: Connections to impacts from ENSO events and dams. *Glob. Planet. Chang.* **2006**, *50*, 212–225.
4. Fu, G.; Charles, S.P.; Viney, N.R.; Chen, S.; Wu, J.Q. Impacts of climate variability on stream-flow in the Yellow River.*Hydrol. Process.* **2007**, *21*, 3431–3439.
5. Wang, R.; Ren, H.; Ouyang, Z. China water vision. In *Foreign Language Book*; China Meteorological Press: Beijing, China, 2000.
6. Zhang, Q.; Peng, J.; Singh, V.P.; Li, J.; Chen, Y.D. Spatio-Temporal variations of precipitation in arid and semiarid regions of China: The Yellow River basin as a case study. *Glob. Planet. Chang.* **2014**, *114*, 38–49.
7. Liu, Q.; Yang, Z.; Cui, B. Spatial and temporal variability of annual precipitation during 1961–2006 in Yellow River Basin, China. *J. Hydrol.* **2008**, *361*, 330–338.
8. Yellow River Conservancy Commission. *Assessment Report of Climate Change Impact and Water Resources of the Yellow River Basin*; UNESCO Office Beijing: Beijing, China, 2009.
9. Zhao, G.; Tian, P.; Mu, X.; Jiao, J.; Wang, F.; Gao, P. Quantifying the impact of climate variability and human activities on streamflow in the middle reaches of the Yellow River basin, China. *J. Hydrol.* **2014**, *519*, 387–398.
10. Zhou, Y.; Yang, Z.; Zhang, D.; Jin, X.; Zhang, J. Inter-Catchment comparison of flow regime between the Hailiutu and Huangfuchuan rivers in the semi-arid Erdos Plateau, NW China. *Hydrol. Sci. J.* **2015**, *60*, 688–705.
11. Sui, J.; He, Y.; Karney, B.W. Flow and high sediment yield from the Huangfuchuan watershed. *Int. J. Environ. Sci. Technol.* **2008**, *5*, 149–160.
12. Zhou, Y.; Shi, C.; Fan, X.; Shao, W. The influence of climate change and anthropogenic activities on annual runoff of Huangfuchuan basin in northwest China. *Theor. Appl. Climatol.* **2015**, *120*, 137–146.
13. Abdul Aziz, O.I.; Burn, D.H. Trends and variability in the hydrological regime of the Mackenzie River Basin. *J. Hydrol.***2006**, *319*, 282–294.
14. Liang, L.; Li, L.; Liu, Q. Precipitation variability in Northeast China from 1961 to 2008. *J. Hydrol.* **2011**, *404*, 67–76.
15. Westmacott, J.R.; Burn, D.H. Climate change effects on the hydrologic regime within the Churchill-Nelson River Basin. *J. Hydrol.* **1997**, *202*, 263–279.
16. Li, E.; Mu, X.; Zhao, G.; Gao, P.; Shao, H. Variation of runoff and precipitation in the hekou-longmen region of the yellow river based on elasticity analysis. *Sci. World J.* **2014**, *2014*.
17. Ward, P.J.; Jongman, B.; Kummu, M.; Dettinger, M.D.; Sperna Weiland, F.C.; Winsemius, H.C. Strong influence of El Nino Southern Oscillation on flood risk around the world. *Proc. Natl. Acad. Sci. USA* **2014**, *111*, 15659–15664.
18. Zebiak, S.E.; Orlove, B.; Muñoz, Á.G.; Vaughan, C.; Hansen, J.; Troy, T.; Thomson, M.C.; Lustig, A.;

Garvin, S. Investigating El Niño-Southern Oscillation and society relationships. *Wiley Interdiscip. Rev. Clim. Chang.* **2015**, *6*, 17–34.

19. Bellenger, H.; Guilyardi, E.; Leloup, J.; Lengaigne, M.; Vialard, J. ENSO representation in climate models: From CMIP3 to CMIP5. *Clim. Dyn.* **2014**, *42*, 1999–2018.
20. Dai, A. The influence of the inter-decadal Pacific oscillation on US precipitation during 1923–2010. *Clim. Dyn.* **2013**, *41*, 633–646.
21. Donders, T.H.; Punyasena, S.W.; de Boer, H.J.; Wagner-Cremer, F. ENSO signature in botanical proxy time series extends terrestrial El Niño record into the (sub) tropics. *Geophys. Res. Lett.* **2013**, *40*, 5776–5781.
22. Zebiak, S.E. Air-Sea interaction in the equatorial Atlantic region. *J. Clim.* **1993**, *6*, 1567–1586.
23. Philander, S.G.H. El Niño and La Niña. *J. Atmos. Sci.* **1985**, *42*, 2652–2662.
24. Feng, S.; Hu, Q. Variations in the teleconnection of ENSO and summer rainfall in Northern China: A role of the Indian Summer Monsoon. *J. Clim.* **2004**, *17*, 4871–4881.
25. Lau, K.M.; Weng, H. Coherent modes of global SST and summer rainfall over China: An assessment of the regional impacts of the 1997–98 El Niño. *J. Clim.* **2001**, *14*, 1294–1308.
26. Wang, W.-C.; Li, K. Precipitation fluctuation over semiarid region in northern China and the relationship with El Nino/Southern Oscillation. *J. Clim.* **1990**, *3*, 769–783.
27. Ropelewski, C.F.; Halpert, M.S. Global and regional scale precipitation patterns associated with the El Niño/Southern Oscillation. *Mon. Weather Rev.* **1987**, *115*, 1606–1626.
28. Hui, G.; Yongguang, W.; Jinhai, H. Weakening significance of ENSO as a predictor of summer precipitation in China.*Geophys. Res. Lett.* **2006**, *33*, L09807.
29. Tong, J.; Qiang, Z.; Deming, Z.; Yijin, W. Yangtze floods and droughts (China) and teleconnections with ENSO activities (1470–2003). *Quat. Int.* **2006**, *144*, 29–37.
30. Ronghui, H.; Yifang, W. The influence of ENSO on the summer climate change in China and its mechanism. *Adv. Atmos. Sci.* **1989**, *6*, 21–32.
31. Ding, Y. *Monsoons over China*; Springer Science & Business Media: Dordrecht, The Netherlands, 1994; Volume 16.
32. Zhang, R.; Sumi, A.; Kimoto, M. A diagnostic study of the impact of El Niño on the precipitation in China. *Adv. Atmos. Sci.* **1999**, *16*, 229–241.
33. Kiem, A.S.; Geogievsky, M.V.; Hapuarachchi, H.P.; Ishidaira, H.; Takeuchi, K. Relationship between ENSO and snow covered area in the Mekong and Yellow River basins. In Proceedings of the International Symposium on Regional Hydrological Impacts of Climatic Variability and Change with an Emphasis on Less Developed Countries, Held during the 7th Scientific Assembly of the International Association of Hydrological Sciences, Foz de Iguaço, Brazil, 3–9 April 2005; Volume 2, pp. 255–264.
34. Wang, G.X.; Shen, Y.P.; Liu, S.Y. On the characteristics of response of precipitation and runoff to the ENSO events in the source regions of the Yellow River. *J. Glaciol. Geocryol.* **2001**, *23*, 16–21.
35. Sang, Y.-F. A review on the applications of wavelet transform in hydrology time series analysis. *Atmos. Res.* **2013**, *122*, 8–15.
36. Maheswaran, R.; Khosa, R. Comparative study of different wavelets for hydrologic forecasting. *Comput. Geosci.* **2012**,*46*, 284–295.
37. Sang, Y.-F. Improved Wavelet Modeling Framework for Hydrologic Time Series Forecasting. *Water Resour. Manag.***2013**, *27*, 2807–2821.
38. Hao, Y.; Liu, G.; Li, H.; Li, Z.; Zhao, J.; Jim Yeh, T.-C. Investigation of karstic hydrological processes of Niangziguan Springs (North China) using wavelet analysis. *Hydrol. Process.* **2012**, *26*, 3062–3069.
39. Grinsted, A.; Moore, J.C.; Jevrejeva, S. Application of the cross wavelet transform and wavelet coherence to geophysical time series. *Nonlinear Process. Geophys.* **2004**, *11*, 561–566.
40. Jevrejeva, S.; Moore, J.C.; Grinsted, A. Influence of the Arctic Oscillation and El Niño-Southern Oscillation (ENSO) on ice conditions in the Baltic Sea: The wavelet approach. *J. Geophys. Res.* **2003**, *108*.
41. Lin, Y.-C.; Yu, H.-L. The cross wavelet and wavelet coherence analysis of spatio-temporal rainfall-groundwater system in Pingtung plain, Taiwan. In Proceedings of the EGU General Assembly 2013, Vienna, Austria, 7–12 April 2013; Volume 15.
42. Shanahan, T.M.; Overpeck, J.T.; Anchukaitis, K.J.; Beck, J.W.; Cole, J.E.; Dettman, D.L.; Peck, J.A.; Scholz, C.A.; King, J.W. Atlantic forcing of persistent drought in West Africa. *Science* **2009**, *324*, 377–380.
43. Yin, Y.; Xu, Y.; Chen, Y. Relationship between flood/drought disasters and ENSO from 1857 to 2003 in the Taihu Lake basin, China. *Quat. Int.* **2009**, *208*, 93–101.
44. Xu, H.; Taylor, R.; Kingston, D.; Thompson, J.; Todd, M. Impact of climate scenarios on water resources in River Xiangxi and Huangfuchuan basin. In Proceedings of the Quest-GSI Water Resources Workshop, Reading, UK, 30 March–3 April 2009.
45. Xu, H.; Taylor, R.G.; Xu, Y. Quantifying uncertainty in the impacts of climate change on river discharge in sub-catchments of the Yangtze and Yellow River Basins, China. *Hydrol. Earth Syst. Sci.* **2011**, *15*, 333–344.
46. Li, T.; Wang, G.; Huang, Y.; Fu, X. Response of soil erosion and sediment yield to the temporal variation of precipitation in the Loess Plateau. In Proceedings of the 15th International Congress of the International Soil Conservation Organization (ISCO), Budapest, Hungary, 18–23 May 2008.
47. Tian, P.; Zhao, G.; Mu, X.; Wang, F.; Gao, P.; Mi, Z. Check dam identification using multisource data and their effects on streamflow and sediment load in a Chinese Loess Plateau catchment. *J. Appl. Remote Sens.* **2013**, *7*.
48. NOAA/OAR/ESRL PSD. Multivariate ENSO Index

(MEI). Available online: http://www.esrl.noaa.gov/psd/enso/mei/(accessed on 15 October 2014).

49. Wolter, K.; Timlin, M.S. El Niño/Southern Oscillation behaviour since 1871 as diagnosed in an extended multivariate ENSO index (MEI.ext). *Int. J. Climatol.* **2011**, *31*, 1074–1087.

50. Alexandersson, H. A homogeneity test applied to precipitation data. *J. Climatol.* **1986**, *6*, 661–675.

51. Buishand, T.A. Some methods for testing the homogeneity of rainfall records. *J. Hydrol.* **1982**, *58*, 11–27.

52. Pettitt, A.N. A non-parametric approach to the change-point problem. *Appl. Stat.* **1979**, *126*, 126–135.

53. Von Neumann, J. Distribution of the ratio of the mean square successive difference to the variance. *Ann. Math. Stat.***1941**, *12*, 367–395.

54. Wijngaard, J.B.; Klein Tank, A.M.G.; Können, G.P. Homogeneity of 20th century European daily temperature and precipitation series. *Int. J. Climatol.* **2003**, *23*, 679–692.

55. Kang, H.; Yusof, F. Homogeneity tests on daily rainfall series in Peninsular Malaysia. *Int. J. Contemp. Math. Sci.* **2012**,*7*, 9–22.

56. Chiew, F.; Siriwardena, L. *Trend User Guide 2005*; CRC for Catchment Hydrology: Canberra, Australia, 2005.

57. NOC/NERC. Crosswavelet and Wavelet Coherence. Available online: http://noc.ac.uk/using-science/crosswavelet-wavelet-coherence (accessed on 3 September 2014).

58. Mann, H.B. Nonparametric tests against trend. *Econom. J. Econom. Soc.* **1945**, 245–259.

59. Kendall, M.G. *Rank Correlation Measures*; Charles Griffin: London, UK, 1975; Volume 202.

60. Mondal, A.; Kundu, S.; Mukhopadhyay, A. Rainfall trend analysis by Mann-Kendall test: A case study of north-eastern part of Cuttack District, Orissa. *Int. J. Geol. Earth Environ. Sci.* **2012**, *2*, 70–78.

61. Onoz, B.; Bayazit, M. The power of statistical tests for trend detection. *Turkish J. Eng. Environ. Sci.* **2003**, *27*, 247–251.

62. Chiew, F.H.S.; McMahon, T.A. Detection of trend or change in annual flow of Australian rivers. *Int. J. Climatol.* **1993**,*13*, 643–653.

63. Kundzewicz, Z.W., Robson, A., Eds.; *Detecting Trend and Other Changes in Hydrological Data*; WCDMP-46, WMO/TD-No.1025; World Meteorological Organization (WMO): Geneva, Switzerland, 2000.

64. McGilchrist, C.A.; Woodyer, K.D. Note on a distribution-free CUSUM technique. *Technometrics* **1975**, *17*, 321–325.

65. Karpouzos, D.K.; Kavalieratou, S.; Babajimopoulos, C. Trend analysis of precipitation data in Pieria Region (Greece).*Eur. Water* **2010**, *30*, 31–40.

66. Kim, S. Wavelet analysis of precipitation variability in northern California, U.S.A. *KSCE J. Civ. Eng.* **2004**, *8*, 471–477.

67. Labat, D. Recent advances in wavelet analyses: Part 1. A review of concepts. *J. Hydrol.* **2005**, *314*, 275–288.

68. Morlet, J.; Arens, G.; Fourgeau, E.; Glard, D. Wave propagation and sampling theory-Part I: Complex signal and scattering in multilayered media. *Geophysics* **1982**, *47*, 203–221.

69. Farge, M. Wavelet transforms and their applications to turbulence. *Annu. Rev. Fluid Mech.* **1992**, *24*, 395–457.

70. Gaucherel, C. Use of wavelet transform for temporal characterisation of remote watersheds. *J. Hydrol.* **2002**, *269*, 101–121.

71. Kang, S.; Lin, H. Wavelet analysis of hydrological and water quality signals in an agricultural watershed. *J. Hydrol.***2007**, *338*, 1–14.

72. Nakken, M. Wavelet analysis of rainfall-runoff variability isolating climatic from anthropogenic patterns. *Environ. Model. Softw.* **1999**, *14*, 283–295.

73. Torrence, C.; Compo, G.P. A practical guide to wavelet analysis. *Bull. Am. Meteorol. Soc.* **1998**, *79*, 61–78.

74. Zhang, Q.; Xu, C.; Jiang, T.; Wu, Y. Possible influence of ENSO on annual maximum streamflow of the Yangtze River, China. *J. Hydrol.* **2007**, *333*, 265–274.

75. Jevrejeva, S.; Moore, J.C.; Grinsted, A. ENSO signal propagation detected by wavelet coherence and mean phase coherence methods. In *Nonlinear Dynamics in Geosciences*; Tsonis, A.A., Elsner, J.B., Eds.; Springer New York: New York, NY, USA, 2007; pp. 167–175.

76. Torrence, C.; Webster, P.J. Interdecadal Changes in the ENSO–Monsoon System. *J. Clim.* **1999**, *12*, 2679–2690.

77. Torrence, C.; Webster, P.J. The annual cycle of persistence in the El Nño/Southern Oscillation. *Q. J. R. Meteorol. Soc.***1998**, *124*, 1985–2004.

78. Rasmusson, E.M.; Carpenter, T.H. Variations in tropical sea surface temperature and surface wind fields associated with the Southern Oscillation/El Niño. *Mon. Weather Rev.* **1982**, *110*, 354–384.

79. Wu, R.; Hu, Z.-Z.; Kirtman, B.P. Evolution of ENSO-related rainfall anomalies in East Asia. *J. Clim.* **2003**, *16*, 3742–3758.

CHAPTER 14

MICRO RADAR SURFACE VELOCIMETRY FOR HYDROLOGIC SIGNAL PROCESSING USING A BANDPASS FILTERING APPROACH

Jeng-Wen Lin, Chih-Wei Huang and Yin-Sung Hsu

Keywords: Bandpass filter, hydrologic signal processing, micro radar surface velocimetry, fast Fourier transform, wavelet transform

INTRODUCTION

In recent years, Taiwan has been under the influence of global warming because of the frequent extreme climatic conditions and the disasters caused by typhoons and heavy rains. Hydrological observation operations, and the collection and analysis of the related information have been critical in facilitating disaster prevention, and the response, management, and planning of the rivers. The collection and analysis of hydrological data can provide important references for the allocation and utilization of water resources, disaster prevention and warning, and the development of various supply and demand plans for water, as well as the planning and design of hydraulic engineering projects. The topographic conditions of the rivers in Taiwan are slightly unusual, including steep slopes and rapid streams. Floods that are heavy with sand occur, whenever torrential rainfall and typhoons occur. Thus, the erosion and deposition at the cross section of the rivers and the flow path change significantly, making flow observation a difficult task. In addition, most rivers have various river crossing and flood-control constructions, such as bridges, consolidation works, submerged weirs, and river weirs, and the physical change in the water caused by such man-made interferences can affect the precision of the measured data. However, it is worth mentioning that these structures may also represent good, or even optimal, locations for stream gauging because the weirs impose a fixed bed topography and the bridges facilitate the acquisition of data along a cross-section. Therefore, a knowledge of the accurate flow of the rivers is crucial for planning and utilizing water resources and for designing the associated works.

Flow observation includes the observation of the water depth and the flow velocity; they have to be mostly conducted manually and are time-consuming, strenuous, dangerous, and have low precision; however, the precision (and accuracy) is influenced by various other factors. Currently, there are no instruments employed for automatic observation at home and abroad that are suitable for evaluating the environmental conditions of the rivers in Taiwan. Thus, investigations can only be made indirectly by hydrological models. All of the instruments for the water depth observation, such as the conventional sounding rod, heavy punch, and fish lead to a sounding boat equipped with ultrasonic detectors; the acoustic Doppler current profiler (ADCP) and ground penetrating radar (GPR) used for measuring the water depth require manual field operations and are do not support automatic observation. Instruments used for the flow velocity observation, such as those using the conventional float method, the cup-type current

Citation: Jeng-Wen Lin, Chih-Wei Huang, and Yin-Sung Hsu, Micro Radar Surface Velocimetry for Hydrologic Signal Processing Using a Bandpass Filtering Approach, Water 2016, 8(6), 262; doi:10.3390/w8060262.

meter, propeller-type current meter, electromagnetic current meter, acoustic Doppler current profiler, and the radar surface current meter (which is currently used) already have the capability for automatic observation. In addition, the optics-based methods that are rapidly emerging may provide useful velocity benchmark values for future radar investigations. These methods also utilize the images for identifying the river flow, e.g., large-scale particle image velocimetry (LSPIV) [1,2,3].

As all the microwave radar surface current meters are imported, the measurement results are obtained by the instruments directly and the firmware program cannot be changed. Thus, there are only a limited number of adjustable setup parameters in the instruments and the measurement results cannot be processed further in the form of signals. Thus, a continuous wave radar surface current meter with automatic observation was adopted in this study to analyze the radar signals received by the instruments. The surface velocity was estimated using the conventional fast Fourier transform (FFT), wavelet transform (WT), and a bandpass filter to seek an effective method suitable for real-time observations for analyzing the signals and the uncertainties related to the radar-based flow velocity estimations.

LITERATURE REVIEW

The radar surface current meter includes a pulsed Doppler radar and a continuous microwave radar. Melcher *et al.* [4] mounted a pulsed radar on a helicopter and Cheng *et al.* [5] mounted a pulsed radar on a riverbank to measure the surface velocity of the entire cross-section at different locations and distances by adjusting the tilt angle; however, it still required manual operations. The continuous microwave radar can be placed on cross-river constructions. A single such instrument can measure an elliptical footprint, while multiple instruments can measure the surface velocity of the entire cross section and it can be used directly for automatic measurements without manual operation; e.g., Costa *et al.* [6] mounted a pulsed radar on the riverbank and a continuous wave radar was deployed from a cableway or installed under a bridge.

The continuous microwave radar surface current meter calculates the surface velocity of the river by using the Bragg scattering and the Doppler shift of the reflection of the radar waves, caused by tiny ripples on the stream surface. Therefore, the setup of the tilt angle (the angle between the horizontal plane and the direction of the radar wave transmission) and the yaw angle (the angle between the mainstream direction and the direction of the radar wave transmission) affect the measurement results of the surface velocity apart from the influence of the stream conditions of the rivers. Limited by environmental factors, the measurement direction is not guaranteed to be in parallel with the mainstream direction; thus, the reflected waves received by the microwave radar may not be reflected by the ripples moving in the mainstream direction. A Doppler radar instrument enables the calculation of the component of the velocity vector of the target along the beam. Unless the beam is parallel to the main flow velocity (tilt angle and yaw angle, both equal to zero), this component differs from the actual flow velocity. While the drift angle can be zero (for example, if the instrument is placed on a bridge), the depression angle is always larger than zero for a non-contact water survey setup but, in both the cases, their effect can be calculated using trigonometric relationships.

When a yaw angle (horizontal included angle) exists between the measurement direction and the mainstream direction, the calculation formula for the surface velocity can be revised by the yaw angle; when the angle is considerably large, the measurement results may not give an exact value for the surface velocity along the mainstream direction. The tilt and the yaw angle have the same type of effect on the measurements as they represent the angle between the radar beam and the flow velocity on two different planes. Trigonometric corrections can be applied either on-the-fly or during data post-processing in both cases.

Hsu *et al.* [7] conducted measurements on the surface velocity with a propeller-type current meter and a 10.525 GHz microwave radar current meter having a tilt angle range of 20–60 degrees and a yaw angle range of 0–40 degrees. The results indicated that the standard deviation of the measurement results obtained by the radar wave current meter were larger than those obtained by the propeller-type current meter. Therefore, when measuring the surface velocity with the radar wave current meter, the signals received by the radar waves of this instrument should be analyzed in addition to the settings according to the original correction parameters and the measurement direction should be in line with the positive flow direction as much as possible in order to avoid the yaw angle correction. If correcting the yaw angle becomes necessary, then the corrected angle should be controlled within 20 degrees to improve the precision of the observation results of the surface velocity [7]. Lee and Julien [8] compared some of the laboratory tests with the field tests and demonstrated satisfactory agreements over a wide range of tilt and yaw angles. Laboratory testing showed comparable results between the electromagnetic wave surface velocimeters and other measurement techniques. Field testing at three different locations show that the optimal operation conditions are at a tilt angle of approximately 30 degrees and a yaw angle less than 13 degrees.

Previous studies seldom focus on the noise removed by the radar echo signals and most of them discuss the measurement results with different radar wave transmitting antennae or erections of instruments. Costa *et al.* [9] utilized van-mounted, pulsed Doppler (10 GHz) radar collected surface-velocity data across the 183-m wide Skagit river in Washington at a USGS stream gaging station using Bragg scattering from the short waves produced by turbulent boils on the surface of the river. Melcher *et al.* [4] indicated that the United States Geological Survey and the University of Washington collaborated in a series of initial experiments on the Lewis, Toutle, and Cowlitz Rivers during September 2000 and in a detailed experiment on the Cowlitz River during May 2001 for determining the feasibility of using helicopter-mounted radars for measuring the river discharge (surface velocities were measured using a pulsed Doppler radar). Tamari *et al.* [10] tested a handheld radar to measure the water velocity at the surface of the channels, while Kim *et al.* [11] analyzed the error in the electromagnetic surface velocity and discharge measurement in a rapid mountain stream flow. Therefore, rapid and effective processing, and the method used for analyzing the surface current meter signals, are key issues. The optimal linear filtering theory, also called the Wiener filtering theory, was derived from the study of Kolmogorov [12] and Wiener *et al.* [13].

The greatest disadvantage of the Wiener filtering theory is that it requires data from the infinite past and it is not applicable for time series processing. To overcome this disadvantage, Kalman [14] introduced a state-space model in the filtering theory and deduced a recursive estimation algorithm called the Kalman filtering theory. Kalman filtering adopts the minimum mean square error as the optimum criterion to obtain a recursive estimation algorithm. It updates the estimation of the state variables by the estimated value at the last instance in time and the observed value at present with the state-space model of the signal and noise to obtain the estimated value at present. Another method similar to the Kalman filtering is the adaptive filter [15,16,17]; however, its filter coefficient can vary depending upon certain requirements at any given point in time. The adaptive filter is often adopted when the optimal coefficient of the digital filter cannot be determined. If an adaptive filter is adopted along with the noise elimination system, noise will inevitably be mixed with the signals received by the sensors [18]. Therefore, sometimes two sensors may be adopted for reducing the noise interference, with one mainly detecting the target signals and the other detecting the noise. When a surface current meter is placed on site, the space for storing the component information is limited; all of the above filters perform a significant number of calculations; hence, they are suitable for systems with a large storage space (such as computers). Therefore, this study compares the errors in the various methods used for estimating the surface velocity, via wavelet transform and a bandpass filter that calculate rapidly and do not depend upon previous measurement information, apart from the fast Fourier transform method, necessary for the calculations.

HYDROLOGIC SIGNAL PROCESSING APPROACH

The principle for measuring the surface velocity using a microwave radar surface current meter is derived from the Bragg scattering and the Doppler effect. When the wavelength of the radar waves incident on the water surface are 2n times that of the waves on the water surface, the radar waves generate a Bragg scattering (as shown in Figure 1) owing to interaction with the waves on the water surface, causing resonance. The Bragg resonance yields a Doppler shift when the ripples move on the water surface; thus, the movement speed of the ripples, namely, the surface velocity of the water body, can be obtained using the Doppler shift.

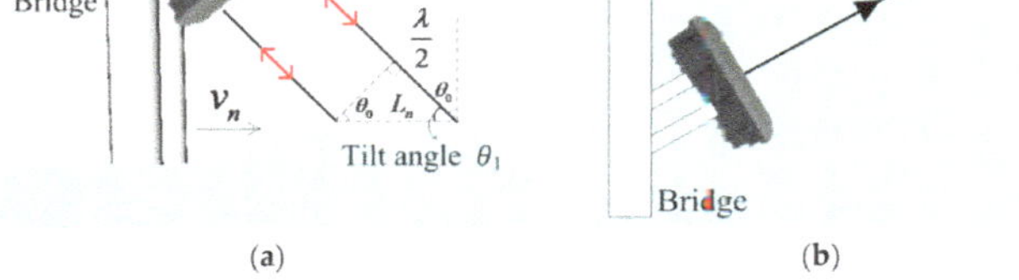

Figure 1: Bragg scattering and surface velocity measured by a radar wave current meter: (**a**) lateral view; and (**b**) top view.

Bragg scattering is shown in Figure 1. L_n is the Bragg resonance wavelength, λ is the wavelength of the radar waves, and θ_0 is the incidence angle of the radar waves. Their relationship can be expressed as follows [19]:

$$L_n = \frac{\lambda}{2\sin\theta_0} \tag{1}$$

According to the Doppler effect, v_n is the speed of the surface ripples moving towards/away from the microwave radar surface current meter; thus, the Doppler peak frequency of the echo Doppler spectrum, f_D, Can be expressed as follows [19]:

$$f_D = \frac{2v_n}{\lambda} \tag{2}$$

In the above formula, v_n is the line-of-sight velocity, *i.e.*, the component of the scattered velocity along the antenna direction. Figure 1 shows the method for measuring the surface velocity of a river, when a microwave radar surface current meter is placed at a fixed location (such as bridge) and the formula for calculating the surface velocity is as follows:

$$v = \frac{c}{2f_o\cos\theta} f_D,\ \cos\theta = \frac{\cos\theta_1 + \cos\theta_2}{2} \tag{3}$$

In the above formula, v is the surface velocity of the water body, c is the speed of the radar waves transmitting in air (3×10^8 m/s), f_o is the frequency of the microwave radar antennae (10.525 GHz), f_D is the Doppler peak frequency caused by the movement of the surface ripples, θ_1 is the tilt angle, θ_2 is the yaw angle, and θ is the angle between the radar waves and the flow direction. The transmission speed of the radar waves, c, the frequency of the radar antennae, f_o, and the angle between the radar waves and the flow direction, θ, are known constants; hence, the surface velocity, v, can be estimated by measuring the receiving frequency of the radar echo, f_D. Therefore, the precision of the radar echo frequency, f_D, is very important in estimating the surface velocity of a river.

In practice, the influence of the ripples generated by the wind and rain on the water surface on the radar echo signals cannot be determined using only conventional fast Fourier transform. To discuss the applicability and feasibility of the different analyses methods for the flow velocity signals, trolley tests were conducted in this study at the Laboratory of Current Meter Calibration of the Water Resources Planning Institute, Ministry of Economic Affairs. The original signals of the microwave radar surface current meter were measured and analyzed by fast Fourier transform, wavelet transform, and a bandpass filter. Further, the precision in estimating the surface velocity by these three analytical methods was compared to determine an effective method that would be suitable for real-time observations for identifying the radar wave frequency.

Fast Fourier Transform

In the past, the corresponding frequency spectrum (frequency-amplitude) of the signals was obtained by discrete Fourier transform (DFT) for analyzing signals to obtain their important parameters. As the number of calculations were significant, the discrete Fourier transform required considerable time for the calculations; hence, it was difficult to obtain results in real-time. Cooley and Tukey [20] simplified the discrete Fourier transform and employed the fast Fourier transform that improved the calculation speed significantly and it was widely adopted in various fields; however, it has certain limits:

(1) periodic signals are required; (2) the sampling period must be an integral multiple of the signal period; (3) the sampling frequency must be at least two times higher than the maximum frequency of the signals; and (4) the length of the sampling information must be in powers of two.

Conventionally, the estimation of the receiving frequency of the radar echo, f_D, is conducted using the fast Fourier transform for analyzing the original radar echo signals (as shown in Equation (4)) and for determining the resonant peak frequency of the power spectral density (PSD) to obtain the Doppler peak frequency, f_D (as shown in Equation (5)).

$$Y(\omega) = \int_{t_1}^{t_2} y(t)e^{-i\omega t}dt,\ \omega = 2\pi f \tag{4}$$

$$PSD(\omega) = |Y(\omega)|^2 \tag{5}$$

In the above formula, y(t) represents the original measured signals, f represents the sample frequency (Hz), the range t_1–t_2 indicates the measurement time, Y(ω) represents the amplitude, PSD(ω) represents the power spectral density, and f_D is the maximum amplitude of PSD(ω), corresponding to the frequency. As radar waves transmit rapidly and receive signals sensitively, echo signals often carry noise that directly affect the estimated value of the Doppler peak frequency, f_D. Further, as the flow velocity signal information is a time-varying signal, the information on only a part of the time domain only can be obtained if it is analyzed by fast the Fourier transform and the power spectral density (PSD). This is unsuitable for the nonlinear and unsteady information on the rivers because the noise signal cannot be distinguished from the master signal. Therefore, the wavelet transform was adopted in this study for signal analysis to filter the noise.

Wavelet Transform

The basic wavelet function is called the mother wavelet and the wavelets generated by extension, contraction, and translation are defined as baby wavelets [21,22,23,24,25,26,27]. The mother wavelet function must meet the following two requirements:

1. The energy of the mother wavelets must be finite, as shown in Equation (6):

$$E_\phi = \int_{-\infty}^{\infty} |\varphi(s)|^2 dt < \infty \tag{6}$$

In the above formula, E_ϕ stands for the energy of the mother wavelet and φ(s) stands for the mother wavelet function.

2. Mother wavelets must meet the admissible conditions, as shown in Equation (7):

$$C_g = \int_0^{\infty} \frac{\left|\widehat{\varphi}(\omega)\right|^2}{\omega} d\omega < \infty,\ \widehat{\varphi}(\omega) = \int_{-\infty}^{\infty} \varphi(s)e^{-i\omega t}dt \tag{7}$$

In the above formula, C_g stands for the admissible constant that varies slightly depending upon the wavelets. Equation (7) is called the admissibility condition and $\widehat{\varphi}(\omega)$ is the Fourier transform of φ(s).

From the signal analysis perspective, any signal can be considered as a multi-frequency signal function and a compound of signals with different resolutions by linear additivity. The conventional Fourier transform can only present a single signal in the time domain or frequency domain, while the signals can extract the required information from the original signals after being processed by the wavelet transform. Multi-scale analysis by the scale extension and translation operation functions is conducive for the enhancement and identification of the signal characteristics, and the analysis of nonlinear and unsteady information. This implies that the wavelet transform decomposes an original signal into a combination (as shown in Figure 2) of approximation signals (represented by A) and detail signals (represented by D), as shown in Equation (8):

$$y(t) = D_1 + D_2 + \cdots + D_n + A_n \tag{8}$$

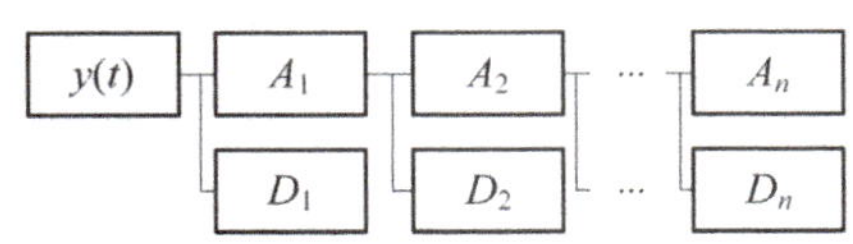

Figure 2: Schematic diagram of the wavelet transform signal analysis.

In the above formula, y(t) represents the original signal, $D_1, D_2, \ldots, D_n$ represent the detail signals from the decomposition of the original signal by the wavelet transform, An represents the approximation signals, and n represents the order of the wavelet decomposition.

Signal composition follows the decomposition of the original signal into detail signals and approximation signals with different composition methods corresponding to different properties of the original signal that can then be divided into detail signal compositions (as shown in Equation (9)) and approximation signal compositions (as shown in Equation (10)). Donoho [28] and Misiti *et al.* [25] adopted the detail/approximation signal composition infinitely to filter signals with a set threshold value. Taswell [29] proposed the elimination of the noise in the original signal by composing only the approximation signals; however, the tests in the study indicated that the approximation signals were only suitable for low flow velocity (under 1.0 m/s) analyses, while the detail signals were required for those above 1.0 m/s. In practical applications, it is difficult to set the filter for automatically compounding the detail/approximation signals, given that the onsite flow velocity cannot be determined. By considering the importance of the flow velocity information, in this study, detail signals were adopted for the wavelet signal composition to obtain the velocity information.

$$y_H(t) = D_1 + D_2 + \cdots + D_n \tag{9}$$

$$y_L(t) = A_n \tag{10}$$

where $y_H(t)$ is the composition of the detail signals and $y_L(t)$ is the composition of the approximation signals. The Daubechies wavelet function used in this study, was adopted as the wavelet transform filter and the dmey function (discrete approximation of the Meyer wavelet) of MATLAB was employed for decomposing the wavelet into five detail signals and one approximation signal. It is difficult for the wavelet signals to set the filter for an automatic switchover between the compositions of the detail/approximation signals; moreover, the different numbers of the decomposed signals can affect the differences in the composed signals. Therefore, a bandpass filter was adopted to conduct an additional signal analysis to filter the noise.

Bandpass Filter

A filter discards the unnecessary low frequency (or high frequency) parts of a complex signal and ensures that most of the passed high-frequency (or low-frequency) signals are useful; the definitions of the low-frequency and high-frequency are determined by the cut-off frequency selected by the designer of the filter. A pure filter can be divided into a high-pass filter, low-pass filter, and a bandpass filter. A high-pass filter allows the passage of high-frequency signals but weakens (or reduces) signals whose frequency is lower than the cut-off frequency; on the contrary, a low-pass filter allows the passage of low-frequency signals but weakens (or reduces) the signals whose frequency is above the cut-off frequency. However, both low-frequency and high-frequency signals (low flow velocity and high flow velocity) can be encountered during the identification of the radar wave surface velocity owing to the interference of the external environmental conditions. Therefore, in this study, a bandpass filter that integrates both high- and low-pass filtering should be applied to the microwave radar surface current meter instead of a pure high-pass filter or a pure low-pass filter.

An ideal bandpass filter has a completely flat passband (*i.e.*, with no gain/attenuation throughout) and will completely attenuate all the frequencies outside the passband. Additionally, the transition out of the passband will have brick-wall characteristics [30], as shown in Figure 3. The resonance frequency is the difference between the cut-off frequency, F_{cL}, of a lower frequency band and the cut-off frequency, F_{cH}, of a higher frequency band; the bandpass width of the filter is the difference between F_{cL} and F_{cH}.

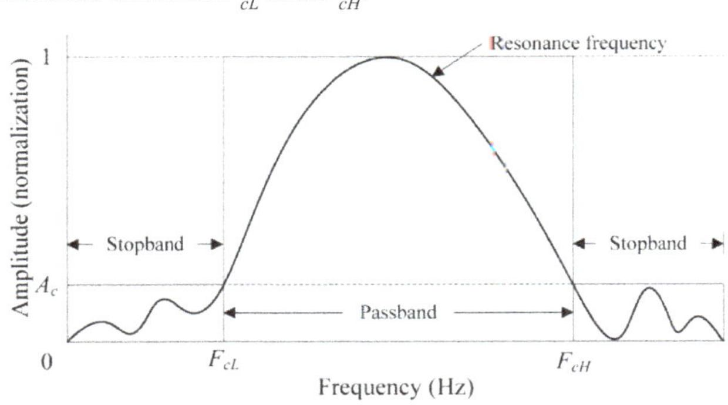

Figure 3: Illustration of the design map of a bandpass (A_c = 0.1) filter for radar surface velocimetry.

To extract each modal signal from a given signal, a digital bandpass stopband filter that contains one to several modal frequencies for a signal was designed. Figure 3 illustrates the design map of a filter that uses a cutoff frequency corresponding in amplitude to 10% of the peak value (*i.e.*, A_c = 0.1) of the signal in the frequency spectrum [31,32]. The signal was only allowed oscillation between the stopband cutoff frequencies (*i.e.*, F_{cL}–F_{cH}), and the amplitudes in the other intervals (*i.e.*, 0–F_{cL} and higher than F_{cH}) were set to zero to filter out the lower and higher portions of the signal, arising from the measurement errors.

EXPERIMENTATION AND ANALYSIS

A 10.525 GHz microwave radar surface current meter (test model before the product development, resolution: 0.01 m/s, voltage: 12–36 VDC, temperature: −35–60 °C, ingress protection: 65) was adopted. The experiments were conducted at the Laboratory of Current Meter Calibration of the Water Resources Planning Institute, Ministry of Economic Affairs that is the only current meter calibration laboratory approved by the Taiwan Accreditation Foundation (TAF). The calibration trolley in the laboratory has a calibration speed of 0.3–6.0 m/s, minimum uncertainty of 2%, and a minimum speed of 0.2 m/s. The length and width of the experimental channel are 120 m and 1.6 m, respectively. The trolley initially advances slowly, accelerating gradually and moves at a constant speed when it reaches the set location (start of the stable movement section). It then decelerates and ultimately stops after moving for 65 m in the stable movement section. The instrument is fixed on the stainless steel shelf on the trolley for the measurement (as shown in Figure 4). The water surface is still during the measurement, and ripples appear on the water surface as the trolley moves; therefore, the radar wave reflected signals could be measured for calculating the errors incurred by the different analytical methods, while identifying the surface velocity and the actual movement speed of the trolley, as shown in Equation (11):

$$RE = \frac{v_i - v_r}{v_r} \times 100\% \tag{11}$$

where RE is the relative error (%), v_i is the identification result of the surface velocity (m/s), and v_r is the actual movement speed of the trolley (m/s).

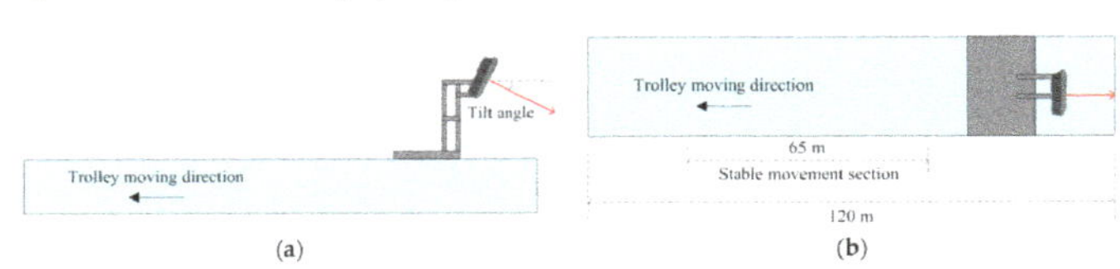

Figure 4: Schematic diagram of the experimental setup in the laboratory for the current meter calibration: (**a**) lateral view; and (**b**) top view.

Herein, the incident direction of the radar waves and the flow direction are mostly parallel (namely, yaw angle = 0 degrees); therefore, this study discusses the differences between the different analytical methods for the identification of the surface velocity under varying flow velocities and incident tilt angles. The tilt angle for the experiments in this study was set as 20, 30, 40, 50, and 60 degrees; the movement speed of the trolley was 0.2, 1.0, 1.5, 2.0, 2.5, 3.0, 4.0, 5.0, and 6.0 m/s; the incident direction of the radar waves was opposite to the movement direction of the trolley, *i.e.*, the simulated incident direction of the radar waves was the same as the stream direction. A total of 45 sets of experiments were conducted.

The surface current meter in this study was set to a continuous measurement and a group of data was obtained every 2 s; each group contained 640 data points and each data point had a measurement interval of 4.27×10^{-4} s; the frequency of the samples was 2.34 kHz. The range of the radar emission angle was 10 degrees and the radar emission position was 1.04 m above the water surface; the width of the illuminated area was approximately 0.11–0.50 m (tilt angle 20–60 degrees) that was 1.6 m narrower than the width of the channel; it is necessary to ensure that the sample is not affected by the presence of the side walls (as shown in Figure 5).

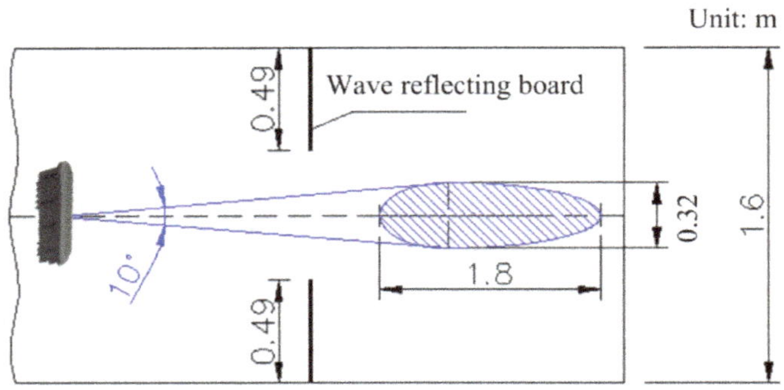

Figure 5: Projection range of the radar waves (tilt angle = 30° and yaw angle = 0°).

For example, the original signals of the radar waves are shown in Figure 6 and the original signals (by conventional FFT only), wavelet analysis, and the signal frequency spectrum after the bandpass filter analysis are shown in Figure 7, when the tilt angle is 30 degrees, and the speed of the trolley is 0.2 m/s and 1.0 m/s. According to Figure 7, there is considerable noise (several peak frequencies) in the original signals. The noise that can be filtered by the wavelet transform is negligible, often resulting in interference and misjudgment of the peak frequency, while the bandpass filter can filter the redundant noise for representing the frequency spectrum chart as a smooth curve. Thus, the peak frequency can be easily identified. Figure 8shows the results of the flow velocity obtained using the three methods. The calculation of the flow velocity using the original signals was unstable and fluctuated because of considerable noise. The calculation of flow velocity using the wavelet analysis was slightly more stable, but it often misjudged the peak frequency, while the calculation of the flow velocity using the bandpass filter was stable and the identification results were closer to the actual values.

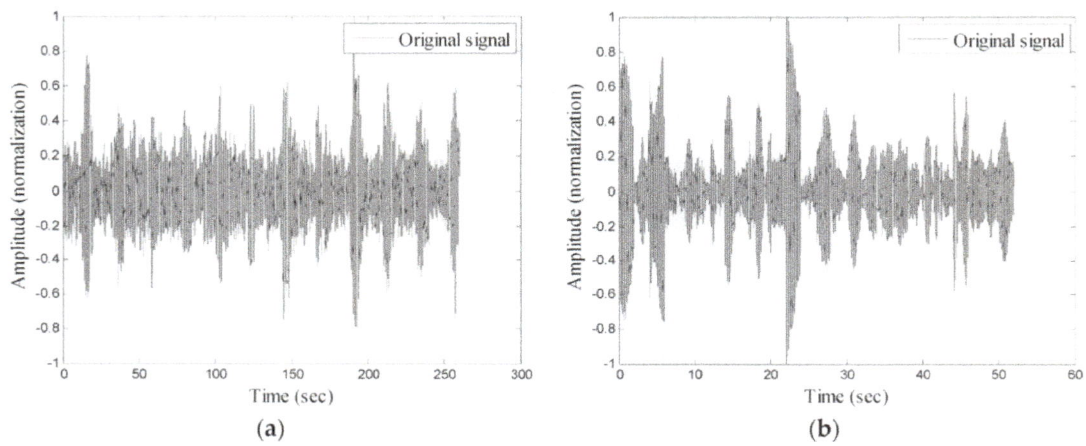

Figure 6: Charts of the original signals of radar waves at a tilt angle of 30°: (**a**) speed = 0.2 m/s; and (**b**) speed = 1.0 m/s.

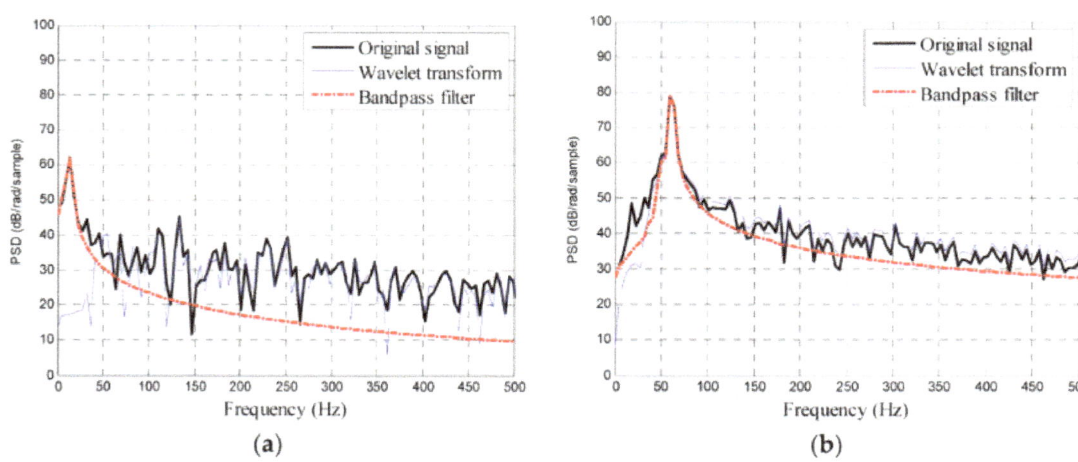

Figure 7: Frequency spectrum chart of the radar signal analysis at a tilt angle of 30°: (**a**) speed = 0.2 m/s; and (**b**) speed = 1.0 m/s.

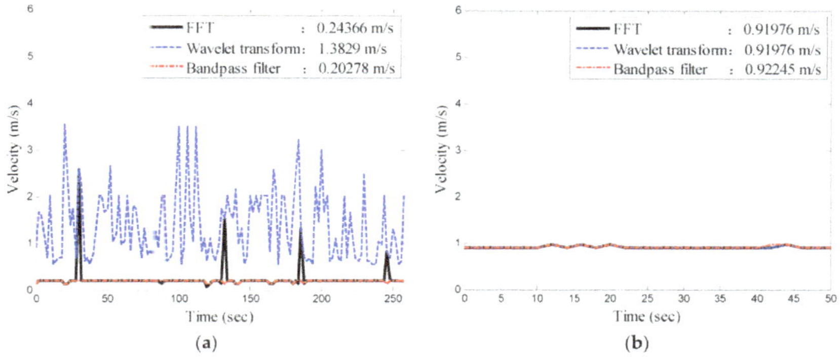

Figure 8: Comparison of the calculated results obtained using different analytical methods at a tilt angle of 30°: (**a**) speed = 0.2 m/s; and (**b**) speed = 1.0 m/s.

Table 1 and Table 2, and Figure 9 and Figure 10, show the comparison of the identification results and the relative errors under different tilt angles and speeds of the trolley. When the velocity is less than 1.0 m/s (low frequency), the relative error of the wavelet analysis is larger because the selected detail signal is suitable for a high frequency signal, which is unsuitable for a low frequency velocity, in general. When the velocity is greater than 1.0 m/s for the high frequency signal, the results of the wavelet analysis and FFT are similar, possibly because of discarding the approximation signal (low frequency part) because it is negligible in the wavelet signal analysis process. In general, the signal processing of the bandpass filter was more stable and the computed flow velocity had the lowest relative error (as low as 2.84% for a tilt angle of 20 degrees). This is because the proposed bandpass filter could remove the white noise effectively. The analysis results show that all the methods tend to be less accurate as the velocity increases owing to the restrictions in the laboratory (*i.e.*, the stable movement section in Figure 4 is 65 m for the measurement). In the 65 m section, when the moving speed of the trolley is higher, fewer original signals could be measured. The length of the measured original signal will affect the analysis results. The optimal range of the tilt angle was between 20 and 40 degrees and the measurement results would be imprecise if it was larger than 40 degrees (the relative error will become higher). This is caused by the limitation of the 10.525-GHz radar wave antennae.

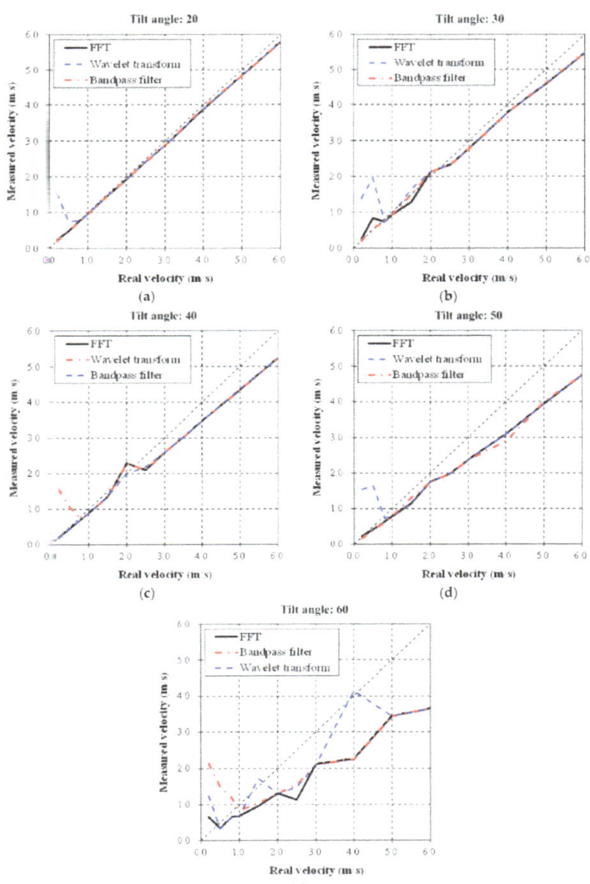

Figure 9: Comparison of the results of the flow velocity experimental analysis using the FFT, bandpass filter, and the wavelet transform. (**a**) Tilt angle = 20°; (**b**) Tilt angle = 30°; (**c**) Tilt angle = 40°; (**d**) Tilt angle = 50°; (**e**) Tilt angle = 60°.

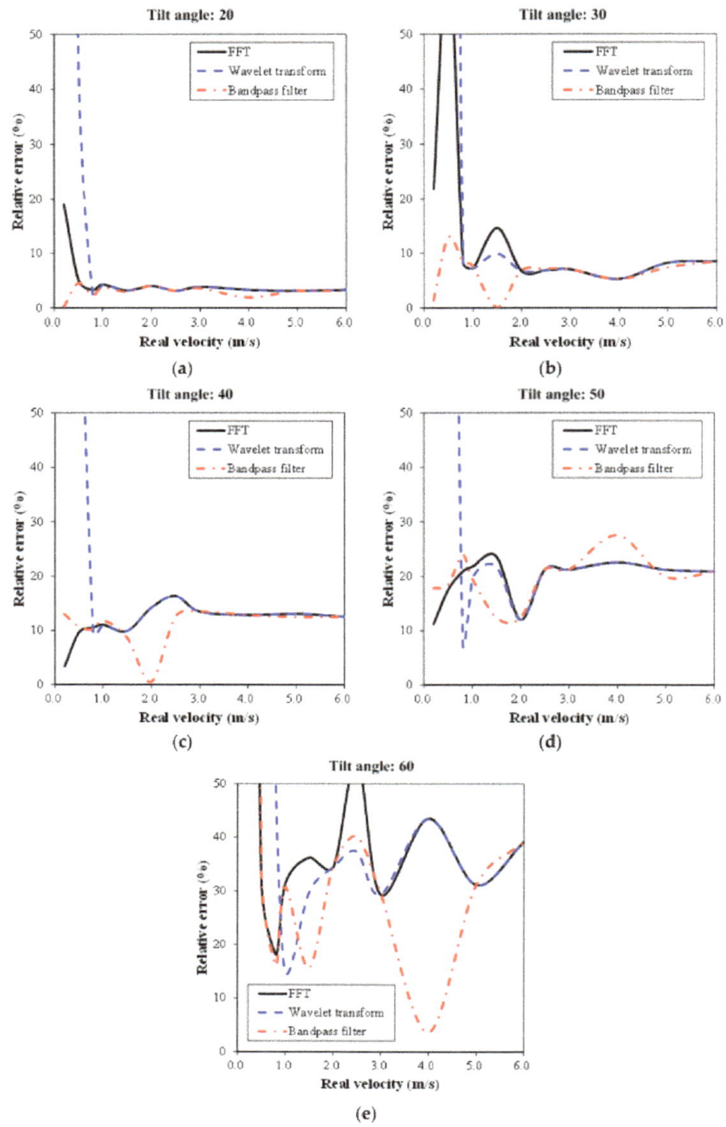

Figure 10: Comparison of the relative errors in flow velocity experimental analysis using the FFT, bandpass filter, and the wavelet transform. (**a**) Tilt angle = 20°; (**b**) Tilt angle = 30°; (**c**) Tilt angle = 40°; (**d**) Tilt angle = 50°; (**e**) Tilt angle = 60°.

Table 1: Comparison of the identification results under different tilt angles and movement speeds of the trolley.

Tilt Angle (Degree)	Method	Velocity (m/s)								
		0.2	1.0	1.5	2.0	2.5	3.0	4.0	5.0	6.0
20	FFT	0.24	0.96	1.45	1.92	2.42	2.88	3.87	4.84	5.80
	Wavelet transform	1.52	0.96	1.45	1.92	2.42	2.88	3.87	4.84	5.80
	Bandpass filter	0.20	0.96	1.46	1.92	2.42	2.89	3.92	4.84	5.82
30	FFT	0.24	0.92	1.28	2.14	2.33	2.79	3.79	4.59	5.49
	Wavelet transform	1.38	0.92	1.65	2.14	2.33	2.79	3.79	4.59	5.49
	Bandpass filter	0.20	0.92	1.50	2.14	2.32	2.79	3.79	4.63	5.49
40	FFT	0.19	0.90	1.35	2.28	2.09	2.60	3.48	4.34	5.25
	Wavelet transform	1.56	0.90	1.35	2.28	2.09	2.60	3.48	4.34	5.25
	Bandpass filter	0.17	0.90	1.37	1.99	2.19	2.60	3.48	4.37	5.25
50	FFT	0.22	0.78	1.14	1.76	1.97	2.36	3.10	3.94	4.75
	Wavelet transform	1.55	0.80	1.17	1.76	1.97	2.36	3.10	3.94	4.75
	Bandpass filter	0.16	0.80	1.31	1.75	1.97	2.36	2.90	4.00	4.75
60	FFT	0.66	0.69	0.96	1.31	1.13	2.12	2.26	3.44	3.65
	Wavelet transform	2.15	0.85	1.05	1.31	1.57	2.12	2.26	3.44	3.65
	Bandpass filter	1.26	0.69	1.73	1.31	1.50	2.11	4.15	3.44	3.65

Table 2: Comparison of the relative identification errors under different tilt angles and movement speeds of the trolley (Note that the values in bold indicated the lowest relative error when the tilt angle ranged between 20 and 40 degrees).

Tilt Angle (Degree)	Method	Velocity (m/s)								
		0.2	1.0	1.5	2.0	2.5	3.0	4.0	5.0	6.0
20	FFT	19.01	4.27	3.19	4.01	3.13	3.86	3.30	3.13	3.30
	Wavelet transform	662.20	4.27	3.19	4.01	3.13	3.86	3.30	3.13	3.30
	Bandpass filter	**0.13**	**3.76**	**2.92**	4.01	3.13	**3.58**	**1.90**	3.13	**3.02**
30	FFT	21.83	8.02	14.72	6.77	6.86	7.06	5.31	8.26	8.52
	Wavelet transform	591.44	8.02	9.96	6.77	6.86	7.06	5.31	8.26	8.52
	Bandpass filter	**1.39**	**7.75**	**0.09**	6.77	7.14	7.06	5.31	**7.42**	8.52
40	FFT	3.39	9.92	9.89	14.23	16.37	13.50	12.88	13.11	12.57
	Wavelet transform	678.32	9.92	9.89	14.23	16.37	13.50	12.88	13.11	12.57
	Bandpass filter	13.05	**9.64**	**8.73**	**0.54**	**12.52**	13.50	12.88	**12.52**	12.57
50	FFT	11.20	21.80	23.69	12.02	21.21	21.24	22.56	21.21	20.91
	Wavelet transform	672.56	19.66	21.82	12.02	21.21	21.24	22.56	21.21	20.91
	Bandpass filter	17.83	19.66	12.48	12.33	21.21	21.24	27.52	19.94	20.91
60	FFT	227.87	31.41	36.21	34.43	54.77	29.32	43.46	31.11	39.11
	Wavelet transform	975.61	15.02	29.73	34.43	37.37	29.32	43.46	31.11	39.11
	Bandpass filter	527.97	30.75	15.64	34.43	40.15	29.69	3.66	31.11	39.11

CONCLUSIONS AND SUGGESTIONS

In this study, the limitations of the manual observations were eliminated using a continuous-wave radar-surface current meter for automatic observations. Experiments were conducted on the flow velocity and tilt angle at the Laboratory of Current Meter Calibration of the Water Resources Planning Institute, Ministry of Economic Affairs. The original echo signals of the microwave radar surface current meter were analyzed using signal processing methods, including the fast Fourier transform, wavelet transform, and a bandpass filter for exploring a suitable and effective method for estimating the surface velocity. A series of experiments were carried out in this study and the results indicated that when the tilt angle was set between 20 and 40 degrees and the yaw angle was set to 0 degrees, and compared to the speed of the trolley, the results of the surface velocity obtained using the bandpass filter adopted in this study were the most stable and precise. This is suitable for the real-time observation of the surface velocity without adjusting the set value of the filter. It is suggested that further studies should be conducted using other signal processing methods for determining the applicability of various research methods and the feasibility for practical applications. Onsite experiments can be conducted at natural rivers with stable stream conditions, cross-section erosions, minor changes in the deposition, and appropriate river widths for determining the effects in the onsite applications.

ACKNOWLEDGMENTS

The authors are grateful for the financial support from the Water Resources Agency of Economic Affairs; the experiments were conducted with the assistance of U&U Engineering, Inc. (Taoyuan, Taiwan) and the Resources Planning Institute of the Resources Agency, Ministry of Economic Affairs. Our sincere gratitude is to all the above institutions.

REFERENCES

1. LeBoursicaud, R.; Pénard, L.; Hauet, A.; Thollet, F.; LeCoz, J. Gauging extreme floods on youtube: Application of LSPIV to home movies for the post-event determination of stream discharges. *Hydrol. Process.* **2015**, *30*, 90–105.
2. Muste, M.; Fujita, I.; Hauet, A. Large-scale particle image velocimetry for measurements in riverine environments. *Water Resour. Res.* **2008**, *44*, W00D19.
3. Tauro, F.; Porfiri, M.; Grimaldi, S. Orienting the camera and firing lasers to enhance large scale particle image velocimetry for stream flow monitoring. *Water Resour. Res.* **2014**, *50*, 7470–7483.
4. Melcher, N.B.; Costa, J.E.; Haeni, F.P.; Cheng, R.T.; Thurman, E.M.; Buursink, M.; Spicer, K.R.; Hayes, E.; Plant, W.J.; Keller, W.C.; *et al.* River discharge measurements by using helicopter-mounted radar. *Geophys. Res. Lett.* **2002**, *29*, 41-1–41-4.
5. Cheng, R.T.; Gartner, J.W.; Mason, R.R.; Costa, J.E.; Plant, W.J.; Spicer, K.R.; Haeni, F.P.; Melcher, N.B.; Keller, W.C.; Hayes, K. *Evaluating a Radar-Based, Non-Contact Streamflow Measurement System in the San Joaquin River at Vernalis*; Open File Report of 2004-1015; U.S. Geological Survey: Menlo Park, CA, USA, 2004. Available online: http://pubs.usgs.gov/of/2004/1015/OFR2004_1015_new.pdf (accessed on 1 March 2016).
6. Costa, J.E.; Cheng, R.T.; Haeni, F.P.; Melcher, N.; Spicer, K.R.; Hayes, E.; Plant, W.; Hayes, K.; Teague, C.; Barrick, D. Use of radars to monitor stream discharge by noncontact methods. *Water Resour. Res.* **2006**, *42*, W07422.
7. Hsu, Y.S.; Tung, T.C.; Chou, H.C.; Chang, K.C.; Lee, J.J.; Yu, B.H.; Huang, P.C. Applications of the radar surface velocimeter. *J. Taiwan Water Conserv.* **2006**, *54*, 82–91.
8. Lee, J.S.; Julien, P.Y. Electromagnetic wave surface velocimetry. *J. Hydraul. Eng.* **2006**, *132*, 146–153.
9. Costa, J.E.; Spicer, K.R.; Cheng, R.T.; Haeni, F.P.; Melcher, N.B.; Thurman, E.M.; Plant, W.J.; Keller, W.C. Measuring stream discharge by non-contact methods: A proof-of-concept experiment. *Geophys. Res. Lett.* **2000**, *27*, 553–556.
10. Tamari, S.; García, F.; Arciniega-Ambrocio, J.I.; Porter, A. Testing a handheld radar to measure water velocity at the surface of channels. *La Houille Blanche* **2014**, *3*, 30–36.
11. Kim, D.S.; Yang, S.K.; Jung, W.Y. Error analysis for electromagnetic surface velocity and discharge measurement in rapid mountain stream flow. *J. Environ.*

Sci. Int. **2014**, *23*, 543–552.

12. Kolmogorov, A.N. Stationary sequences in hilbert space. *Mosc. Univ. Math. Bull.* **1941**, *2*, 1–40.
13. Wiener, N. *The Interpolation, Extrapolation and Smoothing of Stationary Time Series*; OSRD 370, Report to the Services 19, Research Project DIC-6037; Massachusetts Institute of Technology: Cambridge, MA, USA, 1942.
14. Kalman, R.E. A new approach to linear filtering and prediction problems. *J. Basic Eng.* **1960**, *82*, 35–45.
15. Woody, C.D. Characterization of an adaptive filter for the analysis of variable latency neuroelectric signals. *Med. Biol. Eng.* **1967**, *5*, 539–554.
16. Thakor, N.V.; Zhu, Y.S. Applications of adaptive filtering to ECG analysis: Noise cancellation and arrhythmia detection. *IEEE Trans. Biomed. Eng.* **1991**, *38*, 785–794.
17. Haykin, S.O. *Adaptive Filter Theory*, 5th ed.; Prentice-Hall Incorporation: Upper Saddle River, NJ, USA, 2013.
18. Lin, J.W. Repetitive model refinement for structural health monitoring using efficient Akaike information criterion.*Smart Struct. Syst.* **2015**, *15*, 1329–1344.
19. Plant, W.J.; Keller, W.C.; Hayes, K. Measurement of river surface currents with coherent microwave systems. *IEEE Trans. Geosci. Remote Sens.* **2005**, *43*, 1242–1257.
20. Cooley, J.W.; Tukey, J.W. An algorithm for the machine calculation of complex Fourier series. *Math. Comput.* **1965**, *19*, 297–301.
21. Mallat, S.G. A theory for multi-solution signal decomposition: The wavelet representation. *IEEE Trans. Pattern Anal. Mach. Intell.* **1989**, *11*, 674–693.
22. Daubechies, I. The wavelet transform, time-frequency localization and signal analysis. *IEEE Trans. Inf. Theory* **1990**, *36*, 961–1005.
23. Daubechies, I. *Ten Lectures on Wavelets*; Society for Industrial and Applied Mathematics: Philadelphia, PA, USA, 1992.
24. Chui, C.K. *An Introduction to Wavelets*; Academic Press Incorporation: Boston, MA, USA, 1992.
25. Misiti, M.; Misiti, Y.; Oppenheim, G.; Poggi, J.M. *Wavelet Toolbox for Use with MATLAB*; The Math Works Incorporation: Natick, MA, USA, 1996; Available online: http://web.mit.edu/1.130/WebDocs/wavelet_ug.pdf (accessed on 1 March 2016).
26. Gurley, K.; Kareem, A. Applications of wavelet transforms in earthquake, wind and ocean engineering. *Eng. Struct.***1999**, *21*, 149–167.
27. Mallat, S. *A Wavelet Tour of Signal Processing*; Academic Press: San Diego, CA, USA, 1999.
28. Donoho, D.L. De-noising by soft-thresholding. *IEEE Trans. Inf. Theory* **1995**, *41*, 613–627.
29. Taswell, C. The what, how, and why of wavelet shrinkage denoising. *Comput. Sci. Eng.* **2000**, *2*, 12–19.
30. Band-Pass Filter. Available online: https://en.wikipedia.org/wiki/Band-pass_filter (accessed on 19 May 2016).
31. Etter, D.M. *Engineering Problem Solving with MATLAB*; Prentice-Hall Incorporation: Upper Saddle River, NJ, USA, 1993.
32. Lin, J.W. Mode-by-mode evaluation of structural systems using a bandpass-HHT filtering approach. *Struct. Eng. Mech.***2010**, *36*, 697–714.

CHAPTER 15 IMPACT OF CLIMATE CHANGE ON GROUNDWATER RESOURCES

Adama Toure, Bernd Diekkrüger, and Adama Mariko

Keywords: Groundwater, EARTH model, PMWIN, climate change, Klela basin, southern Mali

INTRODUCTION

Groundwater resources provide important sources of drinking water throughout the world, especially in developing countries where most human activities (e.g., domestic water use, small-scale irrigation, *etc.*) depend on groundwater. The investigated basin in Mali is an agriculture zone growing cotton and rice during the rainy season and potatoes during the dry season. The surface area used for agriculture has been extended in the basin to assure food security for the growing population. In addition, most of the rural population uses water from traditional wells that are recharged by rainfall and dry out a few months after the recharge period. Moreover, water sustainability in the town of Sikasso's supply system remains a challenge due to groundwater exploitation in the context of increased water demand and climate variability and change. Therefore, better understanding and management of this vital resource are the principal requirements of responsible water use. Because of the growing population and climate change, a scientific tool should be used to assess present and future difficulties related to water supply and demand in the study area. MODFLOW is one of the most popular hydrologic models in the world and can achieve the objectives of this study. MODFLOW is a modular, three-dimensional finite-difference groundwater model that was developed by the United States Geological Survey (USGS) [1]. It uses the groundwater flow equation, which is a combination of the continuity equation and Darcy's law [2]. Over the past few years, the application of MODFLOW to describe and predict aquifer system behavior has increased significantly [3]. MODFLOW has been used by many authors [4,5,6,7] throughout the world to simulate groundwater dynamics; however, few studies have been conducted in West Africa, especially in Mali, where some studies [8,9] have predicted rainfall decreases. For example, Bricquet *et al.* [10] showed a reduction in groundwater storage in the Bani basin (which contains several subbasins, including the Klela basin) due to decreases in rainfall and runoff. Additionally, Bokar *et al.* [11] demonstrated that groundwater levels decreased from 2 to 15 cm per year from 1940 to 2008 in the Kolondieba catchment, a subbasin of the Bani basin. The interaction between surface water and groundwater has been the topic of many discussions [12,13,14] in Mali, which have concluded that the river system drains the aquifer in the southern part of the country. Until now, no study has been performed in the Klela basin to confirm or refute these results or to quantify groundwater storage.

The partitioning of rainfall into groundwater (recharge) and the interaction between surface water and groundwater is unknown in the Klela basin. However, knowledge of these processes is required to facilitate water resources management in the Klela basin in the context of climate change.

Therefore, the purpose of this study is to evaluate the impact of climate change on groundwater resources in the Klela basin using the RCP4.5 climate scenario. EARTH

Citation: Adama Toure, Bernd Diekkrüger, and Adama Mariko, Impact of Climate Change on Groundwater Resources, Hydrology 2016, 3(2), 17; doi:10.3390/hydrology3020017.

(Extended model for Aquifer Recharge and soil moisture Transport through the unsaturated Hardrock) and PMWIN 5.3 (Processing MODFLOW for Windows) developed by Chiang and Kinzelbach [15] is applied to estimate groundwater recharge and simulate groundwater behavior under steady and transient state conditions.

Study Area

The Klela basin is located in the Sikasso Region in southern Mali, with a small part located in western Burkina Faso. Situated between 5°55′58.8″ and 5°16′12″ longitude and 11°40′58.8″ and 10°59′45.6″ latitude, it covers approximately 3680 km². The elevation of the Klela Basin decreases from south to north, from 748 m above sea level (masl; south) to 305 masl (north) (Figure 1). The main river of the area is the Lotio River, which has periodic flows and is a tributary of the Banifing, which is a tributary of the Bani River basin. The only assured water supply is groundwater, and all the inhabitants depend directly or indirectly on this resource. The minimum monthly temperature varies from 12.3 °C to 26.8 °C, the maximum temperature ranges from 28.8 °C to 39.9 °C and the average temperature is 27.4 °C (Figure 2). The climate belongs to the Soudano-Sahelian zone and is characterized by a dry season (November–May) dominated by a dry wind from the Sahara (harmattan) and a rainy season (June–October) with a wet wind from the Guinean Gulf (monsoon) [16]. The mean annual precipitation is approximately 1131 mm/year. The average annual potential evapotranspiration is approximately 2060 mm/year (Figure 2) based on the Blaney-Criddle method. The geology of the study area is mainly constituted by sandstone and dolerite formations (Figure 3). However, the Tabular Infra-Cambrian fractured sandstone formation is predominant. Groundwater can be found in the fractured bedrock at depths of up to 100 m. The recent formations that cover the sandstone bedrock are constituted by clay, sand and laterite, with a maximum covering thickness of 64 m.

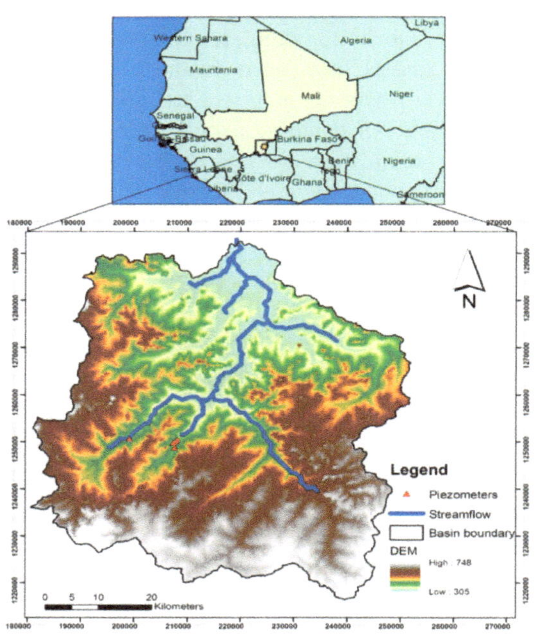

Figure 1: Geographical location of the Klela Basin in southern Mali.

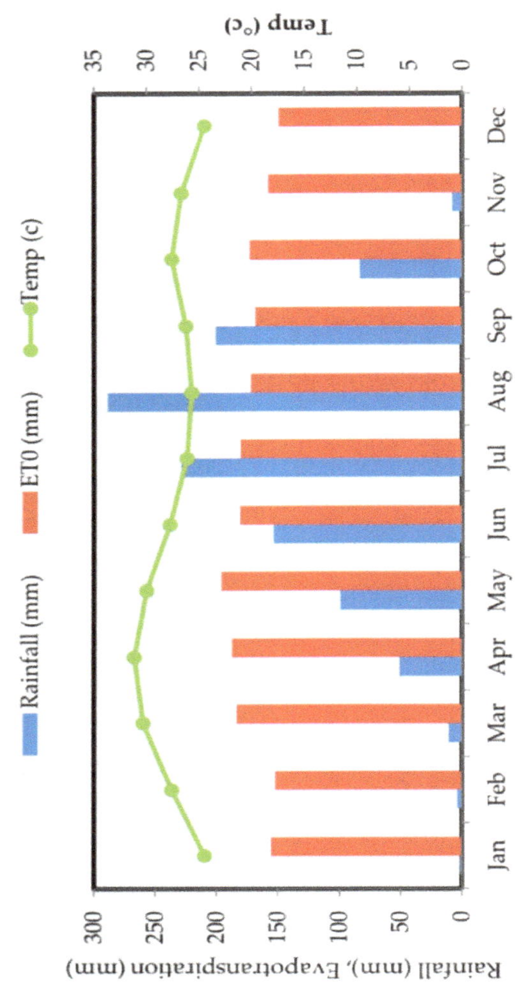

Figure 2: Monthly average rainfall (mm) (blue bars), potential evapotranspiration ET0 (mm) (red bars) and temperature (green line) in the Klela Basin from 1970–2013 (data Source from DNM [17]).

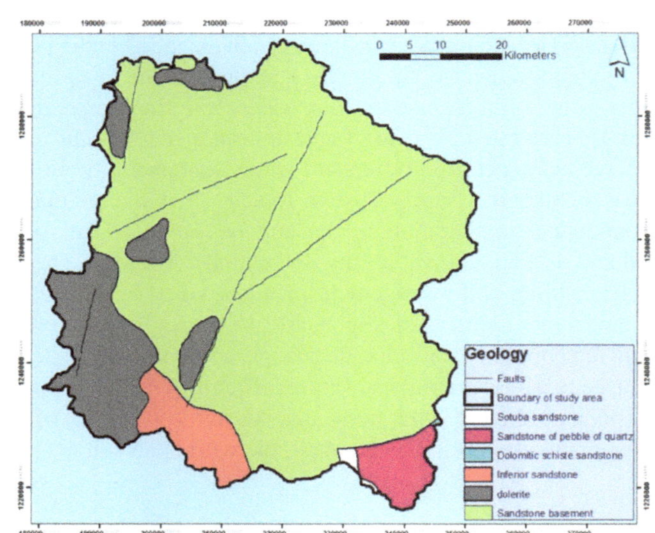

Figure 3: Geologic map of the Klela basin (adapted from DNH [18]).

METHODOLOGY

The data and methods used in this modeling study are summarized in Figure 4.

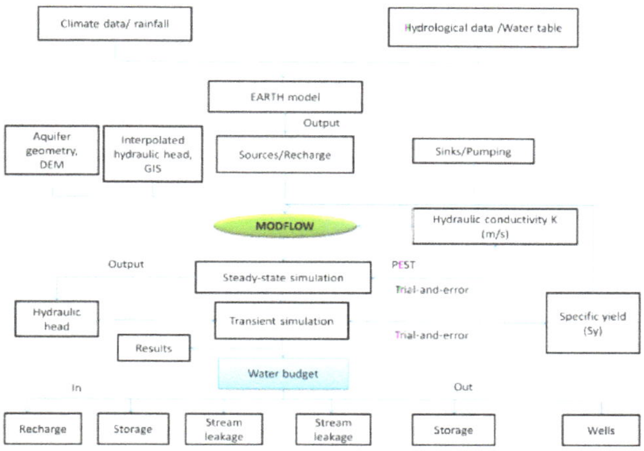

Figure 4: Methodology and data used in this study.

The hydrogeological data (e.g., drilling logs, pumping tests, water levels, *etc.*) were provided by the DRH (Direction Regionale de l'Hydraulique de Sikasso), and climate data are from the DNM (Direction Nationale de la Meteorologie). GIS was used to draw all the maps and interpolate data using spatial analysis. PMWIN version 5.3 was used to model the aquifer behavior in the study area.

EARTH Model

EARTH (Extended model for Aquifer Recharge and soil moisture Transport through the unsaturated Hardrock) is a lumped parametric model that can be used to estimate groundwater recharge. It was developed by van der Lee *et al.* in 1989 in a project in Botswana and tested under different climate conditions [19]. It is a one-dimensional model designed for semi-arid areas. EARTH combines direct and indirect methods. The direct method describes the recharge process from the atmosphere to the soil zone. The indirect method uses groundwater level fluctuations as indicators of the recharge process. The direct method determines recharge using physical processes above the groundwater table, whereas the indirect method calculates the groundwater level that results from recharge estimated by the direct method [19]. The EARTH model has five modules: MAXIL, SOMOS, LINRES, SUST and SAFLOW (Figure 5). MAXIL, SOMOS and LINRES comprise the direct part of the model and simulate recharge in the unsaturated zone. They can be calibrated with measured time series of soil moisture data. SATFLOW is the indirect part of the model and calculates the water level from the recharge simulated by the direct part. SUST calculates the surface runoff.

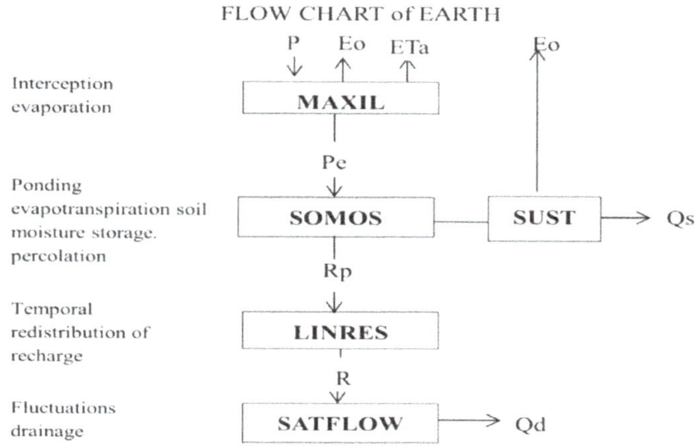

Figure 5: Flow chart of the EARTH model (modified from [19]). E_0 = potential evapotranspiration, ETa = actual evapotranspiration, P = precipitation, Pe = precipitation excess, Rp = percolation, R = recharge, Qs = surface runoff, and Qd = drainage runoff.

Details of the five modules are given by van der Lee and Gehrels [19] and are briefly described as follows:

1. MAXIL (Maximal Interception Loss) estimates the water intercepted by the Earth`s surface.
2. SOMOS (Soil Moisture Storage) describes water storage in the root zone and calculates changes in soil moisture storage by removing actual evapotranspiration, percolation, evaporation and surface runoff to determine the infiltrated precipitation.
3. SUST (Surface Storage) calculates ponding and runoff. It uses SUSTmax, which is the maximum ponding volume that can be stored at the surface. If the amount of ponding is greater than SUSTmax, then runoff occurs.
4. LINRES (Linear Reservoir Routing) redistributes percolation (output of SOMOS) over time in the unsaturated hard rock or soil beneath the root zone using the parametric transfer function.
5. SATFLOW (Saturated Flow) is the last part of the EARTH model. It is a simple one-dimensional parametric model that uses the outputs from the previous modules. SATFLOW calculates the groundwater level using the recharge estimated by the direct part of the model.

A general mathematical explanation of the EARTH model derived from the equation of de Vries 1974 cited in [20] is given by [21], as follows:

$$S\frac{\partial h}{\partial t} = R - \frac{h}{DR} \qquad (1)$$

where $\partial h/\partial t$ is the change in water level head during one month (m/month), h is groundwater level (m), DR is the drainage resistance (a lumped, site-specific parameter), R is the recharge (m/month) and S is the specific yield.

The numerical solution of Equation (1) (linear transfer function) yields Equation (2):

$$h_i = h_{i-1} - \frac{\Delta t \times h_{i-1}}{DR} + \frac{\Delta t \times R_i}{S} \tag{2}$$

where DR = drainage resistance ($L^2/\beta T$), L = length of flow path, β = 2 for radial flow or β = 4 for parallel flow and T = transmissivity.

Monthly observed groundwater levels from two piezometers and rainfall data were used as model inputs. Groundwater levels collected during two years (2012–2013) at wells were used for calibration.

Groundwater Model

In MODLFOW, the three-dimensional movement of groundwater with constant density through porous media is described by the following partial-differential equation [22]:

$$\frac{\partial}{\partial x}\left(K_{xx}\frac{\partial h}{\partial x}\right) + \frac{\partial}{\partial y}\left(K_{yy}\frac{\partial h}{\partial y}\right) + \frac{\partial}{\partial z}\left(K_{zz}\frac{\partial h}{\partial z}\right) + w = S_s\frac{\partial h}{\partial t} \tag{3}$$

where K_{xx}, K_{yy}, and K_{zz} are of the hydraulic conductivities in the *x*-, *y*- and *z*-directions (LT^{-1}); h is potentiometric head (L); w is a volumetric flux per unit volume and represents sources and/or sinks of water (T^{-1}); S_s is the specific storage of the porous material (L^{-1}); and t is time (T).

Spatial Discretization

In numerical groundwater modeling, an aquifer system is replaced by a discretized domain consisting of an array of nodes and associated finite difference blocks (cells). In PMWIN, the model domain is divided into a rectangular mesh comprising columns, rows and layers [15].

In this study, the model area was discretized into 239 columns and 256 rows, resulting in a total grid cell number of 61,184, with a grid size of 300 m × 300 m, taking into account the aquifer extension and hydrogeological data availability. The total number of active grid cells describing the active model domain was 40,992. The aquifer of the study area was modeled under an unconfined condition represented by a single numerical layer.

Aquifer Geometry

The aquifer geometry step establishes the top and bottom of an aquifer. Determination of aquifer geometry is critical in numerical modeling because it influences model calibration results [23]. Drilling log data and an SRTM-based Digital Elevation Model (DEM) from HydroSHED.org were used to estimate the top and bottom of the aquifer and aquifer thickness in the study area. The DEM was used as the aquifer top and the aquifer thickness was arbitrarily set to 300 m.

A scatter plot of surface elevation vs water table (Figure 6) shows that in most wells, the water table is up to 20 m below ground level. This assumes a possible strong interaction between groundwater and surface water bodies within the basin.

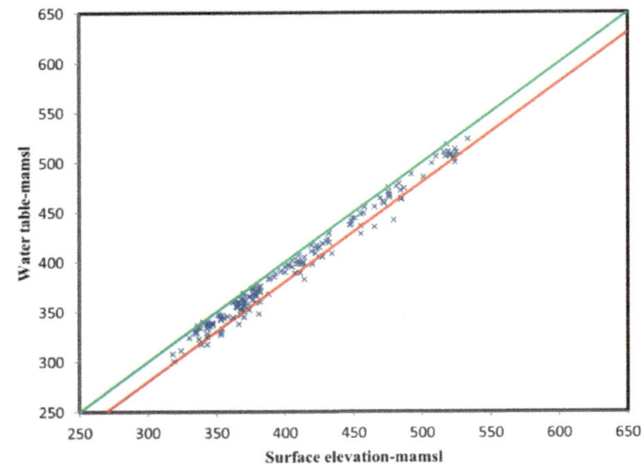

Figure 6: Scatter plot between surface elevation and water table. The blue stars are water table field data, the green line represents surface elevation and the red line corresponds to surface elevation minus 20 m.

Aquifer Properties

The hydraulic properties of the aquifer, including transmissivity, hydraulic conductivity and storativity, determine water movement through the aquifer as well as storage in the aquifer. In this study, the horizontal hydraulic conductivity value was calibrated until a good correlation between simulated and observed heads was obtained. For this reason, the model domain was divided into six hydraulic conductivity zones.

Storativity data were not available in the study area. Therefore, specific yield data were estimated using the Cooper-Jacob method [24] and used for transient calibration.

Boundary Conditions

Boundary conditions specify how an aquifer interacts with the environment outside the model domain [25]. Selection of boundary conditions is a key step in model design because they significantly control flow patterns [26,27]. Because there is not enough hydrogeological information in the study area, the topographical divide was considered a no-flow boundary condition. Thus, there was no hydrogeological interaction between the aquifer inside and outside the basin. Drainage of the basin was limited to groundwater–surface water interaction.

Initial Hydraulic Head

Historical water level data from more than 100 boreholes and data interpolated using the Kriging method were assigned to the model as initial hydraulic heads in the steady-state simulation. The head distribution computed using the calibrated steady-state simulation was used as the starting head distribution in the transient simulation.

Groundwater Sources/Sinks

As previously explained, the average recharge value (0.00044 m/day) estimated by the EARTH model was assigned to the entire model domain.

Groundwater abstraction was based on household (population and livestock) water consumption and irrigation water needs from pumping wells.

Streamflow Routing

In this study, the streamflow-routing package [28] was used to simulate the exchange of fluxes (sources and sinks) between groundwater and surface water. The Streamflow-Routing package is designed to account for the amount of flow in streams and to simulate the interactions between surface streams and groundwater [29]. This package requires parameters such as streambed hydraulic conductance, elevation of the streambed top, elevation of the streambed bottom, width of the stream channel, slope of the stream channel, manning`s roughness coefficient, *etc.* For more details, see [15].

In MODFLOW, the flow from streams to an aquifer (losing stream) or from an aquifer to streams (gaining stream) is calculated depending on the groundwater and streamwater tables. When groundwater table is above the streamwater table, the volumetric flux between streams and an aquifer is calculated using Equation (4) [30]:

$$Q = c_{STR} \times (h_S - h) = \frac{KLW}{M} \times (h_S - h) \text{ with } C_{STR} = \frac{KLW}{M} \quad (4)$$

where Q (L^3/T) is the volumetric flux, CSTR (L^2/T) is streambed hydraulic conductance, K (L/T) is hydraulic conductivity, L (L) is the length of the river within a cell, W (L) is the width of the stream, M (L) is the thickness of the streambed, h_S (L) is the head in the stream and h (L) is the groundwater head.

When the groundwater table h is below the elevation of the streambed bottom, the volumetric infiltration flux from the river to the aquifer is calculated using Equation (5).

$$Q = c_{STR} \times (h_S - z) \cdot z > h \quad (5)$$

The value of CSTR was determined to be 500 m^2/day. The hydraulic conductivity of the streambed is unknown; therefore, it was set to 0.5 m/day based on the dominant soil material in the study area. The length L within the cell is 300 m, the width is 10 m, and the streambed thickness is 3 m.

RESULTS AND DISCUSSION

EARTH Model

The calibrated results in both piezometers are shown in Figure 7. The correlation coefficients (R^2) exhibit a good correlation between simulated and observed groundwater levels. A trial and error calibration process was performed as follows. First, the specific yield was fixed and the other parameters, such as drainage resistance and percentage of rainfall (%R), were adjusted to obtain the best estimation between the observed and simulated groundwater levels (Table 1). To evaluate the model quality, three efficiency criteria were used, namely, the correlation coefficient, root mean square error (RMSE) and Nash-Sutcliffe coefficient (Table 2).

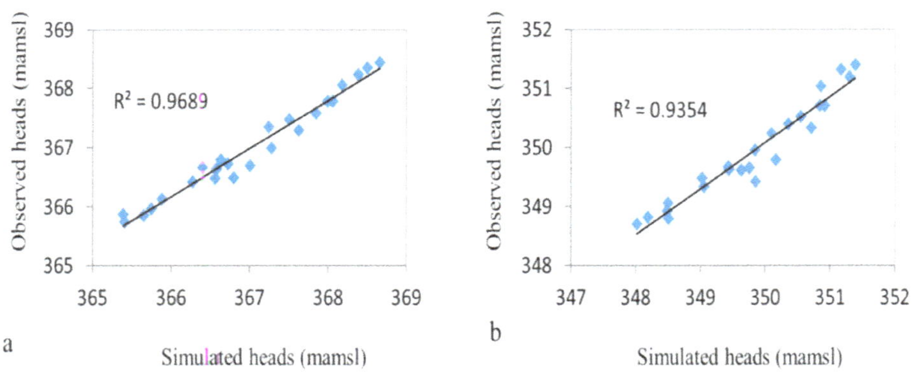

Figure 7: Correlation between observed and simulated groundwater levels in meters above mean sea level (mamsl) at (**a**) piezometer F7 and (**b**) piezometer F15.

Table 1: The values used to calibrate the model (%R: Percentage of rainfall).

Site	Resistance	%R	Mean Annual Rainfall (mm)	Recharge (mm/year)
Piezometer Sko_F7	1241	14	1210.5	169.5
Piezometer Sko_F15	1241	13.3	1210.5	161.0

Table 2: EARTH model performance evaluation (R^2: Correlation coefficient; RMSE: Root mean square error; N_r: Nash-Sutcliffe efficiency).

Sites	R^2	RMSE	N_r
Piezometer Sko_F7	0.968	0.046	0.952
Piezometer Sko_F15	0.935	0.067	0.999

MODFLOW

Calibration

During the calibration process, model input parameters are changed so that the model output matches measured values [31]. Model calibration can be performed under steady-state and transient conditions. In this study, a two-step procedure was applied. In the first step, a steady-state simulation was used to analyze the system and generate consistent initial conditions for the dynamic simulation, which are required to perform the climate change impact study.

Steady-State Simulation

Mean water table data collected by the DRH Sikasso in 1985 (Direction Regionale de l`Hydraulique de Sikasso) and data measured in the field in 12 boreholes in 2014 were used to calibrate the steady-state simulation model. This calibration was first performed via a trial-and-error procedure. Then, the output was optimized using the PEST code integrated in PMWIN. The model performance yielded $R^2 = 0.98$, RMSE = 5.67, and $N_r = 0.86$, exhibiting a good calibration. The scatter plot comparison between simulated and observed groundwater levels is shown in Figure 8. The output heads generated by the model (Figure 9) were used as initial conditions in the transient simulation. The calibration result show that hydraulic conductivity varies from 1.1 m/day to 13.9 m/day.

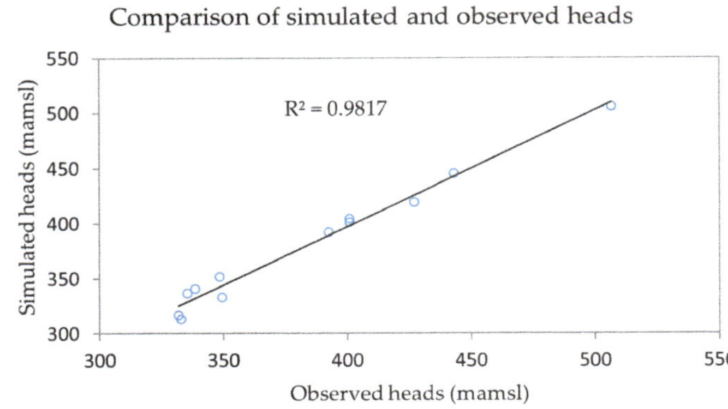

Figure 8: Simulated and observed hydraulic heads in meters above mean sea level based on the steady-state calibration in the Klela basin.

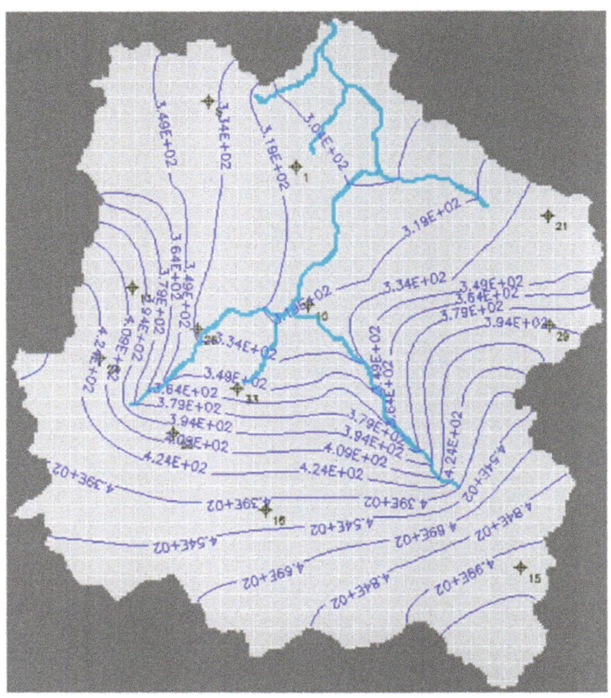

Figure 9: Calculated hydraulic heads (mamsl) in the steady-state simulation, with borehole locations used to calibrate the model.

Transient Simulation

Only the trial and error method was used in the transient simulation due to technical problems with PEST. The groundwater level of the steady-state simulation was used as the initial hydraulic head in the transient simulation (Figure 9). For the transient calculations, the streamflow-routing package was used to simulate the interactions between streams and groundwater. In the same manner, sources (recharge) and sinks (wells) were introduced into the model. Specific yield was set to 0.06 for the whole model domain.

The model was run from November 2010 to November 2014 (48 months). Therefore, the simulation time was divided into 48 stress periods (with 32 dry and 16 wet periods), and each stress period represents one month. The length of a stress period was divided into days, resulting in total time steps over 48 months and a total simulation time of 1,460 days.

Mean monthly groundwater table data from three piezometers (F7, F15 and F18) were used to calibrate the model. In total, five piezometers were monitored weekly in the study area, but only three were chosen for modeling purposes. This choice was based on the data quality and data length. The quality of the calibration results is shown in Figure 10. Four efficiency criteria (Mean Absolute Error (MAE), Root Mean Square Error (RMSE), the Nash-Sutcliffe coefficient (E) and the Correlation Coefficient (R^2)) (Table 3) were used to evaluate the model accuracy. The model efficiency results reflect an acceptable model calibration. However, F7 is not as well simulated as the others.

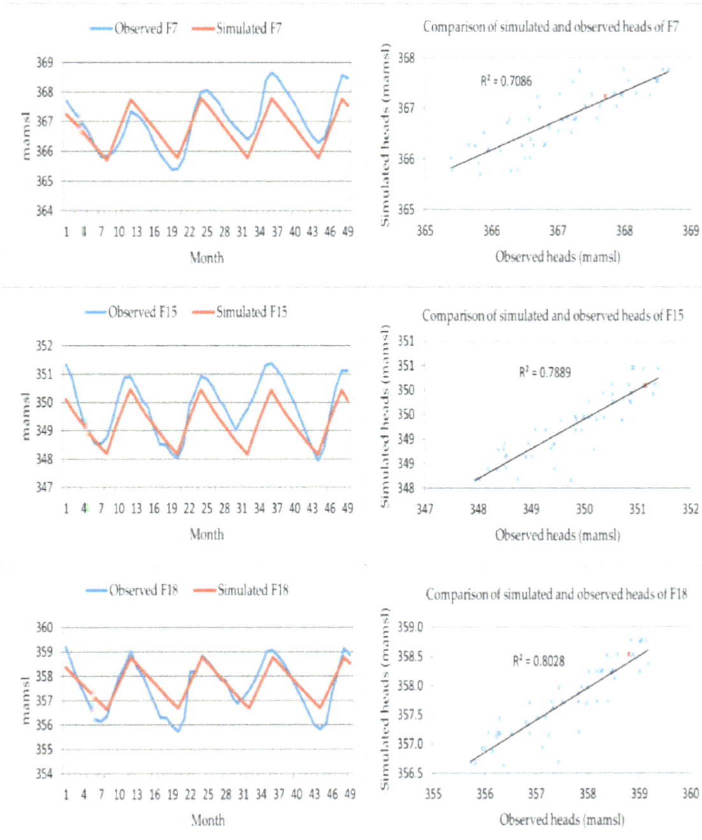

Figure 10: Simulated and observed heads during the transient calibration in the Klela basin from November 2010-November 2014.

Table 3: Summary of calibration errors in the transient simulation (MAE: Mean Absolute Error; RMSE: Root Mean Square Error; E: Nash-Sutcliffe coefficient and R^2: Correlation Coefficient).

	F7	F15	F18
MAE	0.0248	0.0091	0.0101
RMSE	0.1823	0.0670	0.0741
E	−0.4959	0.2141	−0.4439
R^2	0.7137	0.7889	0.8028

Water Budget

One of the best ways to assess model simulation quality is by analyzing the water budget. Although a good water balance cannot guarantee a good simulation, a bad water balance indicates problems in the model [32]. In this study, the percentage discrepancy in all the stress periods of the model is nearly zero. Thus, the model equations have been correctly solved [15]. The average annual water budget is summarized in Table 4.

Table 4: Mean annual groundwater budget in m^3/year in the Klela basin from November 2010 to November 2014.

Flow	Inflow (m3/year)	Outflow (m3/year)
Storage	485,869,302	446,718,545
Recharge	635,293,979	0
Wells	0	57,287,607
Stream Leakage	947,401	618,122,906
Total	1,122,110,683	1,122,129,060

Because the boundary condition is a no flow boundary, the only water source entering the aquifer is recharge. The annual recharge is slightly greater than groundwater seepage (Table 4), *i.e.*, the amount of water that is entering the aquifer is almost the same as the amount exiting the aquifer and entering surrounding streams, which are the discharge points of the study area. According to the boundary condition, groundwater can leave the catchment only via the stream interaction. Inflow from the stream is calculated from the interaction between surface water and groundwater throughout the river system. The mean annual water budget shows that groundwater storage is slightly decreasing by 39,169,133 m^3/year (approximately 10.6 mm/year) (calculated from Table 4).

The seasonal recharge from July to October was constant and represented the important water inflow entering the aquifer. The stream leakage that is leaving the aquifer and entering streams decreases during the dry season and increases during the rainy season (Figure 11). Figure 12 shows that during the dry season, groundwater storage decreases. It then increases during the rainy season. The maximum value of storage occurs in November.

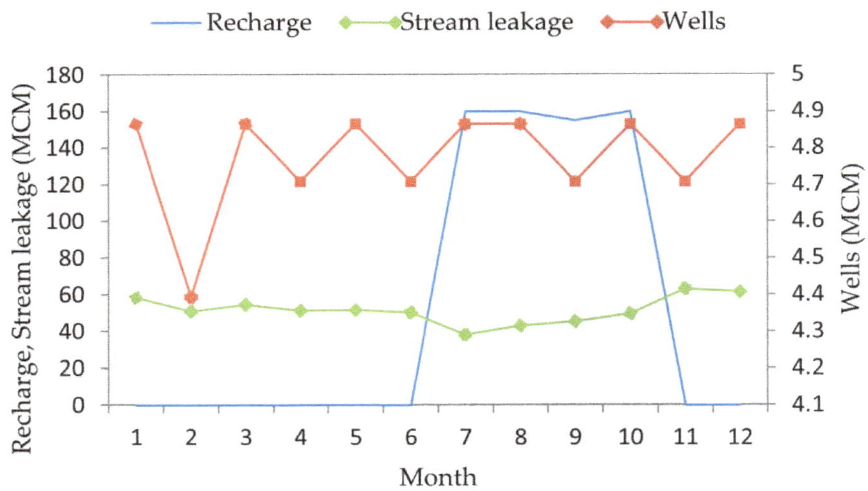

Figure 11: Monthly average water budget in the Klela Basin in Mm^3/month.

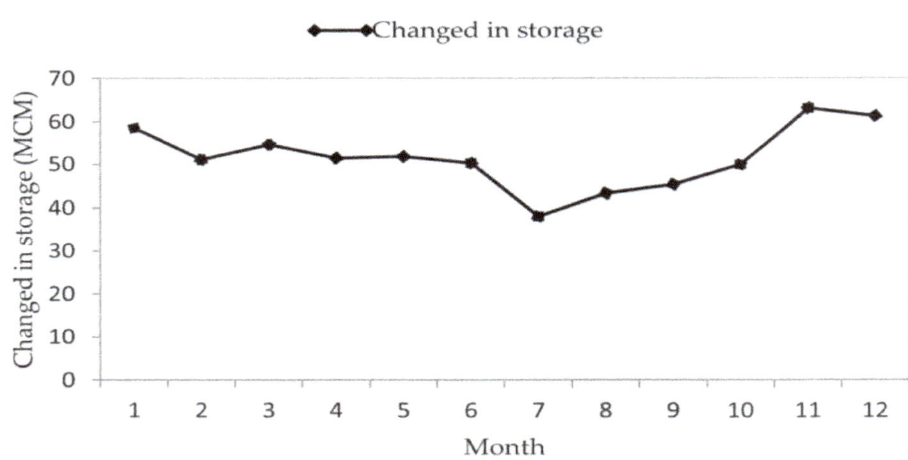

Figure 12: Monthly change in groundwater storage (Mm^3/month) in the Klela Basin.

Scenario Quantification

One of the most important objectives of our groundwater modeling study is to use the scenario data to predict the future behavior of the aquifer system. Thus, the model was used to quantify groundwater levels considering scenario data after model calibration and verification using three different piezometer data sets.

Monthly precipitation data from the RCP4.5 (Representative Concentration Scenario 4.5 W/m^2) scenario were taken from the GCM ECHAM, downscaled to a 0.44° resolution by the Swedish Meteorological Service (Sveriges Meteorologiska och Hydrologiska Institute, SMHI) and provided by the CORDEX (Coordinated Regional Climate Downscaling Experiment) initiative. They were used to calculate the groundwater recharge (Figure 13), which was used as model input to quantify future groundwater levels in the Klela basin from June 2010 to November 2050. RCP4.5 is an intermediate pathway and a stabilization scenario based on the assumption that all countries around the world implement emissions mitigation policies [33,34].

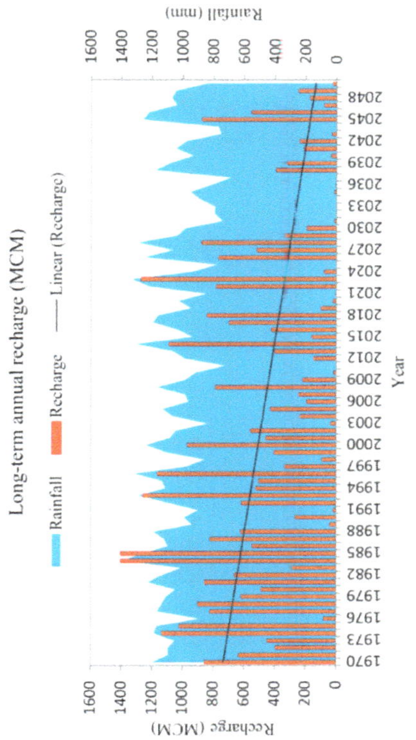

Figure 13: Long-term groundwater recharge in million cubic meters (MCM) calculated using historical and future climate data from the RCP4.5 scenario based on the Thornthwaite Monthly Water Balance and rainfall (mm) from 1970 to 2050.

The Thornthwaite Monthly Water Balance model [35], which is based on monthly temperature and rainfall data, was used for this recharge estimation. The model was calibrated to fit the groundwater recharge previously simulated using EARTH. To calibrate this model, direct runoff was set to zero because local scale surface runoff infiltrates on the way to the river system. There is no direct runoff contribution to the discharge of the basin. The soil moisture storage capacity was calibrated to be 375 mm by comparing the groundwater recharge of the two models (EARTH and Thornthwaite). All other parameters were unchanged. More details on this model can be found in [35]. After calibration, the model was applied to simulate groundwater recharge in the past (1970–2010) and the future (2010–2050). The overall result shows that recharge decreases from 1970 to 2050 and severe droughts occur from 2030 to 2037 (Figure 13).

The results of the simulated MODFLOW model reveal a decrease in groundwater level (GWL) over time (Figure 14). From June 2010 to November 2050, GWL decreases from 365 mamsl to 348 mamsl, or to approximately 17 m, in piezometer F7. During the same period, F18 decreases from 356 mamsl to 338 mamsl, or approximately 18 m. F15 decreases from 348 mamsl to 336 mamsl, or approximately 11 m. The overall mean of all the piezometer drops was approximately 15 m. Furthermore, an important decline in groundwater storage in the 2030s was modeled in all the piezometers considered in this study. Groundwater level reductions may intensify if water extraction increases over time.

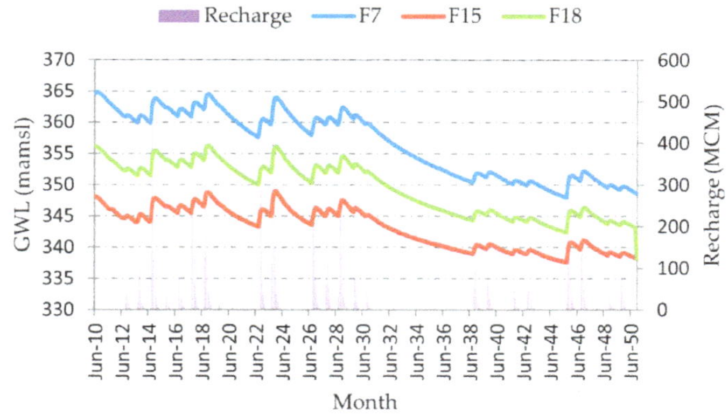

Figure 14; Long-term monthly groundwater level in meters above mean sea level and groundwater recharge (MCM) in the Klela basin calculated using RCP4.5 climate data from June 2010 to November 2050.

Comparing the results of the EARTH model in this study to others carried out throughout the world in arid and semi-arid areas, the results are reasonably similar. For example, Usher *et al.* [36] applied the EARTH model in a fractured sandstone aquifer in South Africa (Kalkveld area) and obtained a recharge rate of 7.4% of the mean annual precipitation of 559 mm. In a similar study, Baalousha [37] applied the same model in Palestine in calcareous sandstone and sand aquifers and computed the recharge to be 36.95% of the average annual rainfall of 321 mm. Values between 7.2% and 26.5% of mean annual rainfall were also noted by Wang *et.al.* [38] in China in unsaturated sand, with rainfall ranging from 511 to 557 mm. No previous studies have been conducted in the Klela basin using the EARTH model; however, Henry [39] applied the HELP (Hydrologic Evaluation of Landfill Performance) model in southern Mali, including the Klela basin, and obtained an annual recharge of 12.6% of the mean rainfall of 1049 mm.

Regarding the MODFLOW results, despite a lack of hydrogeological data, they appear to be acceptable based on the model performance evaluation results. One of the most important uncertainties in this modeling is the establishment of boundary conditions. In this study, the boundary conditions were set as no flow boundaries. In reality, this is very difficult to prove. For example, Belay [40] applied the PMWIN model to establish a water balance for Lake Beseka in Ethiopia and concluded that the choice of boundary conditions has a larger effect on the dynamics than does precipitation. Therefore, neglecting lateral outflow in groundwater modeling may affect the water balance.

CONCLUSIONS AND RECOMMENDATIONS

The behavior of groundwater in the Klela basin has been evaluated in the context of climate change. This evaluation was performed using the RCP4.5 scenario data and applying the groundwater model MODFLOW and the Thornthwaite model for groundwater recharge. In this scenario, groundwater storage will decrease by approximately 10.6 mm/year. Furthermore, hydraulic heads generated by the model displayed a reduced groundwater level, increasing groundwater availability stresses in areas where agriculture is the main activity. By considering the scenario analysis in which the groundwater recharge decreases significantly compared to past recharge, the reduction of groundwater storage is important, especially in the 2030s, when severe drought events are simulated due to climate change effects. These events threaten access to groundwater for irrigation, agriculture and domestic use.

Groundwater recharge, the most important parameter in groundwater modeling, calculated using the EARTH model could have been over- or underestimated in some areas of the basin because the piezometers were not spatially distributed in a way that adequately allowed determination of spatial recharge patterns. Moreover, the true boundary of the basin was not known, potentially increasing the uncertainties in the results of this study. Despite these shortcomings, the results are reliable and can serve as a basis for watershed management.

In conclusion, a focus should be placed on determination of the hydrogeological situation in future studies. More piezometers should be implemented in the Klela basin to permit permanent monitoring of groundwater resources in the context of global change.

This study also shows that MODFLOW, which is widely used to model groundwater resources, can be applied in the Klela basin to quantify the influence of climate change on groundwater resources and can support water resources management.

ACKNOWLEDGMENTS

This work was funded by the German Federal Ministry of Education and Research (BMBF). It is part of the WASCAL (West African Science Service Center on Climate Change and Adapted Land Use) project, which is available online atwww. wascal.org. The authors thank all the National Directions in Mali for providing hydrological and climate data in the DRH of Sikasso, DNM. The authors also want to thank Wolfgang Kinzelbach and Lamine Baba Sy for helping to understand the PMWIN software.

REFERENCES

1. McDonald, M.G.; Harbaugh, A.W. A modular three-dimensional finite-difference groundwater flow model. *Techniques of Water-Resources Investigations*; Book 6, Chapter A1; U.S. Geological Survey: Reston, VA, USA, 1988; p. 586.
2. Council, G.W. *A Lake Package for MODFLOW (LAK2), Documentation and User's Manual Version 2.2*; HSI GEOTRANS, A Tetra Tech Company: Pasadena, CA, USA, 1999.
3. AL-Fatlawi, A.N. The application of the mathematical model (MODFLOW) to simulate the behavior of groundwater flow in Umm Er Radhuma unconfined aquifer. *Euphrates J. Agric. Sci.* **2011**, *3*, 1–16.
4. Saatsaz, M.; Chitsazan, M.; Eslamian, S.; Sulaiman, W.N.A. The application of groundwater modelling to simulate the behaviour of groundwater resources in the Ramhormooz Aquifer, Iran. *Int. J. Water* **2011**, *6*, 29–42.
5. Valerio, A.M. Modeling Groundwater-Surface Water Interactions in an Operational Setting by Linking Riverware with MODFLOW. Ph.D. Thesis, University of Colorado, Fort Collins, CO, USA, 2008.
6. Banning, R.O.B. Analysis of the Groundwater/Surface Water Interactions in the Arikaree River Basin of Eastern Colorado. Ph.D. Thesis, Colorado State University, Fort Collins, CO, USA, 2010.
7. Sehatzadeh, M. Groundwater Modelling in the Chikwawa district, lower Shire area of southern Malawi. Master's Thesis, University of Oslo, Oslo, Norway, 2011.
8. Taylor, C.M.; Lambin, E.F.; Stephenne, N.; Harding, R.J.; Essery, R.L.H. The Influence of Land Use Change on Climate in the Sahel. *J. Clim.* **2002**, *15*, 3615–3629.
9. Funk, C.; Rowland, J.; Adoum, A.; Eilerts, G.; White, L. A Climate Trend Analysis of Mali. Available online: http://pubs.usgs.gov/fs/2012/3105/fs2012-3105.pdf (accessed on 27 November 2015).
10. Bricquet, J.P.; Bamba, F.; Mahe, G.; Toure, M.; Olivry, J.C. Évolution récente des ressources en eau de l'Afrique atlantique. *Rev Sci Eau.* **1997**. (In French)
11. Bokar, H.; Mariko, A.; Bamba, F.; Diallo, D.; Kamagaté, B.; Dao, A. Impact Of Climate Variability On Groundwater Resources in Kolondieba Catchment Basin, Sudanese Climate Zone in Mali. *Int. J. Eng. Res. Appl.* **2012**, *2*, 1201–1210.
12. Mahe, G. Surface/groundwater interactions in the Bani and Nakambe rivers, tributaries of the Niger and Volta basins, West Africa. *Hydrol. Sci. J.* **2009**, *54*, 704–712.
13. Mahé, G.; Olivry, J.C.; Dessouassi, R.; Orange, D.; Bamba, F.; Servat, E. Relations eaux de surface-eaux souterraines d'une rivière tropicale au Mali. *Sci Terre Planetes.* **2000**, *330*, 689–692. (In French)
14. WWAP. Rapport national sur la mise en valeur des ressources en eau: Mali. 2006. Available online: http://unesdoc.unesco.org/images/0014/001472/147267f.pdf (accessed on 27 November 2015). (In French).
15. Chiang, W.-H.; Kinzelbach, W. Processing Modflow A Simulation System for Modeling Groundwater Flow and Pollution. 1998. Available online: https://www.researchgate.net/publication/238774706_Processing_MODFLOWa_Simulation_System_for_Modelling_Groundwater_Flow_and_Pollution (accessed on 27 November 2015).
16. USAID. Plan de sécurité alimentaire commune rurale de Klela. 2006. Available online: http://fsg.afre.msu.edu/mali_fd_strtgy/plans/sikasso/sikasso/psa_klela.pdf (accessed on 27 November 2015). (In French).
17. Direction Nationale de la Meteorologie du Mali. Departement de gestion des donnees. 2014. Available online: http://www.linguee.com/french-english/translation/direction+nationale+de+la+m%C3%A9t%C3%A9orologie.html (accessed on 27 November 2015). (In French).
18. Direction Nationale de l`Hydraulique du Mali. Projet Mali Sud II. Bibliotheque. 1987. Available online: http://documentation.2ie-edu.org/cdi2ie/opac_css/index.php?lvl=notice_display&id=15175 (accessed on 27 November 2015). (In French).
19. Van der Lee, J.; Gehrels, J. *Modeling Aquifer Recharge, Introduction to Lumped Parameter Model EARTH*; Hydrological Report; Free University of Amsterdam: Amsterdam, The Netherlands, 1990.
20. Nyende, J. Application of isotopes and recharge analysis

in investigating surface water and groundwater in fractured aquifer under influence of climate variability. *J. Earth Sci. Clim. Change* **2013**, *4*.

21. Tonder, G.V.; Yongxin, X. A guide for the Estimation of Groundwater Recharge in South Africa. 2000. Available online: http://www.uovs.ac.za/igs (accessed on 1 September 2014).
22. Harbaugh, A.W. MODFLOW-2005, the U.S. Geological Survey Modular Ground-Water Model—The Ground-Water Flow Process. U.S. Geological Survey Techniques and Methods: Reston, VA, USA, 2005.
23. Middlemis, H. *Groundwater flow Modeling Guideline*; Murray-Darling Basin Commission: Canberra, Australia, 2001.
24. Cooper, H.H.; Jacob, C.E. A generalized graphical method of evaluating formation constants and summarizing well-field history *Am. Geophys. Union Trans.* **1946**, *27*, 526–534.
25. Owais, S.; Atal, S.; Sreedevi, P. Governing Equations of Groundwater Flow and Aquifer Modelling Using Finite Difference Method. In *Groundwater Dynamics in Hard Rock Aquifers*; Ahmed, S., Jayakumar, R., Salih, A., Eds.; Springer: Dordrecht, The Netherlands, 2008; pp. 201–218.
26. Anderson, M.P.; Woessner, W.W. *Applied Groundwater Modeling: Simulation of Flow and Advective Transport*; Academic Press Incorporated: Cambridge, MS, USA, 1992; p. 406.
27. Ahern, J.A. Ground-Water Capture-Zone Delineation: Method Comparison in Synthetic Case Studies and a Field Example on Fort Wainwright, Alaska. Ph.D. Thesis, University of Alaska Fairbanks, Fairbanks, AK, USA, 2005.
28. Prudic, D.E.; Konikow, L.F.; Banta, E.R. *A New Streamflow-Routing (SFR1) Package to Simulate Stream-Aquifer Interaction with MODFLOW-2000*; Open-File Report 2004–1042; U S. Geological Survey: Reston, VA, USA, 2004; p. 95.
29. Prudic, D.E. *Documentation of a Computer Program to Simulate Stream-Aquifer Relations Using a Modular, Finite Difference, Ground-Water Flow Model*; Open File Report; U.S. Geological Survey: Reston, VA, USA, 1989; pp. 88–729.
30. Brunner, P.; Simmons, C.T.; Cook, P.G.; Therrien, R. Modeling Surface Water-Groundwater Interaction with MODFLOW: Some Considerations. *Ground Water* **2010**, *48*, 174–180.
31. Hill, M.C.; Tiedeman, C.R. *Effective Groundwater Model Calibration: with Analysis of Data, Sensitivities, Predictions, and Uncertainty*; John Wiley & Sons, Inc.: Hoboken, NJ, USA, 2007; p. 455.
32. Jackson, J.M. Hydrogeology and Groundwater Flow Model, Central Catchment of Bribie Island, Southeast Queensland. Master's Thesis, Queensland University of Technology, Queensland, Australia, 2007.
33. Moss, R., Babiker, M., Brinkman, S., Calvo, E., Carter, T., Edmonds, J., Elgizouli, I., Emori, S., Erda, L., Hibbard, K., *et al.*, Eds.; *Towards New Scenarios for Analysis of Emissions, Climate Change, Impacts, and Response Strategies: Intergovernmental Panel on Climate Change*; UNISDR: Geneva, Switzerland, 2008; p. 132.
34. Thomson, A.M.; Calvin, K.V.; Smith, S.J.; Kyle, G.P.; Volke, A.; Patel, P.; Delgado-Arias, S.; Bond-Lamberty, B.; Wise, M.A.; Clarke, L.E.; *et al.* RCP4.5: A pathway for stabilization of radiative forcing by 2100. *Clim. Chang.* **2011**, *109*, 77–94.
35. McCabe, G.J.; Markstrom, S.L. *A Monthly Water-Balance Model Driven by a Graphical User Interface*; U.S. Geological Survey: Reston, VA, USA, 2007.
36. Usher, B.; Pretorius, J.; van Tonder, G. Management of a Karoo fractured-rock aquifer system—Kalkveld Water User Association (WUA). Available online: http://www.wrc.org.za (accessed on 4 February 2015).
37. Baalousha, H.M. Risk Assessment and Uncertainty Analysis in Groundwater Modelling. Ph.D. Thesis, Westfälischen Technischen Hochschule Aachen University, Aachen, Germany, 2003.
38. Wang, B.; Jin, M.; Liang, X. Using EARTH model to estimate groundwater recharge at five representative zones in the Hebei Plain, China. *J. Earth Sci.* **2015**, *26*, 425–434.
39. Henry, C.M. An Integrated Approach to Estimating Groundwater Recharge and Storage Variability in Southern Mali, Africa. Ph.D. Thesis, Simon Fraser University, Vancouver, BC, Canada, 2011.
40. Belay, E.A. Growing lake with Growing Problems: Integrated Hydrogeological Investigation on Lake Beseka, Ethiopia. Ph.D. Thesis, University of Bonn, Bonn, Germany, 2009.

CHAPTER 16 USING A MODIFIED LANE'S RELATION IN LOCAL BED SCOURING STUDIES IN THE LABORATORY CHANNEL

Marta Kiraga and Zbigniew Popek

Keywords: Local scour, Lane, dynamic balance, bedload, hydraulics, local scour depth

INTRODUCTION

Numerous approaches to local scour forming studies have been developed [2,3,4,5]. The present research discloses different scientific approaches to the local scouring phenomenon—the Lane's relation [1] application in laboratory studies is performed and used commonly in alluvial rivers morphology forming context in a dynamic balance (equilibrium) conditions. Natural river stays in equilibrium conditions if in a long period of time basic flow parameters, such as even longitudinal river decline, average channel width and depth, and bedload granulation's characteristics remain constant, which is originally described by Lane as the following:

$$Q_s \times d \sim Q_w \times S \tag{1}$$

where: Q_s—bedload's transport discharge, d—particle diameter, Q_w—water discharge, *S*—energy grade line slope.

It was attempted to describe the river morphology forming process more precisely using classic Lane's relation, according to Lane's sentence: "The science of fluvial morphology has developed from two roots which have been largely independent of each other (…)" [1]. Geology and engineering—these two roots have remained largely separate down to the present time. Scientists [6] focused on one of these roots: geology, but the second—engineering—is up to now insufficiently studied and needs a wider description. Others considered the geomorphology parameters of the natural river channel connection with sediment transport discharge, as for example [6,7,8]. Concomitantly, Lane's relation in scope of local scouring in laboratory conditions has not been examined yet. Schumm's formula [8] attempts to forecast river adjustment for the maintenance of a stable equilibrium state following disturbance, and it establishes a connection as it follows:

$$Q_s = \frac{W \times L \times S}{d \times P} \tag{2}$$

where: W—channel width; L—the length of river; S—the channel slope; P—the channel sinuosity index. Schumm's model was further modified by other scientists [7] to derive Lane's modified relation considering the width-to-depth ratio $\left(\frac{W}{H}\right)$:

$$Q_w \times S \sim Q_s \times D_s \times \left(\frac{W}{H}\right)^{-1} \tag{3}$$

where: D_s—the sediment size; W—the channel width, H—the depth of the water.

The present paper consists of a further modification of derived relation, considering width-to-depth ratio in modified relation, developed into an equation in local scour forming process phenomenon in laboratory conditions. To describe equilibrium conditions in scope of scouring phenomenon,

Citation: Marta Kiraga and Zbigniew Popek, Using a Modified Lane's Relation in Local Bed Scouring Studies in the Laboratory Channel, Water 2016, 8(1), 16; doi:10.3390/w8010016.

Chabert and Engeldinger classified local scours into two categories—clear-water scour and live-bed scour. Clear-water scour takes place in the absence of sediment transport by the approaching flow (no sediment feeding adoption), and live-bed scour occurs when the scour hole is continuously fed with sediment by the approaching flow [9]. The time development of clear-water and live-bed scour is shown in Figure 1.

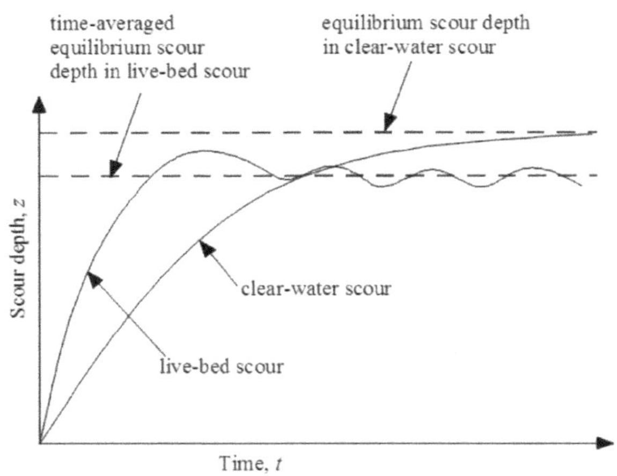

Figure 1: Time development of clear-water and live-bed scour [10].

The original form of Lane's relation is not an equation, but a qualitative expression which cannot be directly used to estimate the influence of a change in one parameter on the magnitude of others. Consequently, it is not possible to estimate the changes in the morphological characteristics of a river which result from its seeking to retrieve the dynamic balance lost due to the change in water flow conditions and/or bedload's transport. Lane's relation, despite its qualitative and simplified character, serves well to describe the nature of the process of forming alluvial stream channels. Results of many field studies, as well as analyses of causes and effects of the river's loss of dynamic balance, support the assumptions of Lane's relation, which can predict tendencies in changes of hydraulic and morphological characteristics of stream channels [8,11,12,13,14].

MATERIALS AND METHODS

The Modified Lane's Relation

The original form of Lane's relation (1) is not applicable to the functional description of the bedload's dynamic balance conditions due to the lack of consistency of parameters on the left and right side. In order to eliminate this inconvenience, the relation modification is suggested, which relies on the replacement of bedload's *d* design grain size with dimensionless grain parameter, described by Bonnefille with the following relation [15]:

$$D_* = d_{50} \times \left[\frac{(s-1)\,g}{v^2}\right]^{1/3} \tag{4}$$

where: d_{50}—median grain diameter for 50% of sieve curve (m), s—specific density of solid particles (-); $s = \rho_s/\rho_w$, where $_s$—bed material specific density (kg·m^{-3}), ρ_w—water specific density (kg·m^{-3}), g—acceleration of gravity (m·s^{-2}), v—kinematic viscosity parameter (m^2·s^{-1}).

Once Lane's relation, in its original form, describes the natural river's character, not local scouring phenomenon in laboratory conditions, there exists the possibility of crediting the local scour distance energy grade line into hydraulic gradient in researched reach.

The modified Lane's relation with the [7] adjustment (Equations (1), (3) and (4)) eventually takes the following form with D_* parameter included:

$$Q_s \times D_* \times \left(\frac{W}{H}\right)^{-1} = Q_w \times S \tag{5}$$

where: Q_s—mean bedload's transport discharge (m^3·s^{-1}), D_*—dimensionless grain parameter of laboratory material (-), W—the channel width (m), H—the depth of the water (m), Q_w—water flow discharge (m^3·s^{-1}), S—mean energy grade line slope in local scour distance (-).

Aim and Scope of Research

The aim of this research is to examine the possibility to use the modified Lane's relation (Equation (5)) in the local scouring phenomenon description in laboratory channels with a rectangular cross-section, in which the solid bottom transforms into sandy bottom in the intake part. Due to the increase of flow resistance on the whole length of the bed, resulting from varied roughness of the solid and sandy bottom, the hydraulic gradient increases causing an increase of shear stress on the bottom. After exceeding the critical shear stress, the motion of sediment grains begins and is followed by gradual scour of the bed during the time of experiment until the shape and the parameters of local scours stabilize. Then, the local scour obtains equilibrium depth in clear-water scour conditions, z_{max}, because no sediment feeding system application (Figure 2). The experimental conditions in this study may be compared to a case of the transport continuity being disrupted by the accumulation of bedload material in the retention reservoir located upstream [6,7].

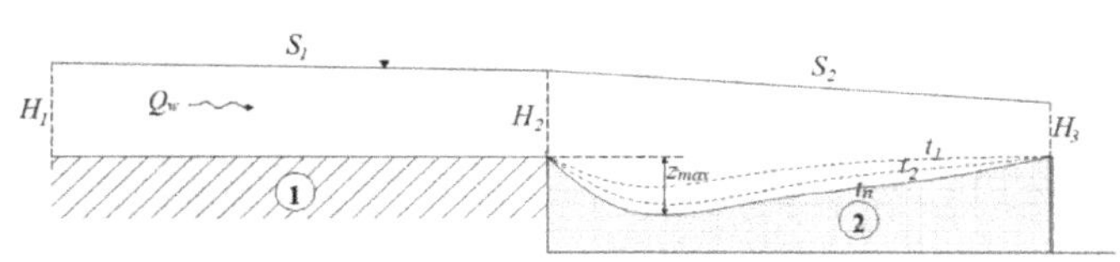

Figure 2. Scheme of local scour forming in washable area of sandy bed. Where: 1—solid bottom; 2—sandy washout bed; H_1, H_2, H_3—water depth; Q_w—water flow discharge; S_1, S_2—energy grade line slope; z_{max}—maximal depth of local scour while achieving stabilization in time t_n (clear-water equilibrium scour depth); t_1, t_2, t_n—shapes of bed while duration of experiment in time.

Description of the Test Stand

Studies of local scours phenomenon were conducted in a laboratory channel of 0.58 m in width (W) and full length (L) of 10.00 m (Figure 3).

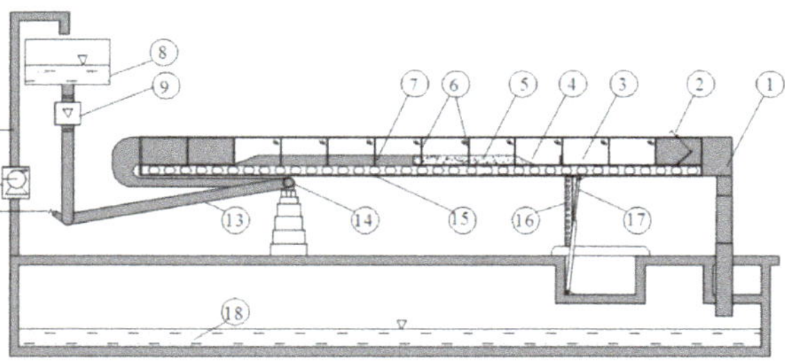

Figure 3. Experimental channel scheme—side view. Where: 1—chute chamber; 2—the regulatory gate; 3—glass side wall; 4—collection chamber; 5—sandy bed; 6—supports for side walls; 7—solid bottom; 8—upper reservoir; 9—electromagnetic flow meter; 10—a pipeline conducting water; 11—the pump; 12—the regulatory valve; 13—the feeding pipeline; 14—the support with joint; 15—the support plate of the channel; 16—the support with adjustable frame elevation; 17—the hydraulic cylinder; 18—the lower reservoir.

The solid bottom, which is 5.82 m in length, precedes the washout part, which is 2.18 in length (Figure 4), that is filled with sand with a mass median diameter of grain d_{50} = 0.62 mm and d_{90} = 1.50 mm [16]. A pin water gauge was used in the intake part in order to measure the water surface elevation, regulated with a gate. In order to measure the ordinate of water surface level, a moving pin water gauge was used and was placed on the trolley pushed on guides along the channel. The level of the sandy bottom within the washout bed was measured with a moving disc probe. The water flow discharge was examined with the use of electromagnetic flow meter. No sediment feeding system was adopted.

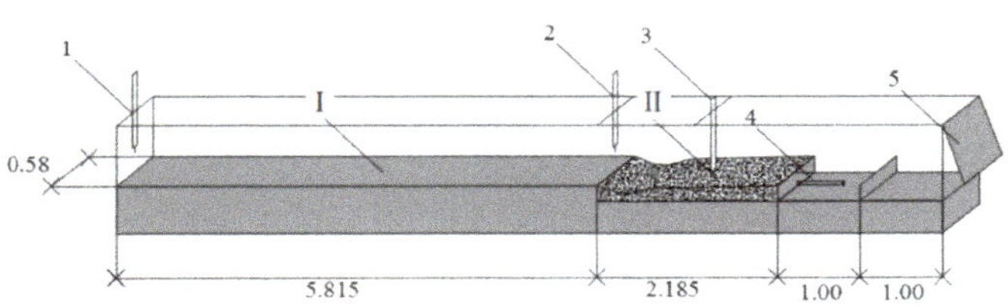

Figure 4: Experimental channel scheme (all dimensions in meters). Where: I—solid bottom; II—washout bed (sandy); 1,2—pin water gauge; 3—disc probe; 4—collection chamber; 5—the regulatory gate.

Methodology and the Scope of the Study

In the studies on local scour phenomenon, which were conducted in changeable water flow discharge and water surface level conditions, the parameters were described for facilitating the establishment of the correlation between the right and the left side of modified Lane's relation, *i.e.*, the product of water flow discharge, water level decline (hydraulic gradient), and the product of bedload discharge and Bonnefille's dimensionless grain parameter.

Water flow discharge was measured by electromagnetic flow meter. The average decline of water level on the washout bed was examined on the basis of the readings of a movable pin water gauge made in the bed's axis in the cross-sections of bed probing. The shape of the sandy washout bed was examined with the results of bed probing measured with a disc probe in five representative points in each of the 11 cross-sections (Figure 5).

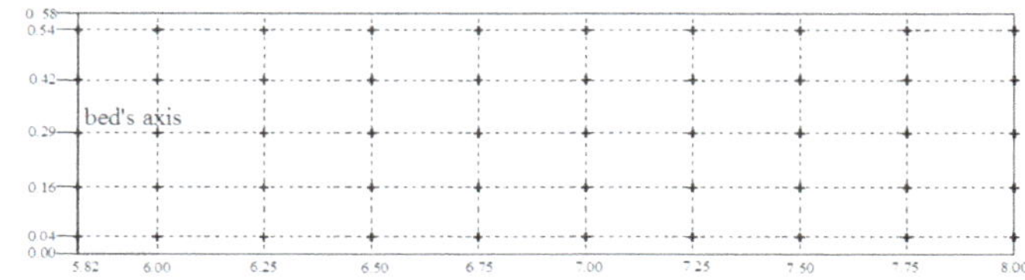

Figure 5: Probing points in the area of sandy bed—plan view; all dimensions in meters; "+"—bed probing points.

The experiment was being conducted until a stabilized scour hole shape (and simultaneously a maximal scour depth z_{max}) was reached and controlled with a time step of 0.5 h or 1 h. The changes in bed shape were described within these time steps under the influence of the set water flow discharge levels. The application of Surfer software enabled conversion data to grid format, computing washed-out bedload volume after each probing with an established time step, plotting a graph of the increase in scours volume, and creating a graphic illustration of the changes occurring in the bedload. It shows the equilibrium depth of clear-water scour. The grid of bed shape was described with coordinates (x,y,z) and then, Surfer was used, also for creation a graphical images of scours (Figure 6). The final volume of local scour was measured firstly by filling calibrated containers with material, gathered in collection chamber, and then, compared with the final scour volume, computed by Surfer as a difference between initial conditions surface (upper surface) and the final coordinates of bed (x,y,z_{max}) (lower surface). Comparison reveals a 5% maximal error of the values measured in laboratory conditions and computed in Surfer; therefore the volume gauged in containers was accepted to further analysis.

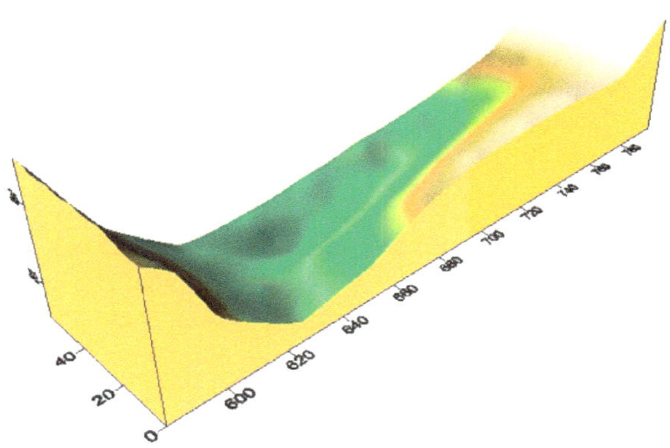

Figure 6: Local scour in sandy washout bed (measurement no. 8 in conditions of Q_w = 0.040 m^3·s^{-1} and H = 0.1 m).

Thirteen experiments were conducted in the research in a scope of water flow discharge Q_w = 0.020–0.045 m^3·s^{-1} and depth H = 0.1–0.2 m (Table 1). Apart from established water flow discharge and water surface level, the water discharge per unit width q, and time t are shown in each measurement. All the tests lasted long enough to reach equilibrium scour depth. Maximum scour depth z_{max} was measured after the equilibrium scour depth in clear-water scouring conditions achievement as a maximal distance between initial and final ordinates (z_{max} and the initial ordinate z, before water flow input into channel).

Table 1: Summary of the test's main parameters.

Number of Measurement	Q_w ($m^3 \cdot s^{-1}$)	H (m)	q ($m^3 \cdot s^{-1} \cdot m^{-2}$)	T (°C)	ν ($1 \times 10^{-6} \cdot m^2 \cdot s^{-1}$)	D_* (-)	V_s (m^3)	t (h)	z_{max} (cm)
1	0.020	0.10	0.003	16.8	1.086	14.72	0.00153	7.25	0.42
2	0.025	0.10	0.004	16.5	1.094	14.65	0.01841	10.50	3.41
3	0.025	0.12	0.004	16.1	1.104	14.56	0.00151	6.50	0.41
4	0.030	0.10	0.005	16.5	1.094	14.65	0.03701	5.00	4.74
5	0.030	0.15	0.005	16.7	1.089	14.70	0.00151	6.00	0.47
6	0.035	0.12	0.006	16.3	1.099	14.61	0.04681	8.50	5.64
7	0.035	0.15	0.006	15.9	1.109	14.52	0.00404	7.50	2.17
8	0.040	0.10	0.007	16.0	1.107	14.54	0.09746	9.25	11.82
9	0.040	0.12	0.007	17.2	1.076	14.81	0.05500	10.50	9.28
10	0.040	0.15	0.007	17.0	1.081	14.77	0.01900	8.00	4.23
11	0.040	0.20	0.007	16.8	1.086	14.72	0.00240	6.00	0.59
12	0.043	0.12	0.007	16.6	1.091	14.68	0.06800	8.50	7.64
13	0.045	0.15	0.008	16.0	1.107	14.54	0.04500	8.50	6.29

Where: Q_w—water flow discharge, H—water depth in the channel in control profile, q—discharge per unit width, T—average temperature of water, —kinematic viscosity parameter, D_*—Bonnefille's dimensionless grain parameter, V_s—total volume of bedload scattered from washout sandy bed in equilibrium conditions (volume of local scour), t—duration of measurement, z_{max}—maximal (equilibrium) scour hole depth.

Mean bedload discharge Q_s was determined as the product of bedload volume (gathered in collection chamber) V_s, and the experiment's duration *t*. The experiment included the measurement of water temperature, which made it possible to determine the kinematic viscosity parameter v and then Bonnefille's dimensionless grain parameter D_* (2). The specific density of sand ρ_s in the washout bed was 2610 $kg \cdot m^{-3}$. Therefore, assuming a water density on the level of 1000 $kg \cdot m^{-3}$, the relative density of bed material was estimated at *s* = 2.61 [-]. Measurements of washed-out bedload volume took place after the achievement of scour stabilization—*i.e.*, no visible particle movement and after dewatering the bed.

To recognize better the hydraulic conditions, Froude number was also calculated for initial conditions. In bedload movement conditions during local scour forming process in sandy bed, dimensionless shear stress in bed region θ exceeds critical shear stress (Shield number) θ_{cr}: $\theta > \theta_{cr}$, where θ for the bed region could be calculated as:

$$\theta = \frac{\tau_b}{(\rho_s - \rho_w) g d_{50}} \tag{6}$$

Mean shear stress for bottom part of flume τ_b is calculated as the product of density of water ρ_w, the gravity acceleration g, hydraulic radius of the sandy bottom part R_b, and hydraulic gradient S [17] ($\tau_b = \rho_w \cdot g \cdot R_b \cdot S$). Because of continued changes of the bed shape during the experiment, the hydraulic radius of the sandy bottom part Rb and energy grade line slope S is changing during the scour forming process. Calculations were conducted for the hydraulic radius of the bottom part R_b in time t=0 (for initial assumed water depth H and initial shape of the bed in flume) and for mean energy grade line slope for each measurement S.

To determine the R_b value in the chosen cross section (located in 5.82 meter of laboratory flume lengh), Einstein's division of velocity field was used [18,19]. The hydraulic resistance and roughness coefficients change alongside the wetted perimeter of the chosen cross section in flume, which influences the shear stress distribution in cross section [20]. Einstein's method predicates the assumption of velocity distribution in the cross section, which is dependent on mutual relation between total value of hydraulic resistance for the cross section and hydraulic resistances for the separated part of wetted perimeter of cross section:

$$\lambda = \frac{\sum \lambda_i O_i}{\sum O_i} \tag{7}$$

where: λ—total hydraulic resistance coefficient for cross section, λ_i—hydraulic resistance coefficient for the separate part of cross section with an absolute roughness k_{si}, O_i—wetted perimeter of separate part of cross section.

It is assumed that the mean velocity in total velocity field V is equal to velocity in separate part of total velocity field, *i.e.*, in zones of water flow influence on bottom V_b and walls of the channel V_g. With the assumption of roughness equality of both walls in the cross section, they are fulfilled following relations:

For mean velocity in each part of the cross section:

$$V = V_b = V_g \tag{8}$$

$$\frac{1}{\lambda}\sqrt{8gRS} = \frac{1}{\lambda_b}\sqrt{8gR_bS} = \frac{1}{\lambda_g}\sqrt{8gR_gS} \tag{9}$$

where: λ, λ_b, λ_g—total hydraulic resistance coefficient for all velocity field; for the sandy bottom part of cross section; for glass walls; R, R_b, R_g—total hydraulic radius for all velocity field; hydraulic radius, that express impact of velocity field on the sandy bottom part; hydraulic radius, that express impact of velocity field on the single glass wall part.

For cross section area: $A=2A_g+A_b$, where: A—area of total velocity field, A_g—area of single velocity field, that impact on glass wall, A_b—area of velocity field, that impact on sandy bottom.

For hydraulic radius: $R = \frac{A}{O}$, $R_b = \frac{A_b}{O_b}$, $R_g = \frac{A_g}{O_g}$, where O, O_b, O_g—total wetted perimeter, wetted perimeter of the bottom part, wetted perimeter of the walls.

Hydraulic radiuses and hydraulic resistance coefficients for the separate part of the cross section were obtained from iterative calculations using the Colebrook-White equation:

$$\frac{1}{\sqrt{\lambda}} = -2log\left(\frac{2.51}{Re\sqrt{\lambda}} + \frac{\frac{k_s}{d}}{3.71}\right) \qquad (10)$$

where k_s is absolute roughness, assumed as: $k_{sb}=3d_{90}=0.0045$ m [15] for the bottom and $k_{sg}=3.0\times10^{-6}$ m for the glass walls [21] and Re is a Reynold's number.

For the R_b value obtained from iterations, it calculated τ_b and finally, dimensionless shear stress θ for the bed area. To perform a comparison with critical shear stress, Shields number for the bed velocity field was calculated [22] as a function of the local Reynold's number Re_*, dependent on median grain diameter d_{50}, dynamic water velocity in velocity field connected with bottom region $u_* = \sqrt{\tau_b/\rho_w}$ and kinematic viscosity parameter v:

$$Re_* = \frac{d_{50} \times u_*}{v} \qquad (11)$$

Then, in accordance with Zanke:

$$\theta_{cr} = 0.432Re_*^{-2} + 0.04\left(1 - 3.32Re_*^{-1}\right) \qquad (12)$$

Attempt to correlate dimensionless shear stress in bed region θ with mean sediment discharge Q_s and with maximal scour depth z_{max} for each measurement were undertaken.

RESULTS AND DISCUSSION

Figure 7 and Figure 8 illustrate the amount of the washout bedload, *i.e.*, the volume of sand which was washed out from the bed and kept in the collection chamber during the experiment. The observation lasted until the bedload motion stabilized, *i.e.*, until it reached a visible equilibrium (maximal) scour depth, defined by Chabert and Engeldinger (Figure 1). Initially, the local scour hole was rapidly increasing (the volume of washed out sand was increasing), and after a specific time, the growth curve of the bedload approached the horizontal position—it reached the dynamic balance condition (dynamic equilibrium).

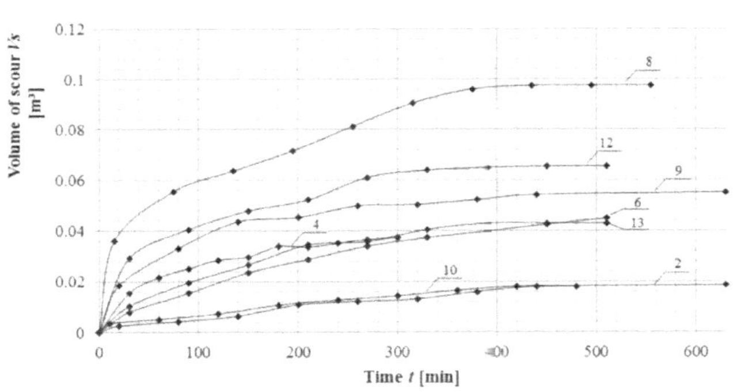

Figure 7: Amount of bedload scattered from the dissipation basin (sandy bed) in variants with visible scour during test runs No 2, 6, 8, 9, 10, 12; Where 1–13 are the numbers of the measurements (compare with Table 1 and Table 2).

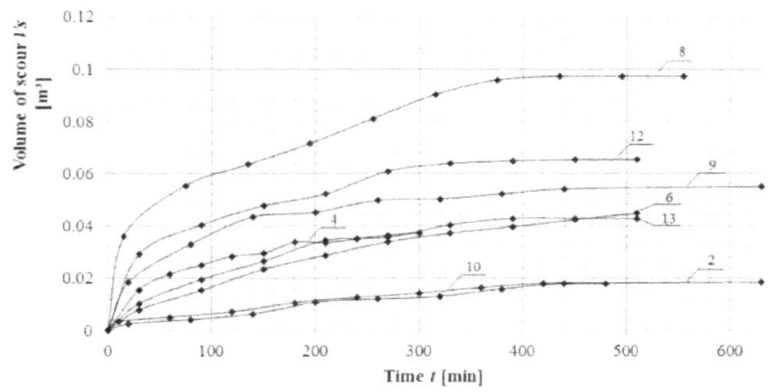

Figure 8: Amount of bedload scattered from the dissipation basin (sandy bed) in variants with invisible scour during test runs No 1, 3, 7, 11; Where 1–13 are the numbers of the measurements (compare with Table 1 and Table 2).

The data collected from each measurement and the calculated parameters of the modified Lane's relation are presented inTable 2. The correlation of the right and left side of the modified Lane's relation is shown in Figure 9. The coefficient of determination R^2 between variables was also calculated. The trend line plotted on the basis of the points indicates exponential measured link between variables.

Table 2: Summary of calculation results.

No of Measurement	Q_w	H	Q_s	D_*	$\left(\frac{W}{H}\right)$	S	$Q_S \times D_* \times \left(\frac{W}{H}\right)^{-1}$	$Q_w \times S$
	$(m^3 \cdot s^{-1})$	(m)	$(1 \times 10^{-6} \cdot m^3 \cdot s^{-1})$	(-)	(-)	(-)	$(1 \times 10^{-6}\ m^3 \cdot s^{-1})$	$(1 \times 10^{-6}\ m^3 \cdot s^{-1})$
1	0.020	0.10	0.059	14.72	5.80	0.0005	0.15	10.00
2	0.025	0.10	0.487	14.65	5.80	0.0007	1.23	17.50
3	0.025	0.12	0.065	14.56	4.83	0.0003	0.19	7.50
4	0.030	0.10	2.056	14.65	5.80	0.0014	5.19	42.00
5	0.030	0.15	0.070	14.70	3.87	0.0004	0.27	12.92
6	0.035	0.12	1.530	14.61	4.83	0.0010	4.62	35.00
7	0.035	0.15	0.150	14.52	3.87	0.0005	0.56	17.50
8	0.040	0.10	2.927	14.54	5.80	0.0013	7.34	52.00
9	0.040	0.12	1.455	14.81	4.83	0.0010	4.46	40.00
10	0.040	0.15	0.660	14.77	3.87	0.0009	2.52	34.46
11	0.040	0.20	0.111	14.72	2.90	0.0002	0.56	8.00
12	0.043	0.12	2.222	14.68	4.83	0.0012	6.75	51.60
13	0.045	0.15	1.471	14.54	3.87	0.0009	5.53	40.50

Where: Q_w—water flow discharge, H—the depth of water in the channel in control profile, Q_s—mean bedload transport discharge, S—mean hydraulic gradient in local scour distance, D_*—Bonnefille's dimensionless grain parameter, z_{max}—maximal scour depth, W—channel width, H—water depth.

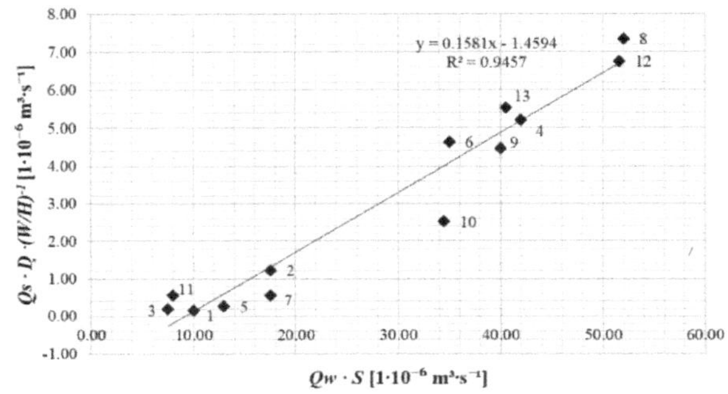

Figure 9: Modified Lane's relation graph; 1–13 are the numbers of the measurements (compare with Table 1 and Table 2).

The acknowledgement of the linear trend line slope of function allows the formulation of the final version of modified Lane's relation on the grounds of relation (Equation (5)), as it follows:

$$Q_s \times D_* \times \left(\frac{W}{H}\right)^{-1} = 0.158 \times Q_w \times S$$

(13)

Verification proposed by authors, a modified Lane's relation is performed on the basis of real, *in-situ* measured data [23]. Measurements concerned local bottom scour downstream of weir, where the solid bottom precedes the sandy part of bed (with median diameter d_{50}=0.42 mm), as it is also designed in the laboratory model described in the present paper. The subject of the studies are located on the Zagożdżonka River at Czarna gauge (the catchment area 23.4 km^2) on the east side of Poland, about 100 km south of Warsaw. Scour was the result of a flood that occurred on 11 June 2013. The water flow and sediment parameters are: S=2.4 ‰,ρ_s=2610 kg·m^{-3},s=2.61, T=16°C, and d_{50}=0.42 mm, D_*=9.85 [–]. Sediments are stemmed in the reservoir and placed in the upper part of the Zagożdżonka River so the scour below the solid bottom is made in clear-water conditions, as laboratory experiment also assumed.

The verification of modified Lane's relation is performed in unsteady conditions of water flow for a flood event, during which, water discharges and inversed width-to-depth ratio $\left(\frac{W}{H}\right)^{-1}$ were variable in time:

- -initial water discharge before the flood occurrence

$$Q_w = 0.50\, m^3 \cdot s^{-1} \text{ with } \left(\frac{W}{H}\right)^{-1} = 0.0375$$

- -peak-flow water discharge

$$Q_w = 5.06\, m^3 \cdot s^{-1} \text{ with } \left(\frac{W}{H}\right)^{-1} = 0.1950$$

During 45 h flood event, local scour was formed of volume V_s=24.8 m^3 and obtained from the sounding results that were done in the area of scour before and during the flood. As assigned in paper, the acknowledgement of scour volume and time allows the calculation of the mean value of bedload transport discharge during flood event:

$$Q_s = \frac{V_s}{t} = 0.000153\, m^3 \cdot s^{-1}$$

To verify the values agreement, the transformation of modified Lane's relation (Equation (13)) below is performed:

$$Q_s = \frac{0.158 \times Q_w \times S}{D_* \times \left(\frac{W}{H}\right)^{-1}} \quad (14)$$

Bedload transport discharge for initial conditions of flood event:

$$Q_s = \frac{0.158 \times 0.5 \times 0.0024}{9.85 \times 0.0375} = 0.0005\, m^3 \cdot s^{-1}$$

And for the peak-flow conditions:

$$Q_s = \frac{0.158 \times 5.06 \times 0.0024}{9.85 \times 0.1950} = 0.0010\, m^3 \cdot s^{-1}$$

Taking into consideration the lack of exact bottom shape data before the flood occurrence, the variability of water surface level during flood conditions, water discharge fluctuations, and variable energy grade line slope in natural conditions, the calculation results appears satisfactory because of the same order of magnitude of obtained values. However, it should be further analyzed in scope of other real objects.

Table 3 presents calculated parameters, representing hydraulic conditions and sediment properties, as the basis for deriving the correlation between the dimensionless shear stress in bed region θ, maximal local scour depth z_{max}, and with medium bedload transport discharge Q_s, shown in Figure 10 and Figure 11. Froude number F_r is lower than one for each measurement, which characterizes the type of movement of the water flow in flume. The subcritical movement was recognized for each measurement.

Table 3: Summary of calculated parameters, connected with hydraulic conditions and sediment properties.

No. of Measurement	Q_w	H	z_{max}	Q_s	R_b	τ_b	θ	u_*	Re_*	θ_{cr}	Fr
	$(m^3 \cdot s^{-1})$	(m)	(m)	$(1 \times 10^{-6} \cdot m^3 \cdot s^{-1})$	(m)	(Pa)	(-)	(ms^{-1})	(-)	(-)	(-)
1	0.020	0.10	0.0042	0.059	0.0923	0.45	0.0451	0.021	12.1	0.0320	0.35
2	0.025	0.10	0.0341	0.487	0.0923	0.63	0.0631	0.025	14.2	0.0328	0.44
3	0.025	0.12	0.0041	0.065	0.1088	0.32	0.0319	0.018	10.0	0.0311	0.33
4	0.030	0.10	0.0474	2.056	0.0923	1.27	0.1263	0.036	20.2	0.0345	0.52
5	0.030	0.15	0.0047	0.070	0.1325	0.52	0.0518	0.023	13.0	0.0323	0.28
6	0.035	0.12	0.0564	1.530	0.1088	1.07	0.1064	0.033	18.5	0.0341	0.46
7	0.035	0.15	0.0217	0.150	0.1325	0.65	0.0648	0.025	14.3	0.0328	0.33
8	0.040	0.10	0.1182	2.927	0.0923	1.18	0.1173	0.034	19.2	0.0343	0.70
9	0.040	0.12	0.0928	1.455	0.1088	1.07	0.1064	0.033	18.8	0.0342	0.53
10	0.040	0.15	0.0423	0.660	0.1325	1.17	0.1166	0.034	19.6	0.0344	0.38
11	0.040	0.20	0.0059	0.111	0.1692	0.33	0.0331	0.018	10.4	0.0312	0.25
12	0.043	0.12	0.0764	2.222	0.1088	1.28	0.1276	0.036	20.3	0.0345	0.57
13	0.045	0.15	0.0629	1.471	0.1325	1.17	0.1166	0.034	19.2	0.0342	0.43

Where: Q_w—water flow discharge, H—the depth of water in the channel in control profile, z_{max}—maximal scour depth, Q_s—mean bedload transport discharge, R_b—hydraulic radius of the sandy bottom part, $_b$—mean shear stress in the bed region, θ—dimensionless shear stress in bed region, u∗—dynamic velocity in velocity field connected with bottom region, Re∗—local Reynolds number, θ_{cr}—Shields parameter, Fr—Froude number.

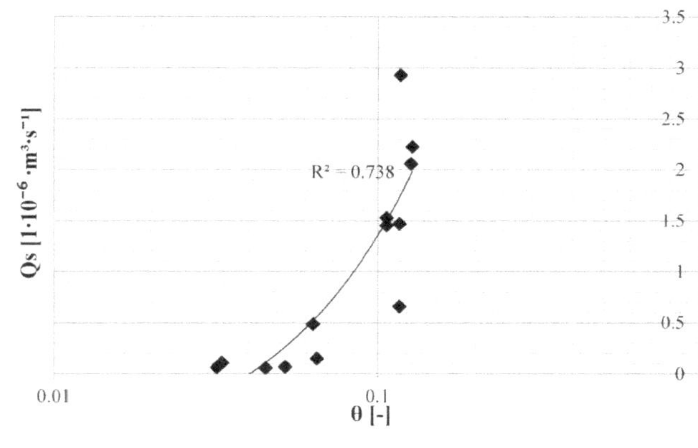

Figure 10: Shields parameter and the sediment discharge correlation graph, where θ—dimensionless shear stress in bed region; Q_s—sediment discharge; R^2—coefficient of determination.

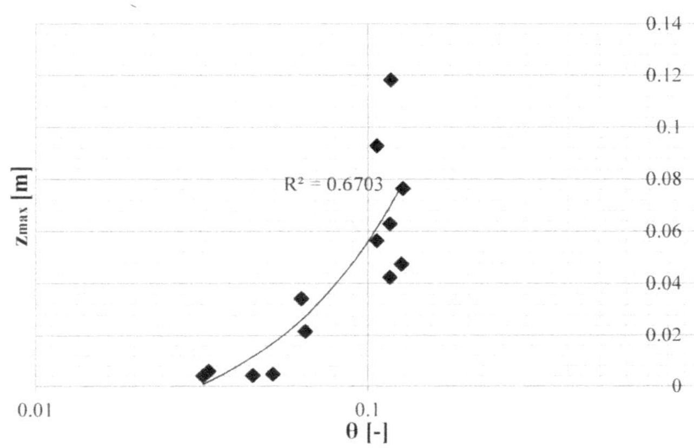

Figure 11: Shields parameter and the maximal scour depth correlation graph, where θ—dimensionless shear stress in bed region; z_{max}—maximal scour depth in clear water scour forming process; R^2—coefficient of determination.

CONCLUSIONS

For over 50 years, the original form of Lane's relation has provided a unique conceptual model useful for geomorphologists and engineers to connect water flow conditions and sediment properties. Authors have not found the literature in scope of Lane's relation application in local scour process description or trials to recast it into equation. Relation is used commonly to describe the natural river morphology forming process, and, in consequence, the dynamic equilibrium achieving by natural lowland rivers or mountain streams. Additionally, the previous studies of formula modifying were focused to correlate it with geomorphological parameters of the channel [6,7,8]. Concomitantly, the Lane's relation, in scope of local scouring in laboratory conditions, has not yet been examined.

The modified Lane's relation, presented in this paper was examined in the subcritical flow of the water in laboratory conditions in a clear-water scour forming process. The results of the experiments supported the experimental hypothesis, *i.e.*, a mutual correlation of the left and right sides of the modified Lane's relation including width-to-depth ratio. It is possible to build functional relationships between the sides of the modified relation in clear-water scouring conditions. Additionally, it is possible to use the Lane's relation in local scour forming process descriptions, despite the fact that Lane's relation is a principle commonly used in natural, alluvial rivers morphology forming descriptions in dynamic balance (equilibrium) conditions, which could be replaced by equilibrium scour depth conditions for local scour forming phenomenon.

The coefficient of determination R2 of the left and right side of the modified relation is high and equals 0.9457, which indicates a high dependency between the sides, signaling a very good correlation of data. It is possible to state the linear function, basing on trend line slope. The modified form of Lane's relation was derived from laboratory experiment results connected with delimited values of water flow discharge (0.020–0.045 $m^3 \cdot s^{-1}$) and depth (H = 0.1–0.2 m). The formulated relation was examined in praxis for water-flow discharge of 5.06 $m^3 \cdot s^{-1}$ in unsteady flow conditions, before and during a flood event in case of the real object on the grounds of *in situ* measurements. The real objects characteristics were similar to laboratory model—the solid bottom precedes sandy bed, and local scour forming takes place in clear-water conditions.

The verification results could be recognized as satisfactory because of the same value orders of magnitude. Miscalculation could be derived from the lack of exact bottom shape data before the flood occurrence, the variability of water surface level during flood conditions, water discharge fluctuations, and variable energy grade line slope in natural unsteady flow conditions.

Propitious results of verification in scope of other real objects could be a scientific area that should be further analysed. Other real-object verification test results would give data that would widen the scope of modified Lane's relation components.

The volume of washed out bedload from the bottom during the time of experiments increases with time until the stabilization of the scour shape is achieved, where the curve approaches the horizontal position, signalizing the achievement of equilibrium scour depth in clear-water scouring conditions.

Dimensionless shear stress in bed region θ exceeds critical shear stress (Shield number) θcr: θ>θcr, confirming the rule that the particle movement is initiated after exceeding this critical stress value—each measurement assumed lower or higher local scour forming intensity.

Coefficient of determination for dimensionless shear stress in bed region θ and mean sediment discharge Q_s equals 0.7380, which could be described as a satisfactory match between data. Additionally, the coefficient of determination for dimensionless shear stress in bed region θ and maximal local scour depth z_{max} equals 0.6703, also signal a satisfactory match of data. It was possible to sketch the trend lines in graphs, however, some points on the graphs diverge from the trend line. In the range of $\theta > 0.1$, it could be observed that shear stress growth stops despite of continual growth of Q_s and z_{max} values, which leads to the conclusion that further analyses are needed to ascertain the explicit correlation.

ACKNOWLEDGMENTS

This research was supported by Department of Hydraulic Engineering, Warsaw University of Life Sciences, Poland.

REFERENCES

1. Lane, E.W. The importance of fluvial morphology in hydraulic engineering. In Proceedings of American Society of Civil Engineers, New York, NY, USA, July 1955; pp. 1–17.
2. Aderibigbe, O.; Rajaratnam, N. Effect of sediment gradation on erosion by plane turbulent wall jets. *J. Hydraul. Eng.* **1998**, *124*, 1034–1042.
3. Gaudio, R.; Marion, A. Time evolution of scouring downstream of bed sills. *J. Hydraul. Res.* **2003**, *41*, 271–284.
4. Lenzi, M.A.; Marion, A.; Comiti, F.; Gaudio, R. Local scouring in low and high gradient streams at bed sills. *J. Hydraul. Res.* **2002**, *40*, 731–739.
5. Melville, B. The physics of local scour at bridge piers. In Proceedings of the Fourth International Conference on Scour and Erosion, Tokyo, Japan, 5–7 November 2008.
6. Dust, D.; Wohl, E. Conceptual model for complex river responses using an expanded Lane's relation. *Geomorphology* **2012**, *139–140*, 109–121.
7. Huang, H.Q.; Liu, X.; Nanson, G.C. Commentary on a "conceptual model for complex river responses using an expanded Lane's diagram by David Dust and Ellen Wohl". *Geomorphology* **2014**, *209*, 140–142.
8. Schumm, S.A. *Fluvial System*; Wiley-Interscience: New York, NY, USA, 1977.
9. Barbhuiya, A.K.; Dey, S. Local scour at abutments: A review. *Sadhana* **2004**, *29*, 449–476.
10. Chabert, J.; Engeldinger, P. *Etude des Affouillements Autour des Piles des Ponts*; Laboratoire d'Hydraulique: Chatou, France, 1956.
11. Brookes, A. *Channelized Rivers—Perspectives for Environmental Management*; John Wiley and Sons: Chichester, UK, 1988.
12. Church, M. Channel morphology and typology. In *The River Handbook*; Calow, P., Petts, G.E., Eds.; Blecwell; Scientific Publications: Oxford, UK, 1992; Volume 1.
13. Hickin, E.J. The analysis of river-planform responses to changes in discharge. In *River Channel Changes*; Gregory, K.J., Ed.; John Wiley and Sons: Chichester, UK, 1977.
14. Werburton, J.; Danks, M.; Wishart, D. Stability of an upland gravel-bed stream Swinhope Burn, Northern England. *Catena* **2002**, *49*, 309–329.
15. Van Rijn, L.C. *Principles of Sediment Transport in Rivers, Estuaries and Castal Seas*; Aqua Publications: Amsterdam, The Netherlands, 1993.
16. Nowak, P. *Results of Specific Density Calculations for Sand and Crushed Aggregate*; The Laboratory of Chemistry and Water and Wastewater Technology, Warsaw University of Life Sciences (WULS-SGGW): Warszawa, Poland, 2009. (In Polish)
17. Popek, Z. Bedload transport conditions in small lowland river. Treatises and Monographs. WULS-SGGW, Warsaw (in Polish) Schumm, S.A. 1969: River metamorphosis. *J. Hydraul. Div. American Soc. Civ. Eng.* **2006**, *95*, 255–273.
18. Horton, R. Separate roughness coefficients for channel bottom and sides. *Eng. News Rec.* **1933**, *111*, 652–653.
19. Indlekofer, H. *Überlagerung von Rauhigkeitseinflüssen beim Abfluß in Offenen Gerinnen*; Mitteilungen Institut für Wasserbau und Wasserwirtschaft, RWTH Aachen: Heft, Germany, 1981; Volume 37, pp. 105–145.
20. Kubrak, J.; Nachlik, E. *Hydraulic Basics of River Channels Bandwidth Calculations*; Warsaw University of Life Sciences: Warszawa, Poland, 2003. (In Polish)
21. Bollrich, G.; Preißler, G. *Hydromechanik, Band 1*; Verlag für Bauwesen: Berlin, Germany, 1992.
22. Zanke, U. *Grundlagen der Sedimentbewegung*; Springer: Berlin, Germany, 1982.
23. Urbański, J.; Hejduk, L. The analysis of local scour size formed after flood event. *Monografie Komitetu Gospodarki Wodnej* **2014**, *XX*, 389–399. (In Polish).

CHAPTER 17

WAVE AND HYDRODYNAMIC MODELING FOR ENGINEERING DESIGN OF JETTIES

Lihwa Lin, Zeki Demirbilek, Donald Ward, and David King

Keywords: Coastal modeling, jetty design, Tangier Island, Chesapeake Bay

INTRODUCTION

Tangier Island (75°59.4′ W, 37°49.8′ N) is the southernmost of a string of islands. The shallower Tangier Sound separates the lower Chesapeake Bay on the west from the east bay (Figure 1). The island, approximately 8 km (5 miles) long by 3.2 km (2 miles) wide, is located in the Virginia portion of Chesapeake Bay, 36 km (20 miles) southwest of Crisfield and 112 km (70 miles) north of Norfolk. Tangier Island is comprised of a few low fine-grained sand ridges with intervening marshlands having numerous islets and tidal creeks. The highest elevations of the island are only a few meters above the mean tide level (MTL). The small populated areas are primarily three interconnected ridges on the south-central portion of the island.

Tangier Island boat canal is a narrow light-draft channel that runs east–west across the mid-section of the island (Figure 2). It is approximately 2.3 km (1.5 miles) long, 80 m (265 ft) wide, and 4 m (13 ft) deep for small-boat traffic. Numerous mooring docks and seafood processing sheds along both sides of the canal are the main infrastructure of local fishing and crabbing industries.

Figure 1: Location of Tangier Island (small rectangular box) in Chesapeake Bay (large box).

Citation: Lihwa Lin, Zeki Demirbilek, Donald Ward, and David King, Wave and Hydrodynamic Modeling for Engineering Design of Jetties, J. Mar. Sci. Eng. 2015, 3, 1474-1503; doi:10.3390/jmse3041474.

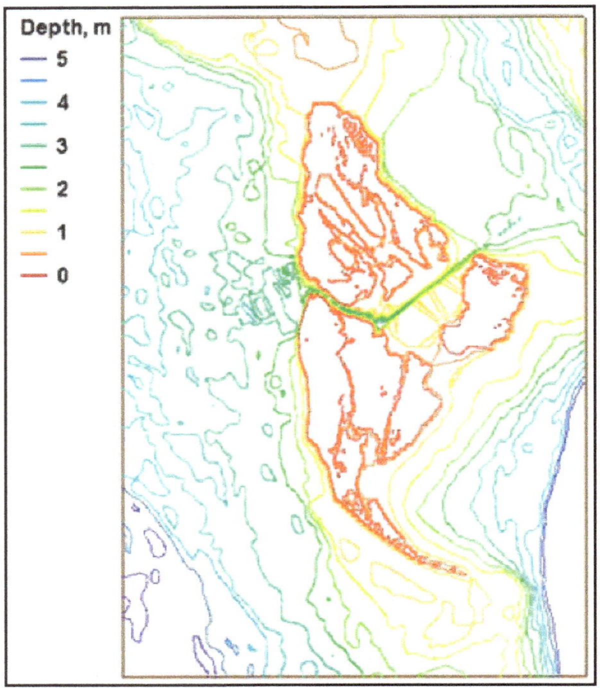

Figure 2: Depth contours in the boat canal crossing through the mid-section of Tangier Island.

The western side of Tangier Island is exposed to large incident waves (up to 2-m wave height) generated during storms in the Chesapeake Bay from the northwest through southwest quadrants. The western shoreline has long experienced progressive flooding and erosion during storms. Due to prevailing wind patterns, the littoral transport along the west shorelines of the island is directed toward the south. During storms, large waves with strong currents and high water frequently enter the western entrance of the boat canal, causing damage to shorelines and structures.

The U.S. Army Corps of Engineers (USACE) Research and Development Center (ERDC) has conducted a numerical modeling study of waves and hydrodynamics for jetty alternatives intended to protect shorelines and reduce wave energy in the western portion of the canal. The primary goal of the study was to develop a quantitative estimate of waves and wave reduction in the canal for alternatives with minimal effects on channel dredging requirements and boat traffic in the channel.

LOCAL ENVIRONMENTAL CONDITIONS

Environmental forces that normally impact the western entrance of Tangier Island boat canal are wind-waves, currents, and water levels. These natural forcings consist of metocean events including summer storms, northeasters, and tropical events, which can impact the Chesapeake Bay and reach Tangier Island from different directions. Seasonal wind patterns vary over the bay. In the winter, the dominant winds are from the north and northwest; they are from the southwest in the summer, with local breezing shifting the wind direction on a daily basis. Larger waves generally occur during northeasters and tropical storms, when high winds blow across the bay.

The west shoreline of Tangier Island is exposed to open water in the lower Bay area, where strong wind can generate large waves. Figure 3 shows two sample wind roses during for 2011 and 2012 from NOAA station 8632837 (37°32.3′ N, 76°0.9′ E) at Rappahannock Light, VA, approximately 35 km (22 mile) south of Tangier Island in the lower Bay. Winds stronger than 10 m/s (~20 kt) mostly follow a longer fetch along the north–south direction in the Bay. During northeasters with sustained winds of 15 to 20 m/s (30 to 40 kt), local wave heights ranging from 1.5 to 2.5 m (5 to 8 ft) can be expected along the west side of Tangier Island.

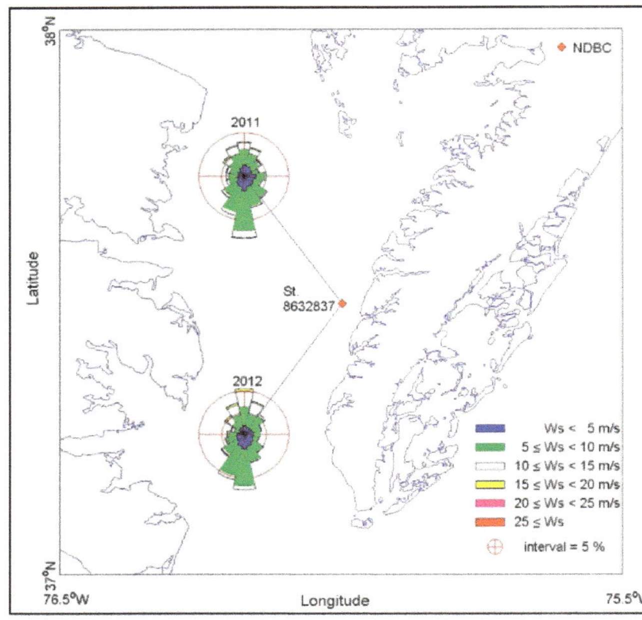

Figure 3: Wind roses for year 2011 and 2012 at NOAA Station 8632837.

Water level fluctuations in the Chesapeake Bay are dominated by oceanic tides interacting with the Bay. Tides at Tangier Island are semi-diurnal, with a 0.6-m mean tidal range. Abnormal water levels or storm surge can occur during tropical events. In the lower Chesapeake Bay, storm surges above the mean water level for 50-year and 100-year recurrence intervals are estimated at 1.5 m and 1.8 m (5 and 6 ft), respectively [1]. The relative sea level rise estimate due to absolute sea level change and land subsidence in the Chesapeake Bay ranges between 3 and 6 mm/year [1].

Historically, the Chesapeake Bay froze more often during the 19th and early 20th centuries, but rarely in the last two decades as a result of regional warming, at approximately +1 °C (+2 °F) per decade. The lower Bay may briefly become covered by thinner ice during a severe winter season. The Bay icing is not considered in the present study.

STRUCTURAL ALTERNATIVES

The primary area of interest in this modeling study is the west channel section of the Tangier Island boat canal shown inFigure 4. This narrow canal is the only navigation route that cuts through the middle of Tangier Island and connects the east and west sides of the island. The average west channel base width is 18.3 m (60 ft), the top width is 30.5 m (100 ft), and the channel depth varies from 2.3 to 4 m (7.5 to 13 ft). The narrowest cross section (bank-to-bank) is 70 m (230 ft).

Figure 4: Western part of Tangier Island boat canal.

Five structural alternatives with a north jetty connecting to the north shoreline were evaluated where the north jetty is either a straight or a dogleg structure. Due to cost constraints, the total length of the north jetty is limited to 200 m (650 ft). Alternatives 1 and 2 consider a north jetty of different length. The north jetty is positioned as close to the channel as possible at a safe (for navigation) distance of 50 m to 100 m (164 ft to 328 ft) from the channel edges. Alternatives 3, 4, and 5 include an additional short spur attaching to the south shoreline.

Figure 5 shows depth color contour maps for the existing western entrance channel (channel position and centerline in black lines) and structural features considered in Alternatives 1 to 5 (Alts 1 to 5). In all five alternatives, both jetty and spur structures have a crest elevation of 1 m (3.3 ft) above MTL (roughly the same as MSL) and crest width of 4 m (13 ft). The selection of jetty/spur crest elevation and crest width is based on the cost-to-benefit ratio and federal funding available for construction of structures in the project. The crest and width of proposed jetty/spur structures may be raised and expanded in the future if a higher crest elevation is required. Alt 1 has a straight north jetty 85 m (280 ft) long that is normal to the north shoreline. Alt 2 has a dogleg north jetty 170 m (560 ft) long with a bayward segment 85 m (280 ft) long that is parallel to the entrance channel. Alt 3 has the same dogleg north jetty as in Alt 2 and a short south spur 25 m (82 ft) long that points north (towards the channel). Alt 4 has the same dogleg north jetty as in Alt 2 and a south spur 25 m (82 ft) long directed towards the northwest (normal to south shoreline). Alt 5 has a straight north jetty identical to Alt 1 and a south spur identical to Alt 4. In addition to guiding the tidal flow along the navigation channel, the second purpose of the north jetty is to protect the shoreline east of the north jetty from wave action. The primary purpose of the south spur is to stabilize the shoreline south of the spur and reduce wave penetration into the western entrance channel. Table 1 presents a summary of existing channel and structural features for the five alternatives.

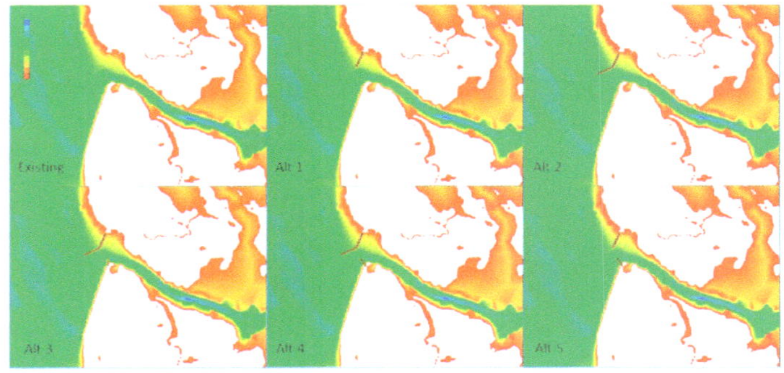

Figure 5: Depth color contours of the existing western entrance channel and Alts 1 to 5.

Table 1: Structural features of alternatives.

Alt	Straight North Jetty (85 m Long)	Dogleg North Jetty (170 m Long)	South Spur (25 m Long) Toward Channel	South Spur (25 m long) Normal to Shoreline
1	X			
2		X		
3		X	X	
4		X		X
5	X			X

NUMERICAL MODELING APPROACH

The Coastal Modeling System (CMS) was used to calculate waves, currents, sediment transport, and morphology change [2]. It is an integrated modeling system that consists of a steady-state spectral wave model (CMS-Wave) and a two-dimensional, time-dependent circulation model (CMS-Flow), which includes sediment transport and bed change capabilities. The CMS uses the Surface-water Modeling System (SMS) interface for grid generation, model setup, and post-processing [3].

The CMS models can run on a grid with variable rectangular cells. To save computational time, large cells can be used in large-area applications away from the area of interest while fine-grid resolution is used in the area of interest. The CMS can also run on nested grids that include many large and small grids [4]. The most commonly applied grid nesting involves two model grids: a large grid (parent grid) and a small grid (child grid). The application of grid nesting can dramatically reduce the computational time as compared to a large grid with fine resolution for the entire model domain. A parent grid may be used to simulate regional processes such as wave generation and propagation in a large domain. A child grid can resolve more complex bathymetry and shoreline geometry in a smaller area for more accurate modeling of nearshore wave processes. Water levels and currents calculated from the parent grid are interpolated to the child grid boundaries. Wave spectra calculated from the parent grid are saved at selected locations along the offshore boundary of the child grid. Conventionally, it suffices to save a single-location spectrum from the large-domain parent grid for wave input and apply it to the entire sea boundary of a comparatively much smaller domain child grid. For a more inhomogeneous wave field, multiple locations of wave spectra may be saved from the parent grid and interpolated for more realistic wave forcing along the seaward boundary of the child grid. The main goal of grid nesting is to minimize computational time while not sacrificing modeling accuracy.

The CMS has been applied in many coastal wave, circulation, and sediment transport studies including open coasts, inlets, bays, estuaries, and lakes [5]. Most recent bay applications include modeling waves, flow, and sediment transport of Braddock Bay in Lake Ontario [6], and Matagorda Bay and Galveston Bay in Texas [7,8].

The CMS uses physics-based theories to calculate wave generation, growth, and dissipation under variable wind/pressure fields in a bay-ocean system and, therefore, can simulate large-scale storm and hurricane events [9,10,11,12,13,14]. The use of the CMS in wave modeling is more realistic for bay applications and more accurate than classical fetch-based empirical curves as the wind wave generation and growth are strongly affected by the complex of bay geometry and varying bathymetry [15,16]. This is particularly true in the Chesapeake Bay because many river tributaries, navigation channels, shoals, islands, and peninsulas coexist in the Bay. The long narrow lake basin with the broader lower bay connecting to the Atlantic Ocean causes more complexity of wave action and flow circulation in the Chesapeake Bay. Because the western channel of Tangier Island is exposed to open water in the lower Bay, strong winds from the northwest and southwest quadrants can generate large waves in the area. During a tropical storm, Tangier Island may be threatened by high water as low atmospheric pressure and strong wind can trap water against the higher ground barriers and land to the east and north of the island. Figure 6 shows the CMS modeling domain for the Chesapeake Bay region and corresponding depth contours. This bay-wide large grid domain, approximately 100 km by 300 km (60 miles by 180 miles), is referred to as the regional grid (parent grid), which has a constant grid cell size of 500 m by 500 m (1600 ft by 1600 ft). The depths in this grid vary from 0 to 45 m (0 to 150 ft). The CMS modeling includes a second domain, referred to as the local grid (child grid) for Tangier Island, which is shown in Figure 1 and Figure 2. This grid has a finer resolution to represent details of Tangier Island shorelines and bathymetry. The domain of the child grid is approximately 5 km by 7 km (3 miles by 4.4 miles), with the local grid cell size varying from 3 m to 50 m (10 ft to 160 ft).

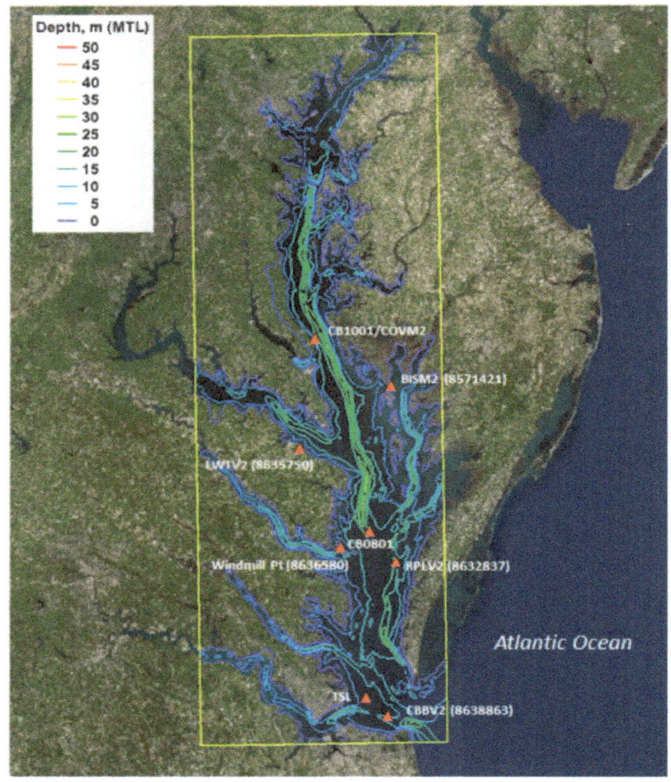

Figure 6: NOAA and VIMS coastal stations (red triangle) and CMS depth contours (m, MTL) for the CMS Chesapeake Bay regional grid domain (yellow box).

Model Calibration

The calibration of the CMS was conducted separately for CMS-Wave and CMS-Flow in the Chesapeake Bay regional grid. CMS-Wave was calibrated using wave spectral data collected at Thimble Shoal Light (TSL) gauge (see Figure 6), maintained by the Virginia Institute of Marine Science (VIMS) from 1988 to 1995 [16,17,18].

The calibration of CMS-Flow was conducted for August 2014. CMS-Flow was forced by hourly water level data, collected from NOAA Coastal Station 8638863 (Chesapeake Bay Bridge Tunnel, VA, USA), along the lower bay entrance boundary for flow exchange with the Atlantic Ocean. Atmospheric input to CMS-Flow was based on hourly wind data collected at NOAA Station 8632837 (Rappahannock Light, VA, USA). August 2014 was selected in CMS-Flow calibration because the wind speed for this month was on average the lowest during the year and so the effect of wind on tidal hydrodynamics is small for the calibration. Figure 7 shows thr wind data collected at NOAA 8632837 and 8638863 in 2014 [19]. A hydro time step of 15 s and spatially constant Manning coefficient of 0.02 were used in the model calibration. Figure 8 shows the model-data comparison water levels at NOAA Stations 8571421 (Bishops Head, MD, USA), 8635750 (Lewisetta, VA, USA), and 8636580 (Windmill Point, VA, USA). Model water levels and data are correlated well. Correlation coefficients between model water levels and data at Stations 8571421, 8635750, and 8636580 are equal to 0.98, 0.97, and 0.93, respectively.

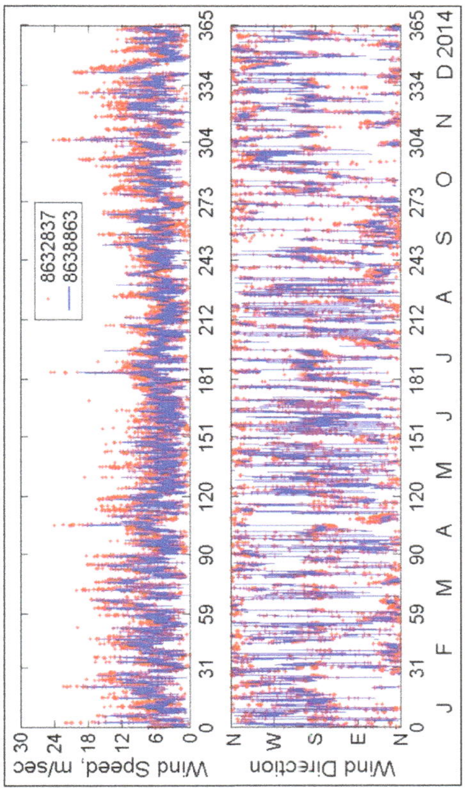

Figure 7: Wind measurements at Rappahannock Light, VA (8632837) and Chesapeake Bay Bridge Tunnel, VA (8638863) for 2014.

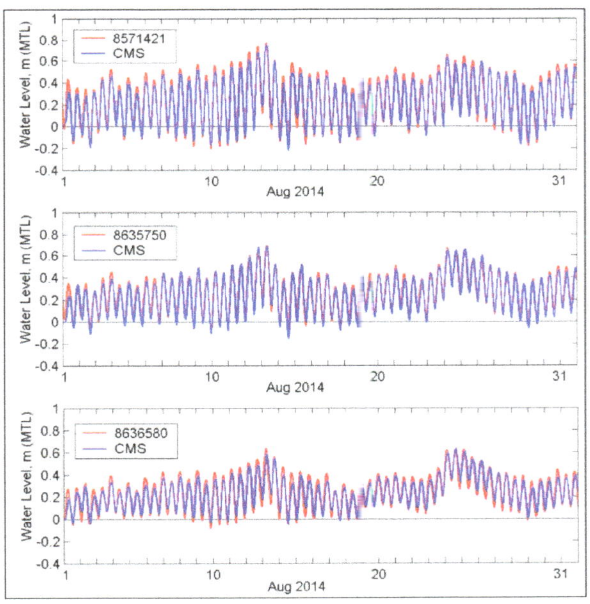

Figure 8: Model and measured water levels at Bishops Head, MD (8571421), Lewisetta, VA (8635750), and Windmill Point, VA (8636580) for August 2014.

Surface current measurements were available in the mid- and south Chesapeake Bay during 2014 from NOAA Stations CB0801 (Rappahannock Shoal Channel, VA, USA) and CB1001 (Cove Point LNG Pier, MD, USA). Figure 9 compares easting and northing current components calculated by CMS-Flow with data collected at CB0801 and CB1001. The positive current speed component along the East–West (E–W) line is directed to the east, and the positive component along the North–South (N–S) line is directed to the north. Because model currents are depth-averaged and the data are for surface currents, the magnitude of the model currents is generally smaller than in the data. Correlation coefficients between model current E–W components and data at CB0801 and CB1001 are equal to 0.27 and 0.88, respectively. Lesser correlation between E–W components and the data at CB0801 is likely due to more wind–wave interactions over weaker E–W components which are not simulated in the calibration. Correlation coefficients between model current N–S components and data at both CB0801 and CB1001 are equal to 0.89.

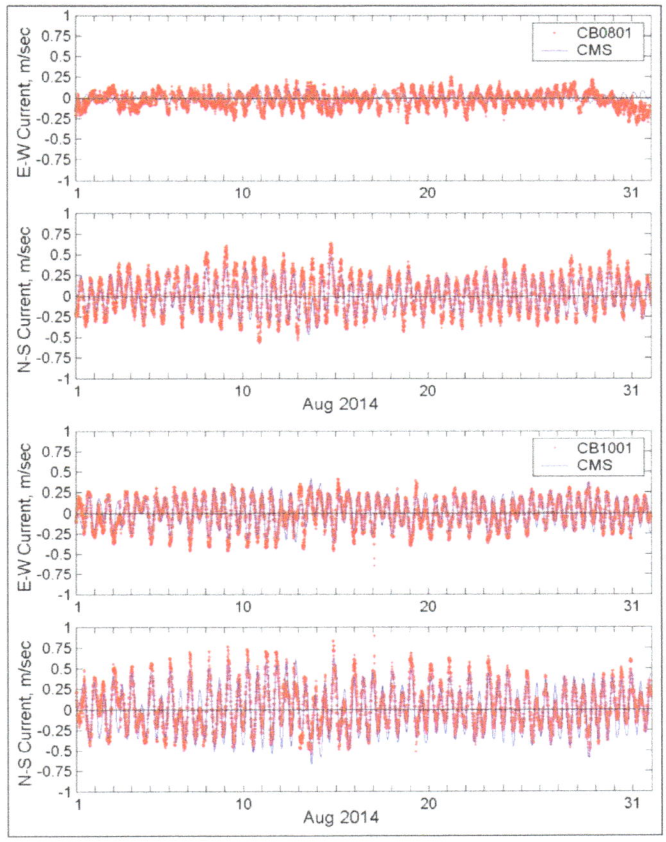

Figure 9: Comparison of calculated depth-averaged current components and measured surface current data at NOAA Stations CB0801 and CB1001 for August 2014.4.2.

Forcing Conditions

After model calibration, the modeling focused on structural design estimates for a 50-year return period for design storm conditions and a historical hurricane. For the 50-year design storm conditions, a constant wind speed of 20 m/s (40 kt) was selected based on a previous study by Basco and Shin (1993) from analysis of 1945–1983 storms at Patuxent Naval Air Station. The storms included both tropical events and northeasters. In the present study, wave generation was simulated for nine constant wind directions covering a westerly half-plane sector from north to south. These nine directions, each covering a 22.5-deg angle, present the design wind from the N, NNW, NW, WNW, W, WSW, SW, SSW, and S directions. Model simulations were conducted for two different water levels representing the observed mean tide level (WL = 0 m) and high water level (WL = 1.5 m or 5 ft) from nearby NOAA coastal stations located in the mid- and lower bay (Figure 6).

Figure 10 shows the example of the water level measurements for year 2012 recorded at three NOAA Stations: 8571421 (Bishops Head, MD), 8636580 (Windmill Point, VA), and 8638863 (Bay Bridge Tunnel, VA). The maximum water level observed at Bay Bridge Tunnel Station was 1.5 m (5 ft), MTL, during Hurricane Sandy. Table 2 lists the 50-year design wind conditions with wind speed of 20 m/s and nine wind directions for two water levels (WL) of 0 and 1.5 m (5 ft). Table 2 also lists the approximate fetch length corresponding to each of the nine wind directions. In general, the wind over longer fetch generates greater wave height at the downwind end, although in reality this also depends on the water depth variation along the fetch and wave refraction over shallower areas before waves reach the oblique shoreline.

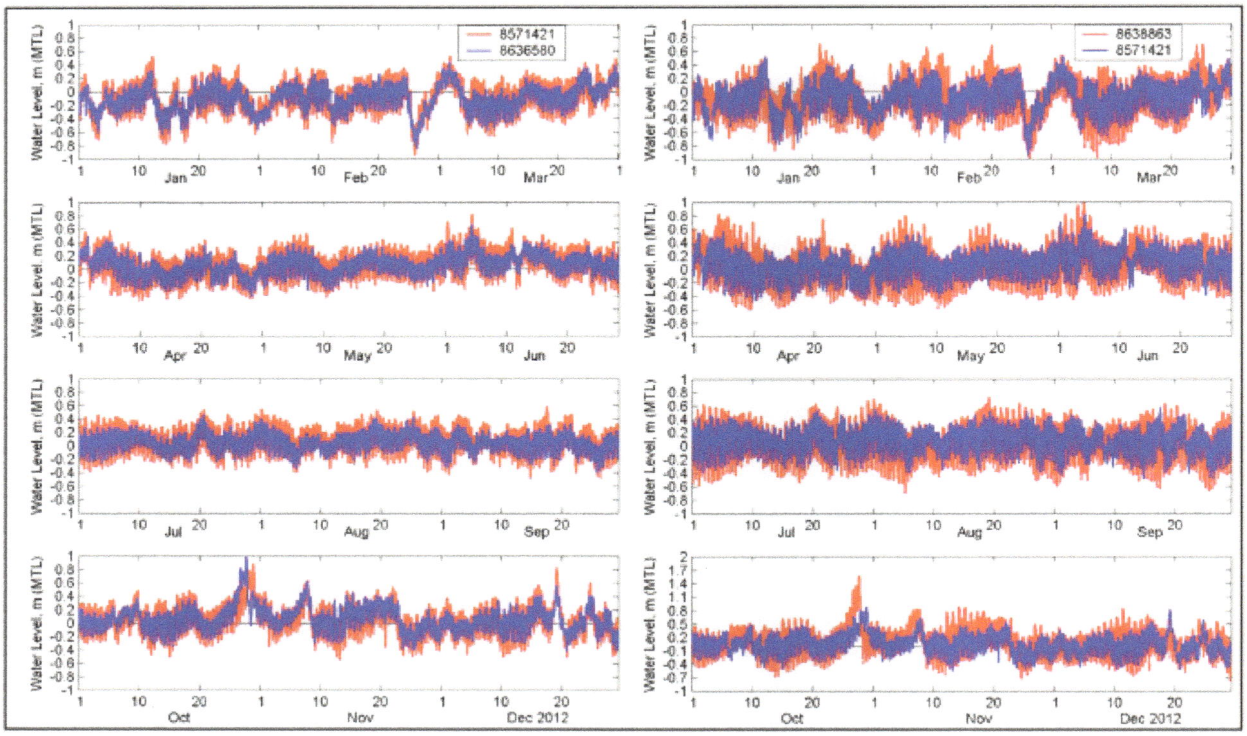

Figure 10: Water levels at Bishops Head (Station 8571421), Windmill Point (8636580), and Bay Bridge Tunnel (8638863) for 2012.

Table 2: Design wind and water level conditions for a 50-year return period northeaster storm.

Wind Dir (deg) *	N (0)	NNW (337.5)	NW (315)	WNW (292.5)	W (270)	WSW (247.5)	SW (225)	SSW (202.5)	S (180)
Fetch ** (km)	30	55	70	25	20	25	30	45	100
Wind Speed					20 m/s				
Mean Tide Level					WL = 0 m				
Mean High Water					WL = 1.5 m				

* Meteorological convention; ** Approximation.

Two water levels (WL) were used in wave simulations for the 50-year design wind conditions: were 0 and 1.5 m (5 ft) with respect to MTL. The WL = 0 m represented the mean water level corresponding to non-tropical storm (e.g., northeasters) design wind conditions, and WL = 1.5 m corresponded to the maximum storm surge for tropical storm (or hurricanes) design wind conditions. Since there is no water level measurement at Tangier Island, the selection of maximum storm surge during a hurricane is based on water level data collected in the last 50 years (after 1965) from three nearby NOAA coastal stations at Bishop Heads (Station ID 8571421), Lewisetta (8635750), and Windmill Point (8636580). These stations show that the maximum storm surge is approximately 1.5 m during Hurricanes Isabel (2003) and Sandy (2012) [13,20]. Therefore, WL = 1.5 m MTL was used as the maximum storm surge for the tropical storm in the 50-year design wind condition.

For a 50-year tropical event, Hurricane Isabel (September 2003) was selected and simulated in the entire Bay [20,21]. The strong east-to-west winds associated with Hurricane Isabel produced higher water levels along the west as well as the south side of the bay and a relatively lower water level along the east side of the bay. Figure 11 shows the examples of water level measurements from Bay Bridge Tunnel (8638863) and Windmill Point, VA (8636580) during Isabel in September 2003. The maximum water levels observed at Stations 8638863 and 8636580 were 1.87 m (6.1 ft) and 1.44 m (4.7 ft) MTL, respectively.

Design Wind and Water Level Simulations

Design wind and wave simulations were performed for nine wind directions and two water levels, 0 and 1.5 m (0 and 5 ft) MTL, listed in Table 2. The design wind conditions, representing the 50-year return period, were used for the existing western channel and Alts 1 to 5 (Figure 5). The 50-year wind condition was based on a previous study by Basco and Shin [22]. The simulations were first conducted with a regional grid, and results were then used as input in the local Tangier Island grid calculations.

A total of 108 simulations (nine wind conditions, two water levels, and six configurations) were performed to develop spatially varying estimates of the wind waves throughout the Chesapeake Bay. As an example, Figure 12 shows the bay-wide wave height fields calculated by the CMS for two wind directions from NW and SW (wind speed of 20 m/s or 40 kt) and two water levels of 0 and 1.5 m (5 ft) MTL.

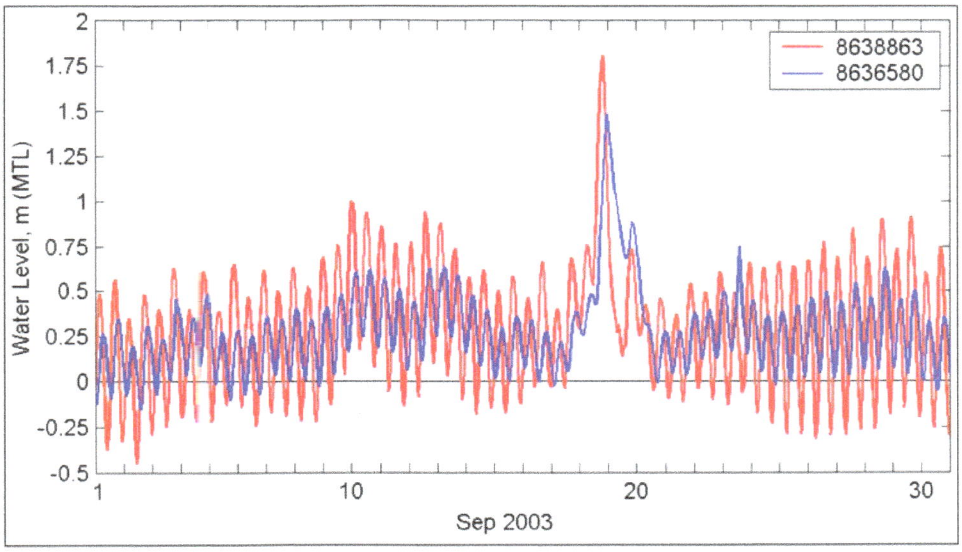

Figure 11: Water levels at Bay Bridge Tunnel (8638863) and Windmill Point (8636580), VA, for September 2003.

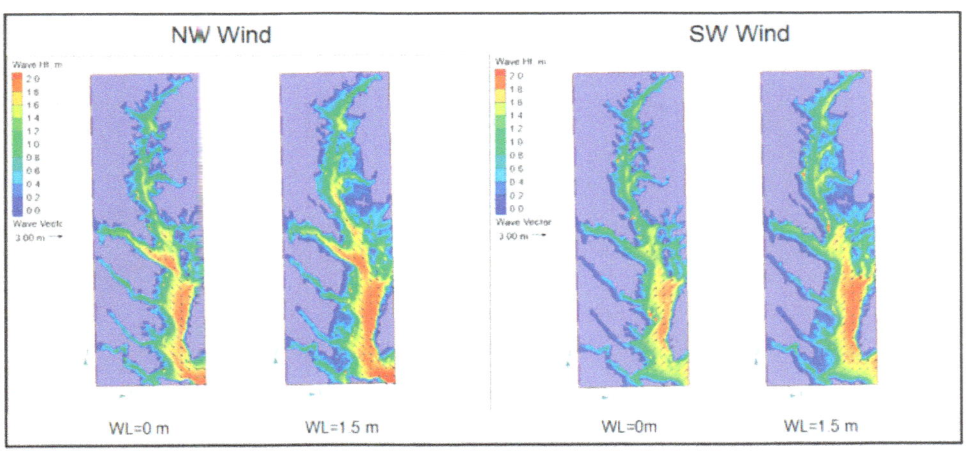

Figure 12: Calculated wave heights in Chesapeake Bay at two water levels (50-year design winds from NW and SW directions).

Wave model results from the regional grid (parent grid) for the entire Bay were used as input in the local grid (child grid) to develop the estimates of waves at the project site. Figure 13, Figure 14, Figure 15 and Figure 16 show contour and vector plots of calculated wave height and direction for Existing, Alts 1, 2, and 3, respectively, for 50-year design wind speed of 20 m/s from NW and SW, and water levels of 0 m and 1.5 m MTL. The extent of wave penetration into the canal, and variation of wave heights along the channel centerline and north and south shorelines are shown in the color-coded plots of wave field in Figure 13, Figure 14, Figure 15 and Figure 16. It should be noted that because CMS-Wave is a steady-state spectral model, the calculated wave field corresponds to a saturated sea state for each input wind condition. The saturated sea is also known as the developed sea that water surface waves cannot grow more under the specified wind input condition. This is possible in open water under constant wind conditions over a sufficiently long time period. In a limited fetch area (e.g., in a bay or lake) for the same wind conditions, wave generation can reach saturated state in a relatively shorter time. Because the Chesapeake Bay has a limited water body, the saturated sea calculated by the CMS-Wave should provide a good approximation of the design wind conditions. The calculated corresponding wave height is the maximum wave field for the design wind. The wave reduction analysis was performed for all simulations by comparing alternatives to the existing channel along three transact lines along the channel centerline and the north and south shorelines (Figure 17). Model wave heights were saved at 103 stations along these three transacts for further analysis (Figure 18).

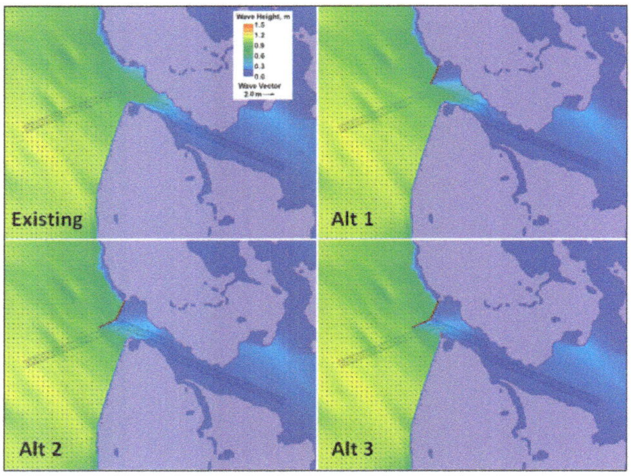

Figure 13: Example of model calculated wave heights in the west channel for the 50-year design winds from NW at 0 m water level.

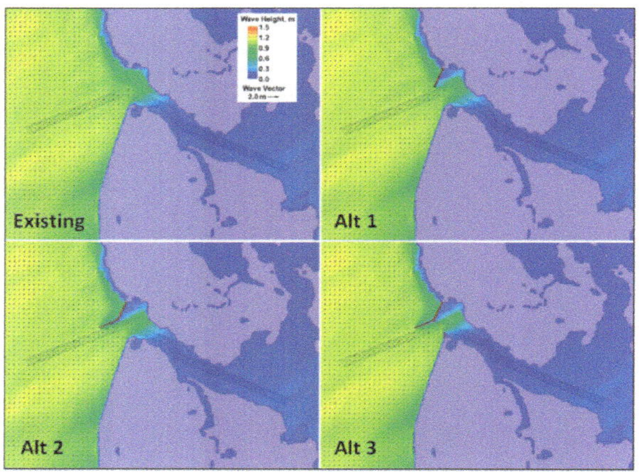

Figure 14: Example of model calculated wave heights in the west channel for the 50-year design winds from SW at 0 m water level.

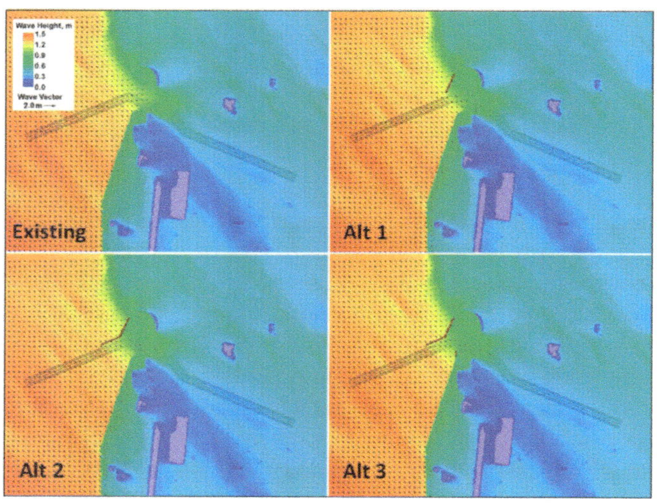

Figure 15: Example of model calculated wave heights in the west channel for the 50-year design winds from NW at 1.5 m water level.

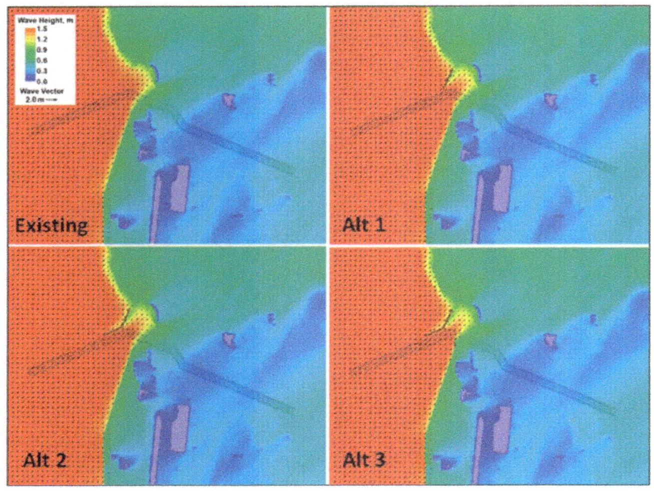

Figure 16: Example of model calculated wave heights in the west channel for the 50-year design winds from SW at 1.5 m water level.

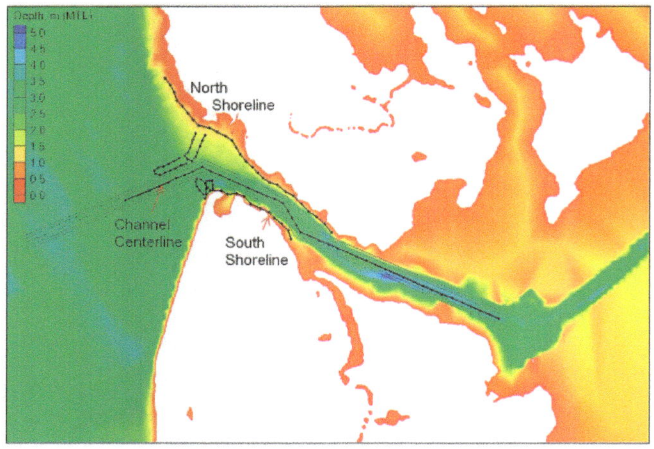

Figure 17: Three transects (lines) used to extract model wave heights.

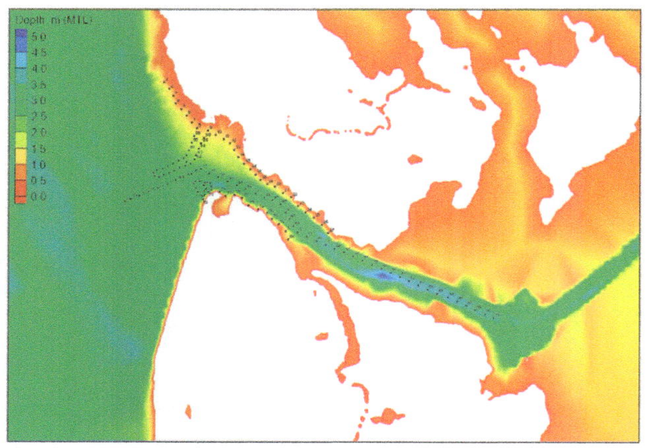

Figure 18: Point locations (stations) used to extract model wave heights.

RESULTS AND DISCUSSION

Fifty-Year Design Wind Waves

The wave reduction analysis was performed for all simulations by comparing model results for the alternatives to the existing channel. Figure 19, Figure 20, Figure 21, Figure 22, Figure 23 and Figure 24 show these comparisons of wave height variation along the west channel centerline for the NW, W, and SW directions. Figure 25 is an example of the calculated wave heights for Alt 4 along the west channel for all directions at 0 m water level. In these figures, incident waves are seen to decrease at Sta 32 (the western channel entrance). The long north jetty in Alts 2–4 has the location between Sta 32 and 38, whereas the short north jetty in Alts 1 and 5 has the location between Sta 36 and 38.

For all wind directions and both water levels investigated in this study, the analysis of wave height reduction from five alternatives is based on the wave height reduction factor calculated as the percentage of wave height reduction to the wave heights in the existing channel without the project condition:

$$\left|\frac{(\text{Wave Height, Alterantive}) - (\text{Wave Height, Existing Channel})}{(\text{Wave Height, Existing Channel})}\right| \times 100\%$$

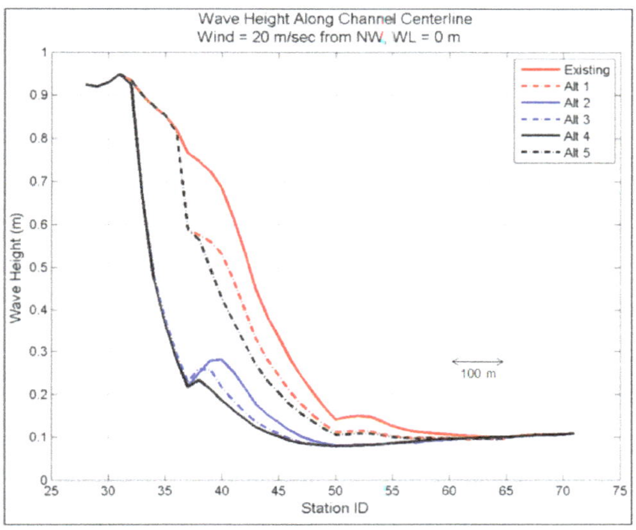

Figure 19: Model wave heights in the west channel for 50-year design wind from NW and WL = 0 m.

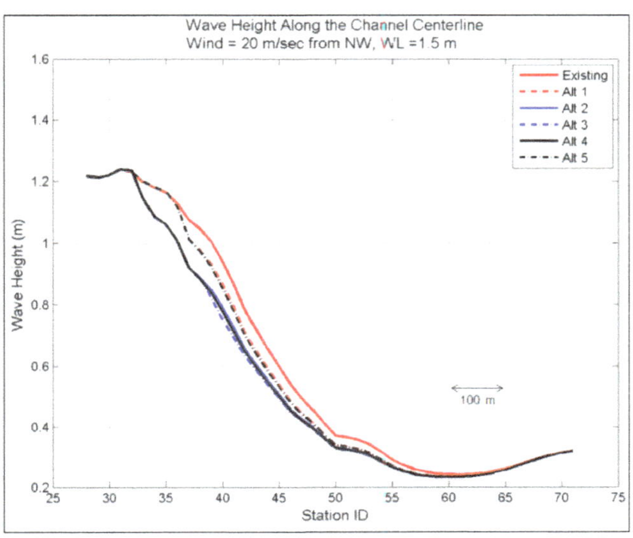

Figure 20: Model wave heights in the west channel for 50-year design wind from NW and WL = 1.5 m.

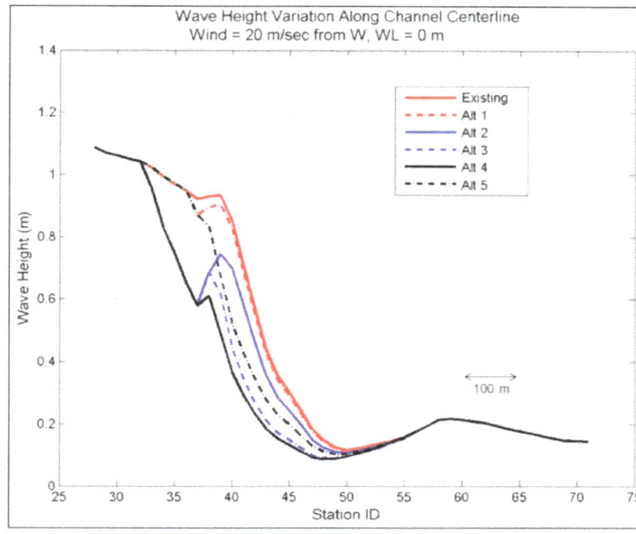

Figure 21: Model wave heights in the west channel for 50-year design wind from W and WL = 0 m.

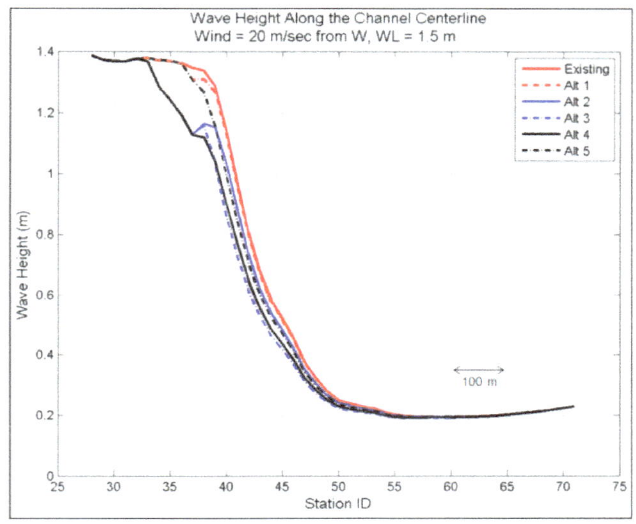

Figure 22: Model wave heights in the west channel for 50-year design wind from W and WL = 1.5 m.

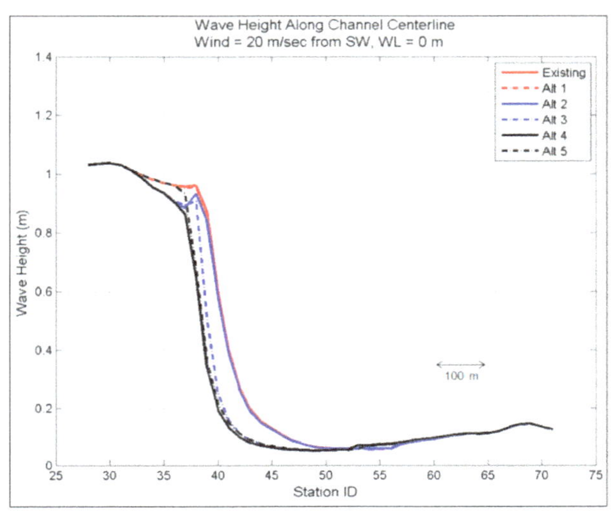

Figure 23: Model wave heights in the west channel for 50-year design wind from SW and WL = 0 m.

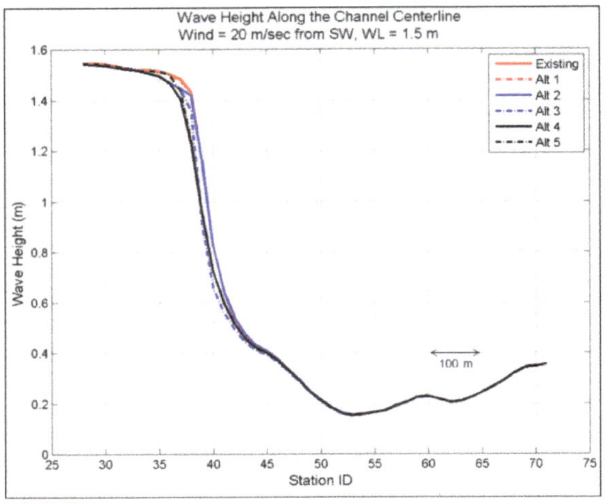

Figure 24: Model wave heights in the west channel for 50-year design wind from SW and WL = 1.5 m.

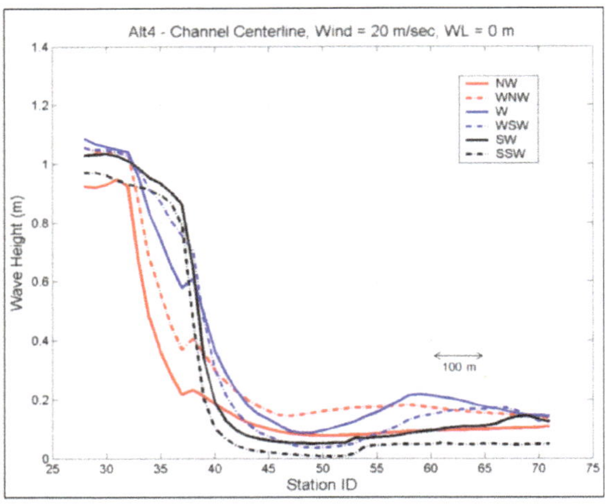

Figure 25: Model wave heights for Alt 4 along the west channel for 50-year design wind from six directions (NW, WNW, W, WSW, SW, and SSW) and WL = 0 m.

For example, Figure 26, Figure 27 and Figure 28 show the wave height reduction factor along the channel centerline for Alternatives 1 to 5 for 50-year design winds from directions of NW,W,SW, respectively, and WL = 0 m.

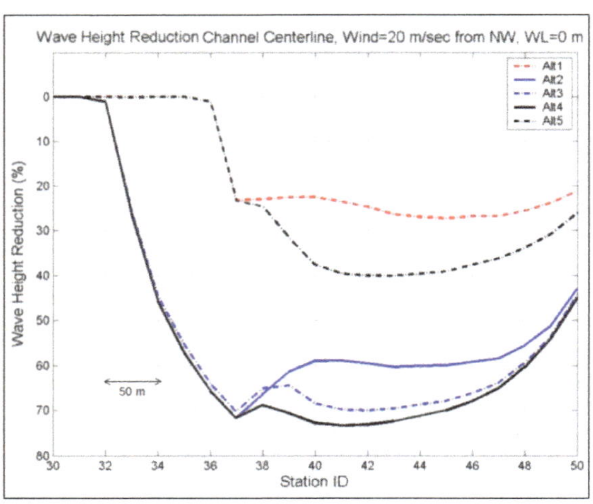

Figure 26: Calculated wave height reduction for Alts 1–5 along the west channel centerline for 50-year design wind from NW and WL = 0 m.

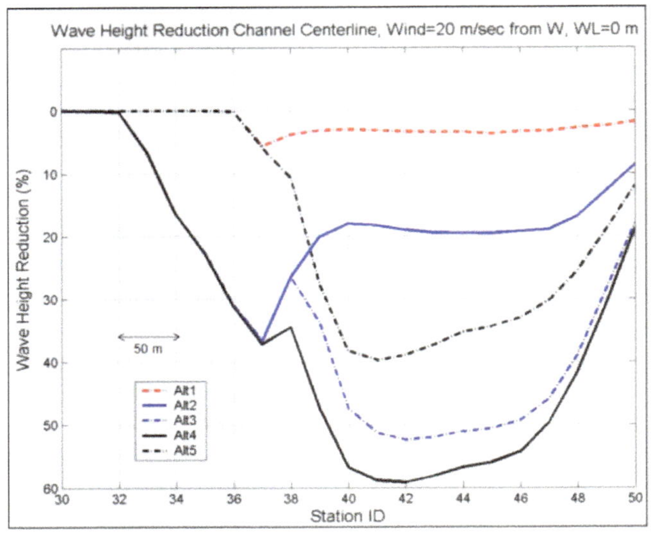

Figure 27: Calculated wave height reduction for Alts 1–5 along the west channel centerline for 50-year design wind from W and WL = 0 m.

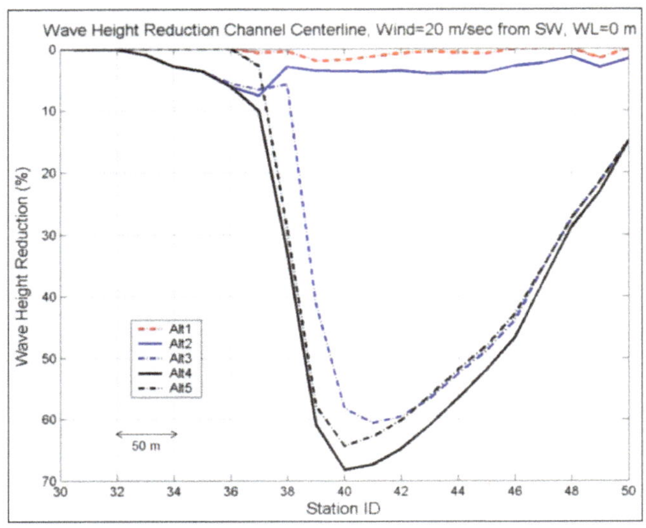

Figure 28: Calculated wave height reduction for Alts 1–5 along the west channel centerline for 50-year design wind from SW and WL = 0 m.

Among all the alternatives, Alts 3 and 4 produced the largest wave reduction for WL = 0 m and 1.5 m, respectively, along the west channel centerline and north and south shorelines. Larger waves are obtained for WL = 1.5 m as compared to WL = 0 m. It is noted that for WL = 1.5 m, the structures used in the alternatives will be submerged, losing much of their effectiveness to intercept and reduce wave energy propagating into the west and mid-sections of the channel. Under such extreme water level conditions, the wave height reduction cannot be used as a measure to rank the alternatives. Consequently, the ranking of alternatives is based on their performance (e.g., wave height reduction factor) calculated for WL = 0 m.

Hurricane Isabel

A similar analysis of waves was performed using Hurricane Isabel for the existing channel and five alternatives. The analysis was to estimate waves and water levels for a 50-year hurricane event. Because Hurricane Isabel approached the Chesapeake Bay from the southeast, the strong easterly winds associated with the hurricane produced elevated water levels along the west side of the bay and lowered the water level along the east side of bay. As a consequence, relatively lower waves had occurred at and around Tangier Island during Isabel. The wind and water level pattern associated with Hurricane Isabel was simulated for 17–23 September 2003. Both surface wind and pressure fields, used as input in the CMS, were generated from a PBL numerical model for tropical storms [21]. Figure 29 shows an example of the calculated maximum wave field in the Bay (regional grid), and at Tangier Island (local grid) for the existing channel configuration during Isabel. Model results indicated comparatively lower waves and water levels at Tangier Island than in the western portion of the bay. Figure 30 compares calculated high water levels (~1.5 m MTL) to data at NOAA Stations 8635750 (Lewisetta, VA) and 8636580 (Windmill Point, VA) during Isabel.

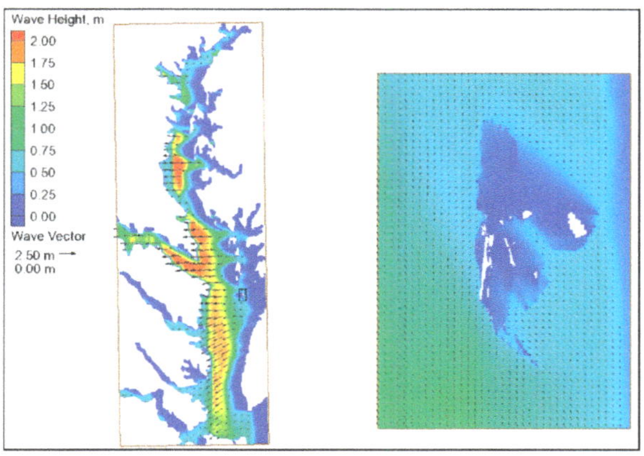

Figure 29: Calculated maximum wave height fields during Hurricane Isabel.

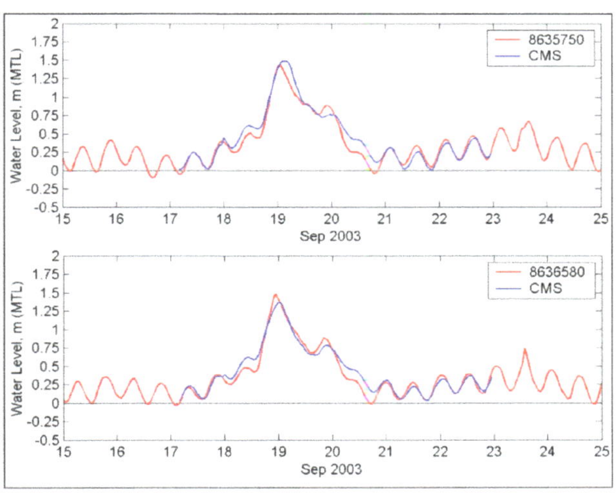

Figure 30: Calculated and measured water levels at NOAA Stations 8635750 (Lewisetta, VA) and 8636580 (Windmill Point, VA) under Hurricane Isabel.

Estimates for Structure Design

Table 3 presents the average of wave height reduction factors at WL = 0 m along the west channel centerline at Sta 30 to 50 (Figure 18) for five alternatives. Table 4 and Table 5 present the average wave height reduction factors at WL = 0 m along the north shoreline (Sta 5 to 25) and along the south shoreline (Sta 74 to 80), respectively. The average of wave height reduction factor is provided for nine wind directions with the average of cases of all nine wind directions for each alternative. Results indicate the wave reduction for Alt 4 was greater than other alternatives for WL = 0 m.

Table 3: Average wave height reduction factors along channel centerline (Sta 30 to 50) at WL = 0 m.

Wind Dir	Alt 1	Alt 2	Alt 3	Alt 4	Alt 5
N	20.3	53.9	57.4	56.5	27.3
NNW	21.5	51.9	55.0	56.7	28.6
NW	16.4	48.6	52.0	54.0	22.9
WNW	5.6	31.2	38.5	40.7	14.4
W	2.2	16.7	31.4	35.1	18.5
WSW	1.1	7.7	30.9	36.1	27.1
SW	0.5	2.9	26.0	30.4	27.4
SSW	0.1	2.8	38.2	43.9	38.4
S	0.2	2.1	25.7	30.6	28.1
Average	7.5	24.2	39.5	42.7	25.9

Table 4: Average of wave height reduction factors along north shoreline (Sta 5 to 25) at WL = 0 m.

Wind Dir	Alt 1	Alt 2	Alt 3	Alt 4	Alt 5
N	38.4	57.5	61.6	61.2	44.9
NNW	39.6	55.9	59.0	60.5	46.2
NW	34.2	51.3	54.6	56.9	40.4
WNW	23.0	40.3	47.2	50.4	31.4
W	15.3	24.8	37.8	42.3	29.3
WSW	12.3	17.3	35.6	40.6	34.4
SW	11.2	15.5	37.2	40.6	36.9
SSW	10.8	15.7	40.5	46.9	40.8
S	11.4	20.0	45.6	50.9	42.1
Average	21.8	33.1	46.6	50.0	38.5

Table 5: Average of wave height reduction factors along south shoreline (Sta 74 to 80) at WL = 0 m.

Wind Dir	Alt 1	Alt 2	Alt 3	Alt 4	Alt 5
N	18.3	56.3	74.0	63.6	35.8
NNW	16.6	48.4	65.7	65.6	35.3
NW	8.7	34.9	52.2	53.2	24.9
WNW	2.4	17.4	36.7	40.3	17.3
W	1.6	11.0	37.7	42.7	25.4
WSW	0.7	4.4	41.6	45.9	36.5
SW	0.3	2.1	45.7	47.1	43.6
SSW	0	2.1	65.1	68.6	61.1
S	0	1.2	59.6	63.6	59.5
Average	5.4	20.0	53.1	54.5	37.7

With a higher water level scenario (WL = 1.5 m), the average wave reduction was less for all alternatives, about 25 percent less than the Existing Channel configuration (without project). Based on model results for 50-year wind forcing and Hurricane Isabel simulation, Alt 4 was more effective alternative overall in reducing wave energy propagation in the canal than the other alternatives at WL = 0 m. It is noted that at the higher water level WL = 1.5 m, the structures evaluated would be either partially or fully submerged, thereby diminishing their effectiveness to intercept and reduce wave energy penetrating into the west and mid-sections of the channel. At this higher water level, wave height reduction cannot be used as a proper measure to evaluate the alternatives.

The bottom lines in Table 3, Table 4 and Table 5 give the

wave reduction factors for each alternative averaged over nine wind directions along the channel centerline, north shoreline, and south shoreline, respectively. By averaging results for the centerline and north and south shoreline, the overall representative wave reduction rating was calculated in Table 6. Among all alternatives, Alt 4 yielded the highest representative wave reduction.

Table 6: Representative wave reduction ratings for Alts 1 to 5.

Alternative	Alt 1	Alt 2	Alt 3	Alt 4	Alt 5
Average Wave Reduction (%)	11.6	25.8	46.4	49.1	34.0

Channel Sedimentation

The proposed alternatives may affect the overall sedimentation pattern in the vicinity of the structures and throughout the channel reaches of Tangier Island. To address this concern, the CMS was used to simulate the sediment transport for Hurricane Isabel. The purpose of the sediment simulation was to provide a quick view of potential shoaling and erosion areas for a 50-year tropical storm condition. It was by no means to model the long-term morphology evolution at Tangier Island. In the absence of sediment grain sized distribution data, a constant sediment median size of 0.15 mm was assumed in this simulation.

Figure 31 shows the calculated spatially varying sediment accretion and erosion field for the existing west channel configuration (without project) in the simulation of Hurricane Isabel. Model results show sediment deposition immediately outside the west entrance channel and bottom erosion inside the west entrance channel. Figure 32 and Figure 33 show calculated erosion and deposition fields in the simulation of Isabel for Alts 2 and 4, respectively. Model results showed sediment scouring in the channel near the tip of the breakwater and getting trapped inside the west entrance channel. More bottom erosion occurred between the north breakwater and south spur structure in Alts 4 and 5. The eroded sediment was carried by the stronger current into the bay, and the sediment deposition was insignificant inside the west channel entrance.

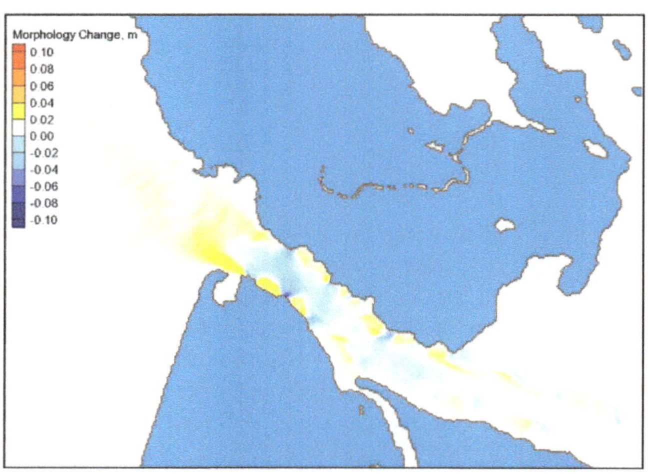

Figure 31: Model sediment accretion and erosion for the existing configuration under Hurricane Isabel.

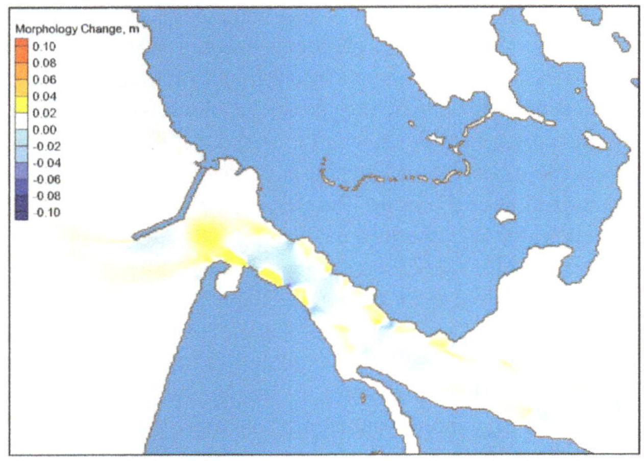

Figure 32: Calculated sediment accretion and erosion for Alt 2 under Hurricane Isabel.

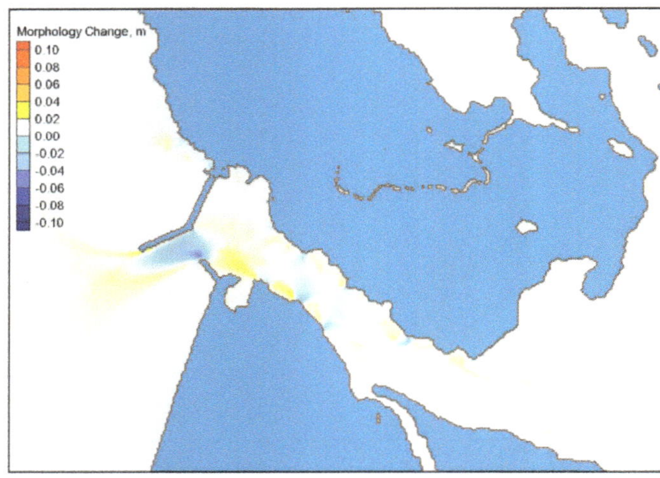

Figure 33: Calculated sediment accretion and erosion field for Alt 4 under Hurricane Isabel.

Overall, sediment transport results for the existing channel configuration and all five alternatives were similar, showing an insignificant morphology change (magnitude of either erosion or deposition less than 2 in. or 0.05 m). While the structures are intended to reduce wave energy in the channel, the currents increase and scour channel near the structures. Some additional settling of suspended sediments occurred away from the channel due to reduced wave-induced currents. Based on the model results for the 50-year return interval tropical storm (Hurricane Isabel), the depth-averaged current magnitude was less than 3 ft/s (1 m/s) in the channel and the maximum channel depth change was less than 2 in. (0.05 m). Therefore, the combined tidally-driven and wave-driven currents in the channel are usually below the threshold for the initiation of sediment motion. No significant effect of structure on channel sedimentation and channel infilling was apparent in the Hurricane Isabel simulation.

SUMMARY AND CONCLUSIONS

Numerical modeling of waves and currents was conducted to assess the impacts of these environmental forces on jetty alternatives to protect a shallow draft navigation channel entrance on Tangier Island, VA, located in the south Chesapeake Bay. The Coastal Modeling System (CMS), an integrated numerical tool that includes a spectral wave model, a two-dimensional depth-averaged hydrodynamic model with sediment transport calculations, is used to investigate the potential effects of waves and hydrodynamics on a relocation and replacement dock.

The existing channel geometry and five alternatives were investigated using the CMS, an integrated wave-hydro-sediment transport modeling system. The five alternatives evaluated consisted of a breakwater system that included a low crest jetty connecting to the north shoreline. Alts 3, 4, and 5 had an optional short structure (spur) joining to the south shoreline. Structural design estimates are based on findings of numerical wave and hydrodynamic modeling conducted for the 50-year design wind speeds, waves, and two water levels. The 50-year wind speed was considered as idealized condition and was based on a previous study by Basco and Shin [22]. Different structure alternatives were evaluated to determine an optimal design, as determined by the level of wave energy reduction in the navigation channel. The hydrodynamic modeling study results (e.g., wave height, period, direction, and water level) along the western side of the proposed structure footprint were used in these preliminary wave control structural design calculations. These calculations include structural stability, run-up/overtopping, and transmission through and over the structure.

Overall, Alt 4 performed better than the other alternatives in its protection of the boat canal entrance. However, some of the other alternatives also provided considerable wave reduction benefits. The comparison of the alternatives shows that Alts 3, 4, and 5 with south spur jetty outperformed Alts 1 and 2 with no south spur for reducing the wave energy in the channel.

It should be noted that the geometry of the channel itself, even without any jetty structure, helps to dampen the propagating waves in the boat canal. For example, at Sta 50, located approximately 300 m (1000 ft) down the channel from the western entrance, the wave energy has dissipated to the extent that wave height is reduced to 10% to 20% of the height in the bay. Thus, while the CMS shows that Alt 4 provides the greatest wave reduction benefits amongst the five alternatives, multiple criteria may be used in the selection of the optimal alternative for the final design. In closing, this modeling study provides estimates of waves and currents necessary for a follow-up structural design study that will determine the final length, orientation, elevation, and width of the structure system and foundation requirements.

ACKNOWLEDGMENTS

The authors would like to thank both of the USACE district, Norfolk, and ERDC Navigation RD & T program for supporting and funding the study. Special thanks go to Julie D. Rosati for her continual support and encouragement towards development and improvement of the CMS. Permission was granted by the Chief, USACE to publish this information.

REFERENCES

1. Boon, J.D.; Brubaker, J.M.; Forrest, D.R. *Chesapeake Bay Land Subsidence and Sea Level Change—An Evaluation of Past and Present Trends and Future Outlook*; Virginia Institute of Marine Science Special Report, No. 425; VIMS, Gloucester Point: Norfolk, VA, USA, 2010.
2. Demirbilek, Z.J.; Rosati, J. *Verification and Validation of the Coastal Modeling System, Report 1: Summary Report*; ERDC/CHL Technical Report 11-10; U.S. Army Corps of Engineers Research and Development Center: Vicksburg, MS, USA, 2011.
3. Zundel, A.K. *Surface-Water Modeling System Reference Manual—Version 9.2*; Brigham Young University Environmental Modeling Research Laboratory: Provo, UT, USA, 2006.
4. Lin, L.; Rosati, J.; Demirbilek, Z. *CMS-Wave Model: Part 5. Full-Plane Wave Transformation and Grid Nesting*; ERDC/CHL CHETN-IV-81; U.S. Army Corps of Engineers Research and Development Center: Vicksburg, MS, USA, 2012.
5. Coastal Modeling System. Available online: http://cirpwiki.info/wiki/CMS (accessed on 8 September 2015).
6. Demirbilek, Z.; Lin, L.; Hayter, E.; O'Connell, C.; Mohr, M.; Chader, S.; Forgette, C. *Modeling of Waves, Hydrodynamics and Sediment Transport for Protection of Wetlands at Braddock Bay, NY*; ERDC TR-14-8; U.S. Army Corps of Engineers Research and Development Center: Vicksburg, MS, USA, 2015.
7. Lambert, S.S.; Willey, S.S.; Campbell, T.; Thomas, R.C.; Li, H.; Lin, L.; Welp, T.L. *Regional Sediment Management Studies of Matagorda Ship Channel and Matagorda Bay System, Texas*; ERDC/CHL TR-13-10; U.S. Army Corps of Engineers Research and Development Center: Vicksburg, MS, USA, 2013.
8. Lin, L.; Reed, C.W. Numerical modeling of mixed sediment transport in GIWW and West Galveston Bay, Texas. In Proceedings of the 8th International Symposium on Coastal Engineering and Science of Coastal Sediment Processes, Coastal Sediments'15, San Diego, CA, USA, 11–15 May 2015.
9. Lin, R.Q.; Lin, L. Wind input function. In Proceedings of the 8th International Workshop on Wave Hindcasting and Prediction, Oahu, HI, USA, 14–19 November 2004.
10. Lin, L.; Lin, R.Q. Wave breaking function. In Proceedings of the 8th International Workshop on Wave Hindcasting and Prediction, Oahu, HI, USA, 14–19 November 2004.
11. Lin, R.Q.; Lin, L. Wave breaking energy in coastal region. In Proceedings of the 9th International Workshop on Wave Hindcasting and Prediction, Victoria, BC, Canada, 24–29 September 2006.
12. Li, H.; Lin, L.; Burks-Copes, K. Numerical modeling of coastal inundation and sedimentation by storm surge, tides, and waves at Norfolk, Virginia, USA. In Proceedings of the 33rd International Conference on Coastal Engineering, Santander, Spain, 1–6 July 2012.
13. Demirbilek, Z.; Lin, L.; Ward, D.L.; King, D.B. *Modeling Study for Tangier Island Jetties, Tangier Island, VA*; ERDC/CHL TR-14-8; U.S. Army Corps of Engineers Research and Development Center: Vicksburg, MS, USA, 2015.

14. Lopez-Feliciano, O.L.; Herrington, T.O.; Miller, J.K. A Morphology change study using the CMS in a groin field during Hurricane Sandy. *J. Am. Shore Beach* **2015**, *83*, 19–26.
15. Lin, L.; Demirbilek, Z.; Mase, H.; Zheng, J.; Yamada, F. *A Nearshore Spectral Wave Processes Model for Coastal Inlets and Navigation Projects*; ERDC/CHL TR-08-13; U.S. Army Corps of Engineers Research and Development Center: Vicksburg, MS, USA, 2008.
16. Lin, L.; Demirbilek, Z. Recent capabilities of CMS-Wave: A coastal wave model for inlets and navigation projects. *J. Coast. Res.* **2011**.
17. Lin, L.; Lin, R.Q.; Maa, J.P.Y. Numerical simulation of wind wave field. In Proceedings of the 9th International Workshop on Wave Hindcasting and Forecasting, Victoria, BC, Canada, 24–29 September 2006.
18. Vimswave-Vims Directional Wave Data. Available online: http://web.vims.edu/physical/research/VIMSWAVE/VIMSWAVE.htm (accessed on 8 September 2015).
19. Turning Operational Oceanographic Data into Meaningful Information for the Nation. Available online: https://tidesandcurrents.noaa.gov (accessed on 8 September 2015).
20. Demirbilek, Z.; Lin, L.; Bass, G.P. Prediction of storm-induced high water levels in Chesapeake Bay. In Proceedings of the Coastal Disaster 2005 Conference, ASCE, Charleston, SC, USA, 8–11 May 2005.
21. Demirbilek, Z.; Lin, L.; Mark, D.J. Numerical modeling of storm surges in Chesapeake Bay. *Int. J. Ecol. Dev.* **2008**, *10*, 24–39.
22. Basco, D.R.; Shin, C.S. *Design Wave Information for Chesapeake Bay and Major Tributaries in Virginia*; Technical Report, No. 93-1; Department of Civil Engineering, The Coastal Engineering Institute, Old Dominion University: Norfolk, VA, USA, 1993.

INDEX

R

S

T

U

V

W